AF885728

Power Systems

Electrical power has been the technological foundation of industrial societies for many years. Although the systems designed to provide and apply electrical energy have reached a high degree of maturity, unforeseen problems are constantly encountered, necessitating the design of more efficient and reliable systems based on novel technologies. The book series Power Systems is aimed at providing detailed, accurate and sound technical information about these new developments in electrical power engineering. It includes topics on power generation, storage and transmission as well as electrical machines. The monographs and advanced textbooks in this series address researchers, lecturers, industrial engineers and senior students in electrical engineering.

Power Systems is indexed in Scopus

More information about this series at http://www.springer.com/series/4622

Yong Tang

Voltage Stability Analysis of Power System

Science Press
Beijing

Springer

Yong Tang
State Key Laboratory of Power Grid Security
and Energy Conservation
China Electric Power Research Institute
Beijing, China

ISSN 1612-1287 ISSN 1860-4676 (electronic)
Power Systems
ISBN 978-981-16-1070-7 ISBN 978-981-16-1071-4 (eBook)
https://doi.org/10.1007/978-981-16-1071-4

Jointly published with Science Press, China
The print edition is not for sale in China Mainland. Customers from China Mainland please order the print book from: Science Press.
ISBN of the Chinese edition: 978-7-03-068105-8

© Science Press 2021
This work is subject to copyright. All rights are solely and exclusively licensed by the Publisher, whether the whole or part of the material is concerned, specifically the rights of reprinting, reuse of illustrations, recitation, broadcasting, reproduction on microfilms or in any other physical way, and transmission or information storage and retrieval, electronic adaptation, computer software, or by similar or dissimilar methodology now known or hereafter developed.
The use of general descriptive names, registered names, trademarks, service marks, etc. in this publication does not imply, even in the absence of a specific statement, that such names are exempt from the relevant protective laws and regulations and therefore free for general use.
The publishers, the authors, and the editors are safe to assume that the advice and information in this book are believed to be true and accurate at the date of publication. Neither the publishers nor the authors or the editors give a warranty, express or implied, with respect to the material contained herein or for any errors or omissions that may have been made. The publishers remain neutral with regard to jurisdictional claims in published maps and institutional affiliations.

This Springer imprint is published by the registered company Springer Nature Singapore Pte Ltd.
The registered company address is: 152 Beach Road, #21-01/04 Gateway East, Singapore 189721, Singapore

Preface

Power system is one of the artificial systems with the widest coverage, the largest scale and the most complex structure in the world. For many years, its security and stability has been a research program with great attention from the power science and technology personnel. For a long time, people often only focus on the stability of rotor angle and frequency. Since the late 1970s, massive power outage of some bulk power systems in the world has occurred many times due to voltage instability, and then the problem of voltage stability attracts our attention, becoming an important branch of power system stability research. For nearly 30 years, China's power system has been in the development trend of long-distance transmission with large units, large power grid and large capacity. Especially since the 1990s, the power system has changed a lot: The load of the receiving-end system is highly concentrated, the scale of the interconnected power grid is expanding, the proportion of hydropower plants, pithead thermal power plants and large-scale renewable energy power plants far away from the load center is increasing, and DC transmission and power electronic control devices are widely used. These changes are of great significance to the rational utilization of resources, the improvement of economic benefits and the protection of the environment, but also bring new challenges to the safe operation of the power system, one of which is the problem of voltage stability. In addition, restricted by environmental factors, the transmission line corridor is increasingly stressful, and the proportion of load that is not conducive to voltage stability is increasing, making the problem of voltage stability more and more outstanding.

Although great progress has been made in the research of the voltage stability of power system in the past few decades, the problem of voltage stability is still one of the challenges perplexing the academic circle of power in the world, and the basic theoretical research of voltage stability is still imperfect. Up to now, no unified definition of voltage stability with clear physical meaning at home and abroad, theoretical and practical judgment index of transient and medium-term and long-term voltage stability, and differentiation method of voltage instability and rotor angle instability are available. Therefore, thorough understanding of the voltage stability mechanism and study of fast and accurate voltage stability analysis methods are major problems to be solved.

This book is a summary of the research results of the Power System Department of China Electric Power Research Institute in voltage stability research, especially the research results of the first sub-project—"Basic Research on the Evaluation of Dynamic Characteristics and Safety of Electric Power Receiving-end System" of the National Key Basic R&D Program (Program 973)—"Basic Research on the Improvement of the Operation Reliability of Large-scale Interconnected Power Grid" (2004CB217901). After 5-year dedicated research and technical research, the research group has carried out basic theoretical research on voltage stability, and made some useful achievements in voltage stability mechanism, criteria and engineering application. Based on the systematic summary of existing research results, this book compiles the latest research results and engineering analysis examples of the program, hoping to be beneficial to the power workers in voltage stability research.

This book was completed under the leadership and support of the Power System Department of China Electric Power Research Institute. The members of the Program 973, Sun Huadong, Zhong Wuzhi, Yi Jun, Lin Weifang, Ma Shiying, Zhang Xin, Song Xinli, Hou Junxian, Zhang Dongxia, doctor Shao Yao, etc., participated in the research, and Sun Huadong, Zhong Wuzhi, Yi Jun and Lin Weifang participated in the preparation. We would like to express our sincere thanks to you hereby!

With limited level and practical experience of the author, the book inevitably has deficiencies or room for improvement. Correction from readers is welcome.

Beijing, China
June 2020

Tang Yong

Brief Introduction

This book comprehensively describes theories and methods in the power system voltage stability in 8 chapters. Chapter 1 introduces the basic theory of the power system and the basic concept and classification of the power system stability; Chap. 2 discusses the basic concepts of voltage stability, including the mechanism of voltage stability, and influencing factors of transient and medium-term and long-term voltage stability; Chap. 3 describes the element characteristics and models of important power system in voltage stability analysis; Chaps. 4–6 discuss the theories and methods of analysis on steady, transient and medium-term and long-term voltage stability analysis respectively; Chap. 7 introduces the measures to improve the voltage stability; Chap. 8 is about two examples of voltage stability engineering application analysis.

This book can be used as a reference for engineers and technicians who are engaged in dispatching operation, planning and design and scientific research of the power system, and teachers and students of electrical engineering major in colleges and universities.

Cataloguing in Publication (CIP) Data

Analysis on Power System Voltage Stability/Prepared by Tang Yong—Beijing: Science Press, 2011

ISBN 978-7-03-032395-8

I. Analysis on Power System Voltage Stability, II. Tang Yong, III. Power System—System Voltage Regulation IV. TM761

Archival Library of Chinese Publications CIP SJHZ (2011) No. 193342

Editor: Pei Yu/Proofreader: Bao Zhihong

Printer: Xu Xiaochen/Cover designer: Chen Jing

Published by Science Press

No. 16 Donghuangchenggen North Street, Beijing

Post code: 100717

http://www.sciencep.com

Printed by Beijing Jinghua Hucai Color Printing Co., Ltd.

Issued by Science Press Distributed in all Xinhua Bookstores

*

October 2011 Edition 1 Format: B5 (720 × 1,000)
Second printing in March 2015 Sheet: 19 3/4
Number of words: 382,000
Price: RMB 108.00
(In case of printing quality problems, the press is responsible for exchange).

Contents

Chapter 1
Introduction

1.1 Development of the Power System

Power is an important secondary energy, playing an irreplaceable role in the development of the national economy. Power safety is directly related to the national economy and social security.

The power system includes the whole process of power generation, power transmission and transformation, and power distribution. In the power system, the power generation equipment produces electric energy, in which the primary energy is converted into electric energy; the transformer and power line transmit and distribute electric power; electric equipment consume electric energy, in which the power is converted into mechanical energy, thermal energy, and light energy, etc.

Different from other energy sources, electric energy cannot be stored on a large scale. In fact, power is produced, transmitted, distributed and consumed at the same time. That is to say, the electric energy produced by the power generation equipment at any time must be equal to the sum of the electric energy consumed transmitted and distributed by electric equipment at that time. The balance of supply and demand must be maintained in real time between power generation and consumption; otherwise, the safety and continuity of power consumption will be endangered.

Safe, reliable and high-quality power supply is one of the bases and important conditions of modern society. The influence of economy, environment, and technology, etc. on the development of power systems constantly puts forward new requirements such as the following:

(1) Clean and sustainable energy with little environmental influence: On the power generation side, to use renewable energy and reduce carbon emissions; increase transmission density of the existing overhead line corridor, design new lines with small electromagnetic and noise impact, environmental requirements and beautiful appearance; gradually replace the overhead line with cables, especially in the city and its surrounding areas.
(2) Availability: To make electricity available and affordable to all.

© Science Press 2021

Y. Tang, *Voltage Stability Analysis of Power System*, Power Systems,
https://doi.org/10.1007/978-981-16-1071-4_1

(3) Applicability: To provide electric energy with the reliability and quality level required by different users.
(4) Advancement: The development of the power industry must depend on the progress of science and technology, and modern control theory, information technology, computer technology, electronic technology, new materials and so on are integrated into all aspects of the electric power industry.
(5) Flexibility and controllability: People can quickly and accurately control the power supply to meet the market demand.

To develop and utilize resources reasonably, the formation of the interconnected power system is the objective requirement and inevitable trend of the development of modern power systems. Part of the power grids in the United States, Canada and Mexico have been interconnected to the North American Power Grid, including four synchronous power grid interconnections, Eastern Interconnection, Western Interconnection, Texas Interconnection and Quebec Interconnection. On the basis of the original Western European Power Grid, the European Power Grid is expanding its scale through interconnection with the neighboring transnational power grid (for example, the Northern Europe Power Grid) and the neighboring national power grid (for example, the power grid in Eastern European countries).

With the great changes of China's economy and energy supply and the rapid development of China's power industry, China will build a strong national power grid taking the UHV AC power grid as the backbone grid frame, layering and zoning UHV, EHV and HV power grids, with a clear grid structure, so as to meet the basic requirements of large-capacity, long-distance, low-loss and low-cost power transmission, adapt to the changes of energy flow in the future and improve the flexibility of power grid operation and scheduling and the scalability of power grid structure. In January 2009, the UHV AC transmission and transformation project with the highest operating voltage level, the largest transmission capacity and the highest level of international power transmission and transformation technology in the world – Southeast of Shanxi-Nanyang-Jingmen 1,000 kV UHV AC Test Demonstration Project was successfully put into operation, which marks a major breakthrough in the localization of the core ultra-high voltage transmission technology and equipment with long distance, large capacity and low loss and is of great significance to optimizing the allocation of energy resources, ensuring the safety of national energy sources and the reliability of power supply.

In order to meet the requirements of long-distance and large-capacity transmission, DC transmission has been widely used in China. China has developed the most complicated AC/DC transmission system in the world, including five synchronous power grids: North China-Central China Power Grid, East China Power Grid, Northwest China Power Grid, Northeast China Power Grid and South China Power Grid. These synchronous power grids have realized the asynchronously interconnected through DC access, and Northeast China Power Grid and South China Power Grid have realized AC/DC parallel operation within the synchronous power grid. By the end of 2010, the DC projects completed and put into operation include: Zhoushan DC Project, Gezhouba-Shanghai Nanqiao DC Project, Three Gorges-Changzhou

DC Project, Three Gorges-Guangdong DC Project, Three Gorges-Shanghai DC Project, Tianshengqiao-Guangdong DC Project, Guizhou-Guangdong Circuit I DC Project, Guizhou-Guangdong Circuit II DC Project, Xiangjiaba-Shanghai UHV DC (±800 kV) Transmission Project and Yunnan-Guangdong UHV DC (±800 kV) Transmission Project. The total length and transmission capacity of DC transmission lines rank first in the world. With the development of the economy and the construction of power grid, it is necessary for the power grid to realize multi-source power supply and multi-point power receiving. However, the traditional two-terminal DC transmission system can only realize point-to-point DC transmission, so the multi-terminal DC transmission system developed from the two-terminal DC transmission system has attracted more and more attention.

Renewable energy and distributed power are the main trend of power grid development in the future. Wind energy, solar energy and other clean energies are new energies that mankind relies on for sustainable development, while distributed power can improve the energy efficiency. The access and variability of new powers make generation resources change fundamentally. The power system is more and more complicated under the influence of variable renewable energy utilization, distributed generation, demand response, access of micro grids and superconducting grids, electric transportation systems and intelligent building management.

In order to improve the energy efficiency and the security of the power system, China has begun to develop a strong smart grid. The strong smart grid is an inevitable result of the economic and technological development, which is to improve the energy conversion efficiency, power utilization rate, power quality and reliability of the power system through advanced technologies. The construction of the strong smart grid has become a good vision of the power industry. It will further promote the revolution and progress of the power industry.

1.2 Operation and Control

1.2.1 Electric System Operation

1. Basic operation requirements

Power interruption may cause production shutdown and chaotic life, and even endanger the safety of people and equipment, resulting in serious consequences. The loss caused by power outage to national economy is far more than that of the power system itself. Therefore, the power system should meet the requirements of reliable and continuous power supply in the operation.

In order to ensure the safe and stable operation of the power system, *Guide on Security and Stability of Power System* in China stipulates that the power system must meet the following requirements [1]:

(1) The system should have sufficient reserves of steady state rotor angle stability and active and reactive reserve capacities to ensure the operating stability of power system and keep the normal frequency and voltage of power grid. Reserve capacities should be allocated reasonably, which can be adjustable. In case of normal load fluctuation and adjustment of active and reactive power flow, there should be no spontaneous oscillation.

(2) A reasonable power grid structure is the basis of safe and stable operation of power system. It is required to take overall consideration and make a reasonable layout in the design stage of power grid. In terms of the operation mode of power grid, the power grid structure should be reasonable. A reasonable power grid structure should meet the following basic requirements:

① To meet the requirements of power flow change under various operation modes, have a certain flexibility and adapt to the requirements of system development;
② To keep the stable operation of the power system and not cause that the loads of other elements to exceed the specified emergency overload and allowable voltage deviation in case of any element fault;
③ To have a strong anti-disturbance capability and meet the relevant standards on security and stability specified in the *Guide on Security and Stability of Power System*;
④ To meet the principle of layering and zoning;
⑤ To reasonably control short circuit current in the system.

(3) Under the normal operation mode (including scheduled maintenance mode), the fault of any element (generator, line, transformer and bus) in the system should not cause the asynchronous operation of the main system, frequency collapse and voltage collapse.

(4) Under the adjusted operation mode after the accident, the power system should still have the specified margin of steady state rotor angle stability and meet the requirements of transient stability and the requirements that the loads of other elements do not exceed the specified emergency overload after another element fault.

(6) There should be predetermined measures to prevent accidents from deteriorating and to reduce the accident loss when the stability of the power system is compromised.

(7) The single fault of any element (including line, bus, and transformer, etc.) in a lower level power grid in any variety should not affect the stable operation of its upper level power grid.

2. Running status

From the perspective of operation conditions, the operation state of a power system can be divided into normal operation, warning, emergency, collapse and recovery. With the change of operation conditions, the operation state of power system will switch among various states, as shown in Fig. 1.1.

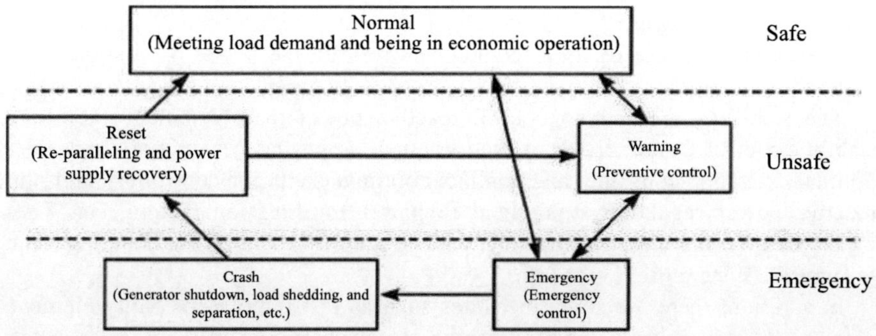

Fig. 1.1 Switching of various operation states of the power system

Under the normal operation, the system supplies balanced and qualified electric energy, meets the above requirements for safe operation with a sufficient reserve, and can realize the economic operation of the system. However, if the safety of the system is lower than a certain value due to the change of operation conditions or fault disturbance, the system will be in the warning state. During this time period, the anti-disturbance capability of the system will be reduced. It is necessary to take preventive control measures to restore the system back to the normal state. In the normal or warning state, when there is a larger disturbance, the system may be in the emergency state. The system may be restored to the warning or normal state after the faulty elements are removed quickly and effective emergency control measures are taken. If the corresponding fault elimination and emergency control measures cannot be taken timely and effectively, the system may be separated by itself, resulting in the whole system crash and large-area power outage, which is called the crash state. After system separation or crash, the system can be in the recovery state when various control measures are taken to stabilize the system. At this time, various measures should be taken to restore the power output and power transmission capacity, so as to restore power supply to users quickly and smoothly, and make the unplanned system separation parallel again. According to the actual situation of the system, the system can be in normal operation or warning state from the recovery state.

1.2.2 Power System Security and Stability Control

The power system security and stability control is to take various measures so that the system runs in a safe and stable operation state whenever possible or restores to a safe and stable operation state as fast as possible. The main reason for power system stability failure is the imbalance of active power or reactive power in the system.

The stability measures to control the active power include: generator shutdown, load shedding, quickly closing of the steam/water valve of the steam/water turbine, dynamic resistance braking (electric braking), automatic access of reserve power, and superconducting energy storage, etc. These measures are mainly used to reduce the accelerating power of the generator and restrain the power oscillation during system transition.

The stability measures to control reactive power mainly include: additional stability control for generator excitation (such as power system stabilizer, linear optimal excitation controller and nonlinear optimal excitation controller), and static reactive power regulating equipment in power transmission system, etc. These measures are usually used to maintain voltage stability and improve damping torque to suppress system oscillation, etc.

In addition, there are some measures such as fast removal of faulty elements, correct selection of reclosing mode and reclosing time, controllable series capacitance compensation control, power modulation of HV DC transmission lines, and fast switching of some new flexible AC transmission system (FACTS) equipment, for example, shunt reactors. When the above measures cannot save the system, selective system separation and load shedding measures can also be adopted.

Power system security and stability control can be divided into preventive control, emergency control and restoration control according to the time when it works. The handling capacity for possible emergency conditions can be improved by determining various preventive controls in advance. These control contents include: adjusting generator output, switching network and load, adjusting power flow, and switching transformer tap, etc., which make the system safety level higher in case of accidents. When the power system is in emergency, it is handled by emergency control. Emergency control measures include relay protection device, various security and stability control devices, etc. The accident is prevented from expanding, active and reactive power is balanced, and the system is restored to normal operation through the correct and rapid action of the automatic response for this equipment.

Generally speaking, power system security and stability control is divided into three lines of defense according to the severity of the fault [2, 3]:

The first line of defense mainly uses relay protection devices to quickly remove faulty elements and prevent the fault from expanding.

The second line of defense is mainly designed with the security and stability control system for automatic control in an emergency to prevent the system from losing stability.

The third line of defense mainly uses automatic control measures such as separation, generator shutdown and load shedding, as well as emergency measures taken by dispatching operators to prevent large-area power outage after the system loses stability.

In order to ensure the safe and stable operation of the power system, the primary system should be provided with a reasonable power grid structure and a reasonable operation mode, equipped with perfect power facilities, and the secondary system should be equipped with a relay protection system of perfect performance as well as security and stability control measures to form a perfect security and stability defense system.

1.3 Definition and Classification of the Power System Stability

1.3.1 Overview

Since the 1920s, electric power workers have recognized the problem of electric power system stability and studied it as an important aspect of the safe operation of the system [4]. In recent decades, there have been many electric power outages caused by power system instability all over the world, which have caused huge economic loss and social impact, and therefore also reflect the importance of studying the power system stability. The basic problem of the study of power system stability is the definition and classification. To clearly understand different types of the stability problems and their interrelations is very necessary for safety planning and operation of electric power system.

Two major international organizations of power systems—the Conseil International Des Grands Réseaux Electriques (CIGRE) and the Institute of Electrical and Electronic Engineers, Power Engineering Society (IEEE/PES)—have respectively given the definition of power system stability [5–8].

With the development of power grid, the form of the instability of electric power system is more complex. Transient rotor angle stability was the main problem of power system stability in the early stage. With the continuous development of power grid interconnection, the continuous application of new technology and new control means, and the increasingly heavy load of system operation, the voltage instability, oscillatory instability and frequency instability have gradually become common phenomena of the instability of electric power systems. The previous definitions given by IEEE/PES and CIGRE are not fully applicable any longer, and their classifications cannot include all the actual stability situations. Therefore, the joint working group on stability definition of IEEE/CIGRE recently gave a new definition and classification report of the power system stability [9].

The definition and classification of power system stability are systematically given in the *Guide on Security and Stability of Power System* (DL 755–2001) on the basis of previous definitions [2], which are supplemented and detailed in references [10, 11], thus forming a complete stability classification and definition of an electric power system. However, the definition and classification of power system stability given by IEEE/CIGRE and China's industrial standard DL755-2001 are not exactly the same. The definition and classification of electric power system stability of IEEE/CIGRE and *Guide on Security and Stability of Power System* (DL 755–2001) are introduced and compared in detail in reference [12], and the basis of different definitions and classifications is analyzed.

1.3.2 Definition and Classification of Electric Power System Stability in China

In China's *Guide on Security and Stability of Power System* (DL 755–2001) and *Code for Calculation of Power System Security and Stability of State Grid Corporation of China*, the stability of electric power system refers to the capacity of the electric power system to maintain stable operation after being disturbed by an accident. Generally, the power system stability is classified into three categories, namely, rotor angle stability, frequency stability and voltage stability, and many subcategories, according to the characteristics of dynamic processes and the components and control systems involved in the action, as shown in Fig. 1.2 [2, 10].

1. Rotor angle stability

The rotor angle instability is that the synchronous generator is no longer running synchronously after being disturbed. According to the magnitude of the disturbance and the leading factors that cause the rotor angle instability (insufficient synchronous torque and insufficient damping torque), the rotor angle stability is classified into the following 4 subcategories: steady state stability, small disturbance dynamic stability, transient stability and large disturbance dynamic stability, as shown in Fig. 1.2.

In practical operation analysis, steady state stability refers to the rotor angle stability of the system without non periodic instability after small disturbance. Its physical property refers to the small disturbance dynamic stability related to

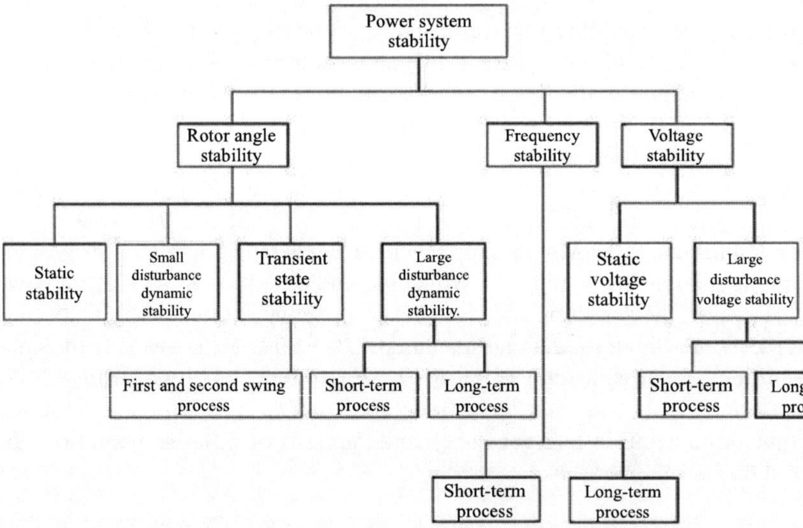

Fig. 1.2 Power system stability classification in the code for calculation of power system security and stability of state grid corporation of China

synchronous torque. It is mainly used to define the steady state stable reserve of the system under normal operation and post-accident operation mode.

Small disturbance dynamic stability refers to the rotor angle stability of the system without periodic oscillatory instability after a small disturbance. Its physical property refers to the small disturbance dynamic stability related to damping torque. It is mainly used to analyze the damping property of the system under normal operation and post-accident operation mode.

Transient stability mainly refers to the stability of the first and second swing of the system after a large disturbance, which is used to determine the transient stability limit and stability measures of the system. Its physical properties refer to the transient stability related to synchronous torque.

Large disturbance dynamic stability mainly refers to the capacity to maintain system stability under the action of system dynamic components and control devices after the system is subjected to a large disturbance, and its physical properties refer to the large disturbance dynamic stability related to damping torque. It is mainly used to analyze the dynamic stability after the system transient stability. In the calculation and analysis, the detailed model of dynamic components and control devices must be considered, such as excitation system and its additional control, prime mover governor, power electronic devices and others.

2. Voltage stability

Voltage stability is the ability of system voltage keeping or resuming the extent permitted so as to avoid voltage collapse when a little or large disturbance occurs to a power system According to the magnitude of disturbance, voltage stability can be classified into steady state voltage stability and large disturbance voltage stability.

Static voltage stability is the ability of system voltage keeping or resuming the extent permitted so as to avoid voltage collapse when a little disturbance occurs to the system. It is mainly used to define the steady state stable reserve of the system under normal operation and post-accident operation mode.

Large disturbance voltage stability includes transient voltage stability, dynamic voltage stability and medium and longterm voltage stability. It refers to the capacity of an electric power system to avoid voltage collapse after large disturbances. Transient voltage stability is mainly used to analyze the problem of fast voltage collapse. Medium and longterm voltage stability is mainly used to analyze the voltage stability of the system responding to the action of slow dynamic components and control devices, such as the secondary control and constant temperature load and others of OLTC, over-current and low excitation limit to generator stator and rotor, operable shunt capacitor, and voltage and frequency.

3. Frequency stability

Frequency stability refers to the capacity of an electric power system to maintain or recover the system frequency to the allowable range without frequency collapse after a sudden active power disturbance. It is mainly used to study the effectiveness and rationality of rotating reserve capacity and low-frequency load reduction configuration of the system, as well as the coordination between units and the power grid.

1.3.3 Definition and Classification of Power System Stability in IEEE/CIGRE

The power system stability refers to the capacity of the system to regain the operation balance point after a physical disturbance under the given initial operation mode, without out-of-limit of most of the system state quantity at the balance point, so as to maintain the system integrity.

The stability definition of the joint working group IEEE/CIGRE classifies the power system stability into three categories, namely, rotor angle stability, voltage stability and frequency stability, and many subcategories, according to the physical properties of the instability of an electric power system, the magnitude of disturbance, and the devices, process and time frame that must be considered when studying the stability, as shown in Fig. 1.3 [9].

1. Rotor angle stability

Rotor angle stability refers to the capacity of synchronous generators in inter-connected system to maintain synchronous operation after disturbances. The rotor angle instability may be caused by insufficient synchronous torque and/or damping torque, the insufficient synchronous torque may lead to non periodic instability, while insufficient damped torque may lead to oscillatory instability.

In order to analyze and understand the stability problems, the rotor angle stability is classified into small disturbance rotor angle stability and large disturbance rotor

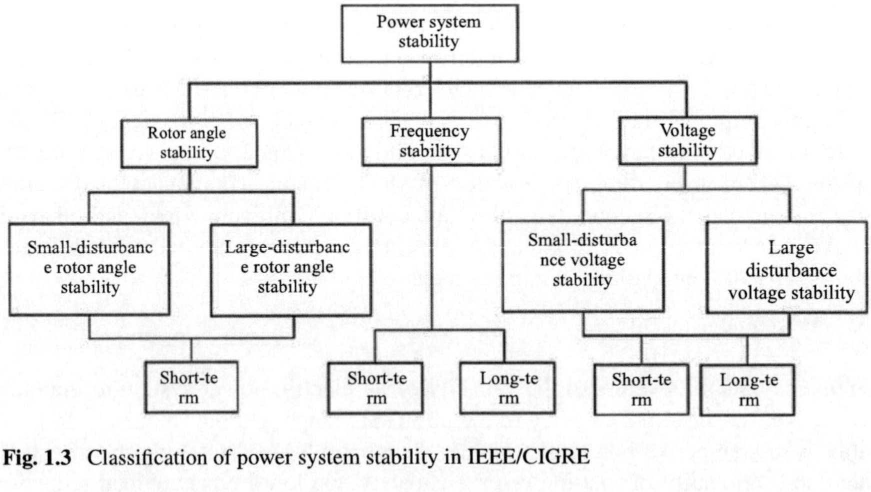

Fig. 1.3 Classification of power system stability in IEEE/CIGRE

angle stability according to the magnitude of disturbance. Since small disturbances can be small enough, for small disturbance stability analysis, the nonlinear differential equation of a power system can be linearized at the balance point, so as to study the stability problem. Large disturbance stabilities must be studied by nonlinear differential equations.

Small disturbance rotor angle stability refers to the capacity of an electric power system to maintain synchronous operation after small disturbances, which is determined by the initial operation status of the system. The small disturbance rotor angle stability can be the non periodic instability caused by rotor insufficient synchronous torque and the rotor increase oscillatory instability caused by rotor insufficient damping torque. The oscillatory instability includes local mode and interconnection mode. The time range of study of small disturbance rotor angle stability is usually 10–20 s after the disturbance.

Large disturbance rotor angle stability, also known as transient stability, refers to the capacity of an electric power system to maintain synchronous operation when it suffers a large disturbance such as short circuit of transmission lines, which is determined by the initial operating state of the system and the severity of the disturbance. In the same way, the large disturbance rotor angle stability can also be expressed in two forms: non periodic instability (first swing instability) and oscillatory instability. For the large disturbance rotor angle stability with non periodic instability, the time frame of study is usually 3–5 s after the disturbance; for the large disturbance rotor angle stability with oscillatory instability, the time frame of the study should be extended to 10–20 s after the disturbance.

In the definition and classification of IEEE/CIGRE, both small disturbance and large disturbance rotor angle stability are regarded as short-term phenomena.

2. Voltage stability

Voltage stability refers to the capacity of all buses to maintain stable voltage in a given initial operating state after the electric power system is disturbed. It depends on the system capacity to maintain/restore the balance between load demand and the power supply to the load. According to the magnitude of disturbance, IEEE/ClGRE classifies voltage stability into small disturbance voltage stability and large disturbance voltage stability, as shown in Fig. 1.3.

Large disturbance voltage stability refers to the capacity of all buses of the electric power system to maintain stable voltage after the system suffers large disturbances, such as system fault, loss of generators or lines. The nonlinear response must be considered in the study of large disturbance voltage stability. The study time period of large disturbance voltage stability can range from a few seconds to dozens of minutes as required.

Small disturbance voltage stability refers to the capacity of all buses of the electric power system to maintain stable voltage after the system subjected to small disturbances, such as load increase. Small disturbance voltage stability may be short-term or long-term.

Voltage stability can be a short-term or long-term phenomenon. The short-term voltage stability is related to the dynamics of fast response induction motor load, power electronic control load and HVDC converter. The study time period is about a few seconds. The dynamic load model must be taken into account in the study of short-term voltage stability. And the analysis of short-circuit fault of adjacent load is very important for the study of short-term voltage stability. The long-term voltage stability is related to the slow dynamic devices, such as OLTC, constant temperature load and generator excitation current limit. The study time period of long-term voltage stability is a few minutes or more. Long-term voltage stability problems are usually caused by interlocking device outages, rather than the severity of the initial disturbance.

In the definition and classification report of power system stability of IEEE/CIGRE, the correct distinction between voltage stability and rotor angle stability is explained: the difference between rotor angle stability and voltage stability is not based on the weak coupling relationship between the active power/rotor angle and the reactive power/voltage amplitude. In fact, for the electric power system under heavy load, there is a strong coupling relationship between the active power/rotor angle and the reactive power/voltage amplitude, and both the rotor angle stability and voltage stability are affected by the active and reactive power flow before disturbance. The two stabilities should be distinguished based on a set of specific opposing force subjected to persistent imbalance and the dominant system variables in the event of subsequent instability.

3. Frequency stability

Frequency stability refers to the capacity of an electric power system to maintain stable frequency, when the system suffers serious disturbances which results in a large imbalance between power generation and load demand. Frequency stability can be a short-term or long-term phenomenon.

1.3.4 Comparative Analysis of Definitions and Classifications at Home and Abroad

The definition and classification of power system stability proposed by IEEE/CIGRE in 2004 is different from that of China in 2001 (DL 755–2001). IEEE/CIGRE and industrial standard DL 755–2001 both consider that the power system stability as a holistic problem, and objectively features only stable and unstable status. However, according to different stability characteristics, magnitude of disturbance and time frames of the system, the system instability can be classified into many different forms. In order to identify the main causes of power system stability, reasonable simplification assumptions, appropriate models, and calculation methods are adopted in the analysis of specific problems, so as to arrange reasonable ways, formulate control strategies to improve system security and stability, and plan and optimize

Table 1.1 Correspondence between the two definitions and classifications

Terms compared	IEEE/CIGRE		Industrial standard DL 755–2001	
Rotor angle stability	Small-disturbance rotor angle stability	Short-term process	Static stability	
			Small disturbance dynamic stability	
	Large-disturbance rotor angle stability	Short-term process	Transient state stability	First and second swing process
			Large disturbance dynamic stability	Short and long-term process
Voltage stability	Small-disturbance voltage stability	Short and long-term process	Static voltage stability	
	Large disturbance voltage stability	Short and long-term process	Large disturbance voltage stability	Short and long-term process
Frequency stability	Short and long-term process		Short and long-term process	

the power grid structure. IEEE/CIGRE and industrial standard DL 755–2001 both classify the power system stability into rotor angle stability, frequency stability and voltage stability. This classification is very necessary to analyze and solve the actual stability problems of power system, and also helps to correctly understand and effectively deal with the problems of power system stability. Table 1.1 shows the comparison and correspondence between the two definitions.

1. Rotor angle stability

IEEE/CIGRE classifies rotor angle stability into small disturbance and large disturbance rotor angle stability from the perspective of mathematical calculation method and stability prediction. Under this classification, the small disturbance rotor angle stability is defined to be small enough to be studied by the small disturbance stability analysis method based on linearized differential equation, while the large disturbance rotor angle stability must be studied based on the nonlinear differential equation with dynamic factors of power system reserved. Small disturbance rotor angle stability can be used to predict and judge the stability of the system by characteristic root analysis, while large disturbance rotor angle stability can be used to predict and judge the stability of the system based on time-domain simulation.

In IEEE/CIGRE, it is believed that the time frame for the study of small disturbances on rotor angle stability is usually 10–20 s after disturbance, while the time frame for the study of large disturbance rotor angle stability of the first pendulum instability is usually 3–5 s after the disturbance, and the time frame for the study of large disturbances on rotor angle stability of oscillatory instability is usually extended to 10–20 s after the disturbance. Therefore, IEEE/ClGRE classifies rotor angle stability (small and large disturbances on rotor angle stability) as a short-term stability problem.

In the definitions of power system stability presented by IEEE and CIGRE earlier, "dynamic stability" is defined as a stable form of rotor angle stability. However, "dynamic stability" represents different phenomena in North America and Europe: in North America, dynamic stability generally refers to a small disturbance stability with control (mainly generator excitation control) considered, to distinguish it from the classical "steady state stability" without generator control considered, which is considered as transient stability in Europe. In order to avoid the confusion caused by the application of the term "dynamic stability", in the new definition of IEEE/CIGRE, the term "dynamic stability" in not used any more.

The *Guide on Security and Stability of Power System* (DL 755–2001) further classifies the rotor angle stability into steady state stability, small disturbance dynamic stability, transient stability and large disturbance dynamic stability from the point of view of the physical characteristics and mathematical calculation method of power system stability. This classification considers both the different properties of instability and the magnitude of disturbances, so it can be studied by different analytical methods.

From the above analysis, it can be seen that IEEE/ClGRE classified the rotor angle stability into small disturbance rotor angle stability and large disturbance power-angle stability according to the magnitude of the disturbance, and the cause of stability problem is not further classified in the subcategories. Considering the differences in physical properties and mathematical calculation methods of stability, the *Guide on Security and Stability of Power System* (DL 755–2001) further classifies rotor angle stability into steady state stability (small disturbance rotor angle stability caused by insufficient synchronous torque under small disturbances), small-perturbation dynamic stability (small disturbance rotor angle stability caused by insufficient damping torque under small disturbances), the transient stability (large disturbance rotor angle stability caused by insufficient synchronous torque under large disturbances) and large disturbance dynamic stability (large disturbance rotor angle stability caused by insufficient damping torque under large disturbances).

2. Voltage stability

As for voltage stability, IEEE/CIGRE classifies voltage stability into small and large disturbance voltage stability from the perspective of mathematical calculation method and stability prediction. In the *Guide on Security and Stability of Power System* (DL 755–2001), people also classify the voltage stability into static voltage stability and large disturbance voltage stability from the point of view of mathematical calculation method and stability prediction. This steady state voltage stability corresponds to the small-perturbation voltage stability in IEEE/CIGRE.

Both IEEE/CIGRE and the *Guide on Security and Stability of Power System* (DL 755–2001) believe that the large disturbance voltage stability can be either a rapid short-term voltage instability caused by fast dynamic load, HVDC, and others, or a long process voltage instability caused by slow dynamic equipment such as OLTC, constant temperature load, and generator excitation current limitation.

For small disturbance voltage stability (steady state voltage stability), IEEE/ClGRE believes that in a given operating point, a power system under continuous load increase, continuous control and discrete control (on-load voltage regulation to recover power) can lead to voltage instability, of which the small disturbance voltage instability may be either a short-term phenomenon or a long-term phenomenon. The definition of *Guide on Security and Stability of Power System* (DL 755–2001) on steady state voltage stability is aimed to investigate the voltage statically stable reserve under normal operation and post-accident operation modes. Therefore, steady state voltage stability is not further classified by time frame.

3. Frequency stability

For frequency stability, both IEEE/CIGRE and *Guide on Security and Stability of Power System* (DL 755–2001) define the stability capacity of frequency under keeping generation and load balance from the perspective of system. In addition, the industrial standard DL 755–2001 defines the allowed frequency range which must be kept or recovered from the point of view of safe operation.

In short, the definition and classification comparison of power system stability in China are summarized as follows:

(1) The power system stability is a holistic problem, and objectively features only stable and unstable statuses. However, according to different stability characteristics, magnitude of disturbance and time frames of the system, the system instability can be classified into many different forms. in order to identify the main causes of instability of the power system, reasonably simplify and select appropriate component models and analysis techniques, so as to arrange a reasonable operation mode, develop the control strategies to improve system security and stability level, and plan and optimize the power grid structure, both IEEE/ClGRE and the *Guide on Security and Stability of Power System* (DL 755–2001) classify the power system stability into rotor angle stability, frequency stability and voltage stability.
(2) As for rotor angle stability, IEEE/CIGRE classifies the rotor angle stability into the small disturbance rotor angle stability and the large disturbance rotor angle stability according to the different magnitude of disturbance, and the causes of instability are not classified any more in the subcategories. Considering the difference in physical properties and mathematical calculation methods of stability, the *Guide on Security and Stability of Power System* (DL 755–2001) further classifies rotor angle stability into steady state stability, small disturbance dynamic stability, transient stability and large disturbance dynamic stability.
(3) As for voltage stability, IEEE/CIGRE classifies voltage stability into small disturbance and large disturbance voltage stability from the perspective of mathematical calculation method and stability prediction. In the *Guide on*

Security and Stability of Power System (DL 755–2001), people also classify the voltage stability into static voltage stability and large disturbance voltage stability from the point of view of mathematical calculation method and stability prediction. This steady state voltage stability corresponds to the small-perturbation voltage stability in IEEE/CIGRE.

(4) As for frequency stability, both IEEE/CIGRE and *Guide on Security and Stability of Power System* (DL 755–s2001) define the stability capacity of frequency under keeping generation and load balance from the perspective of the system. In addition, *Guide on Security and Stability of Power System* (DL 755–2001) defines the allowed frequency range which must be kept or recovered from the point of view of safe operation.

1.3.5 Definition and Classification of Power System Stability

Through the analysis and comparison of the definition and classification presented by the *Guide on Security and Stability of Power System, Code for Calculation of Power System Security and Stability of State Grid Corporation of China,* and Classification of Power System Stability in IEEE/CIGRE, the definition of stability needs to be detailed according to different physical essence and mathematical descriptions. In the definition of IEEE/CIGRE, the rotor angle stability caused by insufficient damping torque and the rotor angle caused by insufficient synchronous torque are not clearly explained in the classification diagram. This book presents the following definitions and classification of power system stability combined with the idioms in power system production practice in our country, and based on physical essence and mathematical description.

Power system stability refers to the capacity of a power system to maintain stable operation after a disturbance. The power system stability can be classified into three categories, namely, rotor angle stability, voltage stability, and frequency stability, and many subcategories, according to the physical properties of the instability of the electric power system, the magnitude of disturbance, and the devices, processes and time frame that must be considered when studying the stability. Figure 1.4 shows the classification of power system stability.

1. Rotor angle stability

Rotor angle stability refers to the capacity of synchronous generators in an interconnected system to maintain synchronous operation after a disturbance. The rotor angle instability may be caused by insufficient synchronous torque and/or damping torque, the insufficient synchronous torque may lead to non periodic instability, while insufficient damped torque may lead to oscillatory instability.

1) Small-disturbance rotor angle stability

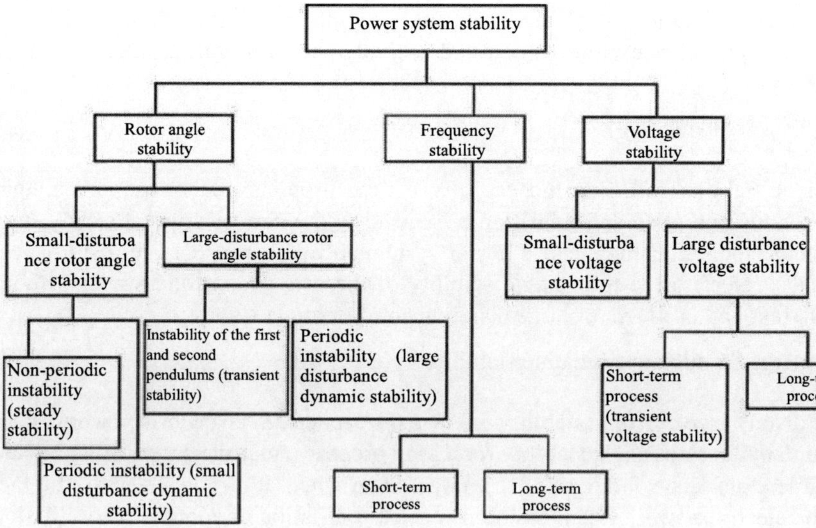

Fig. 1.4 Classification of power system stability

Small disturbance rotor angle stability refers to the capacity of an electric power system to maintain synchronous operation after small a disturbance, which is determined by the initial operation status of the system. The small disturbance rotor angle stability can be the non periodic instability caused by rotor insufficient synchronous torque and the rotor increase oscillatory instability caused by rotor insufficient damping torque.

(1) Steady state rotor angle stability.

Steady state rotor angle stability (steady state stability for short) refers to the capacity of an electric power system to automatically recover to the initial operating state without non periodic out-of-step after a small disturbance, and is a form of small disturbance rotor angle stability.

(2) Small disturbance dynamic stability

Small disturbance dynamic stability refers to the capacity of an electric power system to maintain the operational stability for a long process under the action of automatic adjustment and control device after a small disturbance. It usually refers to that the electric power system has no divergent oscillation or continuous oscillation occurred after the disturbance, which is another form of small disturbance rotor angle stability.

2) Large-disturbance rotor angle stability

Large disturbance rotor angle stability refers to the capacity of an electric power system to maintain synchronous operation when it suffers from serious failure. It is determined by the initial operating state of the system and the severity of the disturbance. The large disturbance rotor angle stability can also be expressed in two forms: non periodic instability (first and second swing instability) and oscillatory instability.

(1) Transient stability

Transient stability refers to the capacity of synchronous generators to maintain synchronous operation and transition or recover to the new or original steady state operating mode after the electric power system suffered from a large disturbance. It usually refers to the rotor angle stability of the first and second swing without out-of-step, and is a form of large disturbance rotor angle stability.

(2) Large disturbance dynamic stability.

Large disturbance dynamic stability refers to the capacity of an electric power system to maintain the rotor angle stability for a long process under the action of automatic adjustment and control device after a large disturbance. It usually refers to the fact that the electric power system has no divergent oscillation or continuous oscillation occurred after the disturbance, which is another form of large disturbance rotor angle stability.

2. Voltage stability

Voltage stability refers to the capacity of all buses to maintain stable voltage in a given initial operating state after the electric power system is disturbed. It depends on the system capacity to maintain/restore the balance between load demand and the power supply to the load. Its essence is that when the power provided by the system to the load increases with the increase of current, the system is in a stable voltage state; otherwise, the system is in an unstable voltage state. When the system is in a wide range of voltage instability, if the load is still continuously trying to obtain more power (active or reactive) by increasing the current, then voltage collapse may occur. According to the magnitude of disturbance, voltage stability can be classified into small disturbance voltage stability and large disturbance voltage stability.

(1) Small-disturbance voltage stability

Small disturbance voltage stability refers to the capacity of all buses of the electric power system to maintain stable voltage after the system suffered a small disturbance, such as load increase, and small disturbance voltage stability is also known as steady state voltage stability.

(2) Large disturbance voltage stability

Large disturbance voltage stability refers to the capacity of all buses of the electric power system to maintain stable voltage after the system suffers large disturbances, such as system fault, loss of generators or lines. Large disturbance voltage stability may be short-term or long-term. Short-term voltage stability is also called transient voltage stability.

3. Frequency stability

Frequency stability refers to the capacity of the system to maintain or recover the system frequency to the allowable range without frequency collapse, when the power system suffers a serious disturbance and unbalance between power generation and load demand. Frequency stability can be a short-term or long-term phenomenon.

References

1. Power Industry Standard of the People's Republic of China (2001) Guide on security and stability of power system (DL 755–2001). China Electric Power Press [CEPP], Beijing
2. National Economy and Trade Commission of the People's Republic of China (2001) (DL 755–2001) Guide on security and stability of power system. China Electric Power Press [CEPP], Beijing
3. National Economy and Trade Commission of the People's Republic of China (2000) (DL 723–2000) Guide on control technology of security and stability of power system. China Electric Power Press [CEPP], Beijing
4. Steinmetz CP (1920) Power control and stability of electric generating stations. AIEE Trans XXXIX(II):1215–1287
5. Crary SB, Herlitz I, Favez B (1948) CIGRE SC32 report: system stability and voltage, power and frequency control. CIGRE, Paris
6. CIGRE Report (1966) Definitions of general terms relating to the stability of interconnected synchronous machine. CIGRE, Paris
7. Barbier C, Carpentier L, Saccomanno F (1978) CIGRE SC32 report: tentative classification and terminologies relating to stability problems of power systems. Electra 56:57–67
8. IEEE TF Report (1982) Proposed terms and definitions for power system stability. IEEE Trans Power Appar Syst PAS-101(7):1894–1897
9. IEEE/CIGRE Joint Task Force on Stability Terms and Definitions (2004) Definition and classification of power system stability. IEEE Trans Power Syst 19(2):1387–1401
10. State Grid Corporation of China (SGCC) (2006) Code for calculation of power system security and stability of state grid corporation of China. State Grid Corporation of China (SGCC), Beijing
11. State Grid Corporation of China (SGCC) (2006) Preparation instructions for code for calculation of power system security and stability of state grid corporation of China. State Grid Corporation of China (SGCC), Beijing
12. Huadong S, Yong T, Shiying M (2006) A commentary on definition and classification of power system stability. Power Syst Technol 30 (17):31–35

Chapter 2
Introduction to Voltage Stability

2.1 History and Current Situation of Voltage Stability Research

Voltage stability is a branch of power system stability that has developed late. In general, the research on voltage stability can be divided into three stages. In Stage I from 1940s when the first voltage stability criterion $\mathrm{d}Q/\mathrm{d}U$ [1] was put forward by Malkovich, a Russian scholar, to the mid-1970s, insufficient attention was been paid to voltage stability. In Stage II from the late-1970s to the mid-1980s, attention was paid to the research on static voltage stability. In Stage III from the late-1980s to the present day, a comprehensive research has been initiated based on the dynamic mechanism discussion which takes modern analysis methods, chaos and catastrophe theory and differential geometry in nonlinear dynamic system as theoretical tools [2–4].

The research on voltage stability is performed often from the relativity or correspondence of the voltage stability to the rotor angle stability with clear physical concepts. In consideration of the long interval between recorded voltage collapse accident and initial failure, researchers generally believe that the voltage stability falls into the static category. Therefore, the mechanism of voltage collapse is researched from the static view, and a large number of analysis methods based on power flow equation or extended power flow equation are proposed. In the early stage, it was generally believed that the rotor angle stability was proposed to research the synchronous operation of generators under various conditions. However, voltage stability becomes a problem once the power system is unable to meet the reactive power demand of the loads. Although static research method cannot well reveal the physical essence of voltage stability and the instability mechanism, it plays an important role in obtaining the limit transmission power of a power grid and guiding the production, dispatching and operation, and also lays a foundation for the later research on dynamic analysis method.

With the deepening of research in various fields, the academic circle gradually realizes that the voltage stability is actually quite complicated. On the one hand,

© Science Press 2021

Y. Tang, *Voltage Stability Analysis of Power System*, Power Systems,
https://doi.org/10.1007/978-981-16-1071-4_2

voltage instability and rotor angle instability are two typical manifestations of instability of the nonlinear dynamic system, which is an important part of stability of a power system. A large number of stability disruption accidents show that in many cases, voltage instability and rotor angle instability are interrelated and interactive; and, it is difficult to distinguish voltage instability and rotor angle instability accurately. Therefore, the mathematical model for researching rotor angle stability of a power system should be consistent with the mathematical model for researching voltage stability. However, the research on specific dynamic phenomenon of voltage stability mainly intends to reveal the influencing factors and main features directly related to voltage stability, which requires special attention to the element model related to voltage stability. Artificially distinguishing the voltage instability and rotor angle instability may result in one-sided conclusions. On the other hand, people gradually realize the necessity of dynamic research. The power system is a non-linear dynamic system. Stability itself falls into the dynamic category. Voltage stability or voltage collapse is also a dynamic process in essence. The influence of dynamic elements of the system cannot be fully considered by the static analysis methods. Therefore, it is impossible to further research the mechanism of voltage instability and its evolution. In addition, when the power system suffers a short circuit or other types of large disturbance, the nonlinear characteristics of power system must be retained in the mathematical description of dynamic behaviors of power system in order to truly reveal the development mechanism of voltage instability.

So far, although many achievements have been made in the research on voltage stability, there is still a lack of comprehensive understanding of the essence of voltage stability and the research methods and theories are inadequate and incomprehensive compared with the mature rotor angle stability. A great deal of thorough and meticulous researches should be still performed by electric power workers with respect to the mechanism of voltage instability, the mathematical model and method for voltage stability analysis, the voltage stability index and voltage stability control, and the relations between voltage stability and rotor angle stability [3–7]

2.2 Voltage Instability Accident and Its Characteristics

Since the late-1970s, many massive voltage collapse accidents have occurred successively to power systems in the world [8–21], as shown in Table 2.1. The common characteristics include the suddenness and concealment of such accidents. It is difficult for operators to identify the voltage collapse accidents and timely take emergency control measures in the initial stage of such accidents. Once the voltage collapse occurs, it is difficult to recover the normal voltage. It often takes several hours or even more than ten hours to recover normal power supply, resulting in huge economic loss and social impact. Such serious consequences have aroused the general concern of the international power industry on voltage stability.

Many power grid collapse accidents show that rotor angle instability and voltage instability often occur in the same time frame and it is difficult to distinguish

Table 2.1 Massive power outage accidents related to voltage collapse since 1970s over the world

No.	Country	Time of occurrence	Accident	Power outage scale/MW	Power failure time
1	U.S.A	1977-07-13	Massive power outage in New York City	5868 (100% load)	25 h 59 min
2	France	1978-12-19	Massive power outage in France	29,000 (75% load)	8 h 30 min
3	Belgium	1982-08-04	Massive power outage in Belgium	2400	1 h 28 min
4	Canada	1982-12-14	Massive power outage in Quebec	15,473	5 h 30 min
5	Sweden	1983-12-27	Massive power outage in Southern Sweden	11,400 (67% load)	5 h 20 min
6	France	1987-1-12	Massive power outage in Western France	1500	6–7 min
7	Japan	1987-07-23	Massive power outage in Tokyo	8168	3 h 21 min
8	Canada	1989-03-13	Massive power outage in Quebec	9450	2 h 45 min
9	U.S.A	1996-07-02	WSCC disturbance in the United States	1.5 to 2 million consumers	1.5–3 h

rotor angle instability and voltage instability accurately according to phenomenon. Defining the distinction and connection between voltage instability and rotor angle instability and analyzing the characteristics of voltage collapse accident are the basis for analysis of voltage collapse mechanism and research on voltage instability criterion.

Some common characteristics of voltage collapse phenomena can be identified by summary of voltage collapse accidents in power system. Analyzing the characteristics of voltage instability or collapse has positive reference and guiding significance for improving the safety and stability level of a power grid and also provides reference for the principles and standards of voltage stability evaluation. The characteristics of massive voltage collapse accidents on power grids at home and abroad are summarized as follows:

(1) Before the voltage collapse, the system is often in a heavy-load operation state, the reserve capacity for operation of the system is tight (especially for reactive power operation), and the power flow of transmission line approaches to the maximum power limit.

(2) The causes of voltage collapse may be different: progressively increase of system loads; massive sudden disturbance; loss of generator set; heavy loads on the line; misjudgment and misoperation of operators in the process of dealing

with abnormal working conditions, thus resulting in propagation of accidents. Sometimes, a small disturbance may also result in propagation of accidents and eventually lead to voltage collapse.

(3) When it is difficult to transmit reactive power from adjacent area due to heavy load line in the system, voltage collapse may occur if the receiving-end system continuously requires the support in increase of reactive power due to load increase or element failure.

(4) Under low voltage, the line distance protection is activated to trip the parallel transmission lines in succession; the generator excitation limiter is activated, thus causing cascaded tripping of the generators; under the low voltage, OLTC is activated to recover the load on secondary side and make the primary system voltage further dropped. Those characteristics are important mechanisms of voltage collapse.

(5) Voltage collapse is generally characterized by slow voltage attenuation, which is the result of the accumulation for actions and interactions of many voltage control devices and protection systems. The voltage collapse process may last on the order of several minutes, and some voltage collapse may, under dynamic condition, last on the order of a few seconds. Such accidents generally result from unfavorable load components (induction motor) or DC converters. The time frame of voltage instability is the same as that of rotor angle instability. In many cases, the phenomena of voltage instability and rotor angle instability are coupled with each other.

(6) The voltage collapse may be deteriorated due to excessive use of shunt capacitor compensation. The most effective reactive power compensation can only be achieved through selection and coordination of shunt capacitor, static var compensator and synchronous condenser.

2.2.1 Overview of Typical Voltage Collapse Accidents

Although the power grid structures are relatively strong, the safety and stability measures are relatively perfect and the transmission technologies are relatively advanced in Europe, North America, Japan and other developed regions, the occurrence of a voltage collapse accident on the whole network is still inevitable. After the accident, various power companies took various positive countermeasures from planning to operation, thus providing valuable experience for the planning, design, dispatching and operation of large power grids. At present, the interconnected power grid in China has entered the era of large power grid and large units. The interconnected power grid in China has begun to take shape. The lessons learned from the foreign massive power grid collapse accidents have given profound enlightenment to the development of the Chinese power grid [21–23].

1. Massive power outage accident on power system in Sweden

On December 27, 1983, a massive power outage accident occurred on the power system in southern Sweden. The accident resulted from the fault of a circuit breaker on a 400 kV bus in a substation at Hamra about 60 km from northwest of Stockholm. Such fault firstly resulted in the separation of the substation from the system and the tripping of two 400 kV transmission lines. After the power outage in this area, the constantly increasing load and current on the transmission line from northern Sweden resulted in constant decrease of the voltage in the Stockholm. Although no important lines were immediately tripped in the following 53 s, all main transmission lines from northern Sweden to southern Sweden were tripped due to heavy load in the following 2 s. After 1 s, the tie line between Swedish and Norwegian lines was separated. During this period, the frequency of the power grid in northern Sweden gradually increased, while the frequency of the power grid in southern Sweden constantly decreased. In addition, all generator sets in this area were tripped due to massive power loss in southern Sweden.

Figure 2.1 shows the voltage curve during this period recorded at a 132 kV substation in Sealand, Denmark. At Point ①, two 400 kV transmission lines in the northern Hamra, Sweden were tripped; at Point ②, four 400 kV main transmission lines, seven 220 kV transmission lines and all 132 kV transmission lines in northern-southern Sweden were tripped, and the voltage curve begun to fluctuate due to such influence; at Point ③, the tie line between Sweden and Norway was out of operation, and the

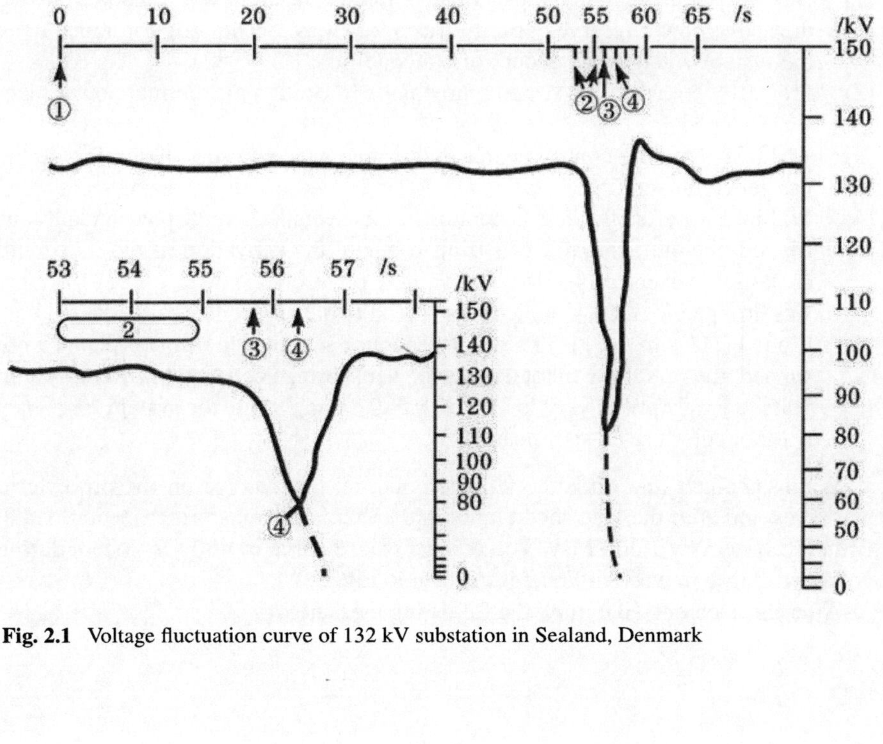

Fig. 2.1 Voltage fluctuation curve of 132 kV substation in Sealand, Denmark

voltage in Sealand begun to drop; at Point ④, one 400 kV transmission line and two 132 kV transmission lines between Sweden and Sealand were tripped, and the voltage dropped to about 0.5p.u. It can be seen from the figure that when t = 53 s, voltage collapse occurred in Sealand within 3 s after the voltage collapse in southern Sweden. Fortunately, when t = 56.3 s, the tie line between the two places was tripped owing to the activation of protection. Therefore, the voltage and frequency of power grid in Sealand were successfully recovered without much impact. While, the load under power outage on the power grid in southern Sweden was about 11,400 MW, accounting for 67% of the total system load, with an impact time of about 7 h.

After the accident, Vattenfall concluded and analyzed the causes of the accident: the fault of substation resulted in the voltage drop and the decrease of system load; however, the activation of OLTC resulted in increase of system load, load increase on transmission line from northern Sweden to southern Sweden, and gradual voltage drop and current increase on main power grid and distribution network; the activation of transmission line protection resulted in cascaded tripping of the line, thus affecting the power supply in surrounding countries and regions. Moreover, the excessive reactive power shortage in the power grid eventually resulted in the occurrence of voltage collapse, with huge losses.

2. Massive Power Outage in Western France

At 10:55 a.m. on January 12, 1987, a voltage collapse accident occurred on the power grid in western France. The sequence of events is as follows:

(1) At 10:30 a.m., the national peak load before the accident was 58,000 MW and the active power reserve capacity was 5,900 MW, accounting for about 10%; the system voltage was under normal condition.
(2) From 10:55 a.m. to 11:41 a.m., three units in Cordemais thermal power plant were tripped.
(3) At 11:28 a.m., the control center in this area gave an order of starting up the gas generator set.
(4) Within 13 s after the 3rd generator in Cordemais thermal power plant was tripped, the 4th generator was tripped due to the activation of the maximum excitation current protection; the voltage on 400 kV power grid in this area was dropped to 380 kV, which was spread to the adjacent area.
(5) From 11:45 a.m. to 11:50 a.m., 9 generator sets in other power plants were tripped, thus resulting in total active power shortage of 9,000 MW. The system voltage was kept at a very low level (0.5–0.8 p.u.), while the system frequency was not deteriorated significantly.

Within about 6 min after the voltage collapse, the voltage on the power grid was recovered after the load shedding of 440 kV/225 kV transformer feeder load in Britanny system by 1,500 MW. The voltage record curve of 400 kV system during voltage collapse in western France is shown in Fig. 2.2.

After the accident, EDF took the following measures:

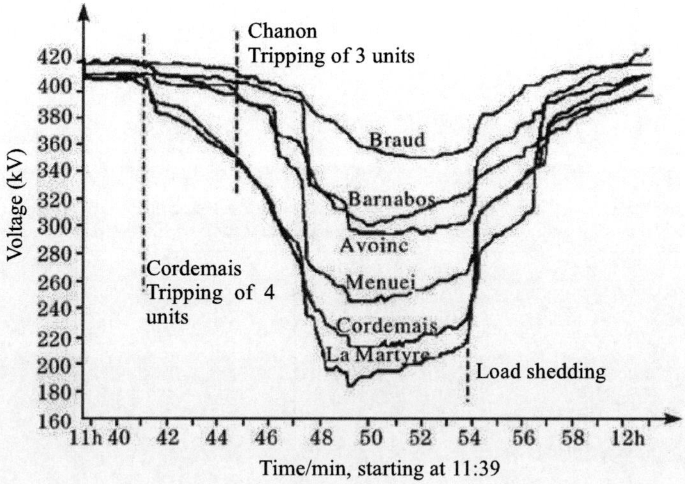

Fig. 2.2 Voltage record curve of 400 kV system during voltage collapse in Western France

(1) They carried out automatic control of secondary voltage, automatic switching of shunt capacitors, and automatic control of reactive power output for regional power generation equipment;
(2) They set real-time voltage instability index;
(3) They took emergency control measures such as automatic locking of OLTC and remote load shedding.

Accident analysis

The accident occurred due to the loss of the generator system. During the accident, the frequency of the system was not deteriorated significantly, while the voltage was dropped to a very low level. Since it was winter in France at the time of the accident, the load power has higher voltage sensitivity. With the voltage drop, the active power consumed by the load also decreased, so that the system could be operated at lower voltage under the condition of massive active power shortage.

3. Massive power outage accident on power system in Tokyo, Japan

On July 23, 1987, a massive power outage accident occurred in Tokyo, Japan, which was a typical voltage collapse accident. At noon on that day, it was hot in Tokyo, with temperature up to 36°. The customer load was increasing constantly. At 13:00 p.m., the maximum increase rate reached 400 MW/min, the total load demand was about 39,300 MW, and the load of whole system approached the limit level. Due to the increase rate being too fast, the reactive power supply controlled by voltage quality controller (VQC) and automatic voltage regulator (AVR) in the system could not be compensated in time, thus resulting in constant drop in voltage on two 500 kV buses in the power grid of western Tokyo. Before the voltage collapse, the voltage was as

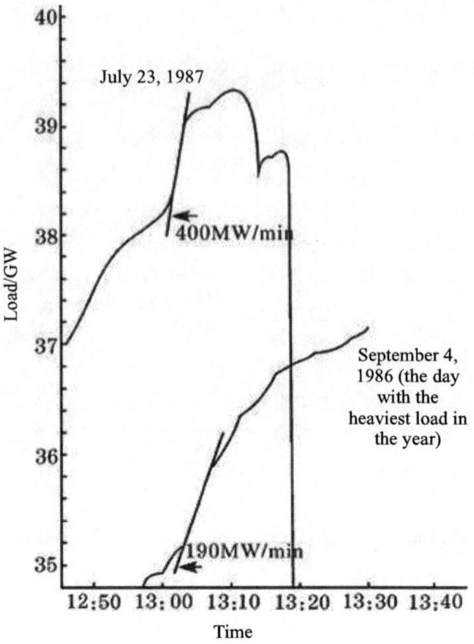

Fig. 2.3 Load variation curve on the day of accident

low as about 370 kV. The activation of bus relay protection resulted in bus tripping. In addition, four 275 kV transmission lines and four 275 kV/66 kV transformers were also out of operation. In consideration of the chain structure of 275 kV and below network in the Tokyo power grid, TEPCO cut back the load by 8,168 MW (accounting for about 21% of the total load) at the terminal of the network at 13:19 p.m. to prevent the accident from expanding, which affected about 2.8 million consumers. It was not until about 16:40 p.m. that the power supply of the system was fully recovered to normal. The power outage lasted up to 3 h and 21 min. The load record curve on the day is shown in Fig. 2.3, and the voltage variation curve of four main 500 kV substations in Tokyo power grid is shown in Fig. 2.4.

After the accident, TEPCO summarized five causes:

(1) Before the accident, the load increased too fast and exceeded the expected value. The lack of understanding of the load characteristics was the main cause of the accident;
(2) Due to the lack of monitoring means for 500 kV bus voltage, the power grid operation managers were unable to understand the operation situation in time;
(3) Before the accident, TEPCO failed to take preventive measures against voltage instability, thus resulting in the lack of operators' knowledge and countermeasures for such accidents;
(4) The shunt reactor and shunt capacitor on the power grid were controlled by VQC. The defect of the control strategy resulted in the spreading of the accident;

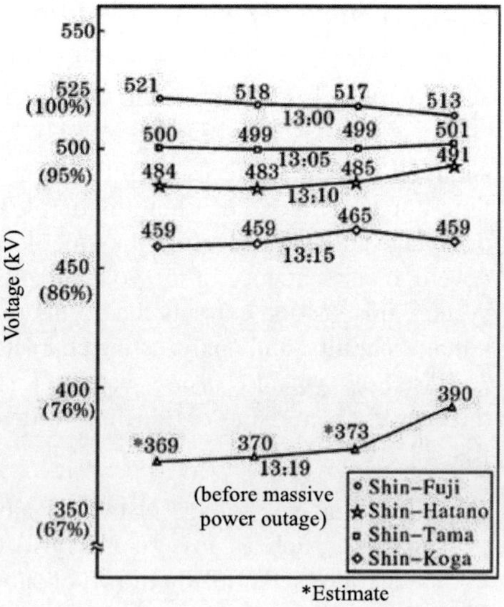

Fig. 2.4 Voltage variation curve of four main 500 kV substations

(5) The uneven distribution of power plants in Tokyo resulted in power flow in the grid frame from the eastern power supply center to the western load center being too heavy, and the shutdown of the two power plants in the central and western areas on the same day for some reason resulted in even larger probability of voltage collapse.

4. Impact of voltage collapse accident on voltage quality control countermeasures

The countermeasures concluded against massive power outage accidents in Sweden, France and Japan are summarized as follows:

(1) Vattenfall recognized the operation problem of on-load tap changer of the transformer and strengthened the coordinated voltage quality control of the power grid;
(2) EDF has paid attention to the coordinated voltage quality control of the power grid, and strengthened the fast cut back measures;
(3) TEPCO has strengthened the safety monitoring of voltage stability and the prevention, and also enhanced the dynamic voltage support for the receiving-end power grid;

In general, various power companies have formulated corresponding countermeasures for the voltage quality planning and control mainly from the perspective of operation. Those countermeasures involve the safety evaluation of voltage stability, which is the premise and foundation of carrying out various tasks.

2.2.2 *Characteristics of Voltage Instability*

1. Characteristics of transient voltage instability

Analyzing the characteristics of transient voltage instability from the perspective of time-domain simulation is helpful for researchers to better understand the mechanism of voltage collapse. Based on the time-domain simulation of the real system, and the development sequence of the voltage collapse process for the typical receiving-end system, this section extracts the events that have important impact on the transient voltage stability, and analyzes the characteristics of transient voltage instability.

Based on typical transient voltage instability accidents and simulated analysis, the main characteristics of transient voltage instability can be summarized as follows [24–27]:

(1) The transient voltage instability almost is always resulted from large disturbances, such as loss of heavy-load transmission lines, loss of large-sized generator sets, full shutdown of substations, and substantial increase in load.
(2) Before the transient voltage instability, the system is often in a heavy-load operation state, the reserve capacity for operation of the system is tight (especially for reactive power operation), and the power flow of transmission line approaches to the maximum power limit.
(3) In the transient process, the dynamic behavior of various load elements and control devices intend to recover the load power, at least to a certain extent, especially the induction motor, the load of which has large impact on the transient voltage stability
(4) The severely insufficient voltage support of the receiving-end system easily results in transient voltage instability. At this time, the system cannot input reactive power from adjoining area due to heavy load. Any additional required active/reactive power support may lead to voltage drop and even transient voltage collapse. This problem becomes even more severe when shunt capacitors are excessively used.
(5) In the receiving-end system with multi-infeed DC, the voltage of most nodes in the AC network constantly dropped due to the increase of reactive power demand after large disturbance at the inverter side, which easily leads to consequential multi-circuit DC commutation failure and, in severe cases, transient voltage collapse of the system.
(6) The transient voltage instability does not always occur in isolation. The transient rotor angle instability and transient voltage instability are often intertwined. In general, one of the two instability modes dominates. However, both modes are not easy to distinguish.

2. Characteristics of medium and long-term voltage instability

The medium and long-term voltage instability phenomenon is characterized by the following aspects: After transient process, the system can keep the stability, while the local area keeps low voltage. With the regulating effect of OLTC, reactive power demand is increased, making the voltage level to constantly and slowly drop. The duration of constant voltage drop is generally long, ranging from a few minutes to several tens of minutes, which is depending on the behavior of the system voltage regulating equipment and the characteristics of load recovery.

Based on typical medium and long-term voltage instability accidents and simulated analysis, the main characteristics of medium and long-term voltage instability can be summarized as follows [24–27]:

(1) With the constant increase of load, a large amount of power transmitted from a remote power supply causes the system to be in a voltage unstable state;
(2) When medium and long-term voltage instability occurs, the system is generally under heavy load state, and a large amount of power is transmitted from a remote power supply, accompanied by sudden large disturbance;
(3) OLTC is activated under low voltage to recover the load at secondary side, which makes the primary system voltage to drop even further;
(4) In the constant low voltage process, the generator excitation limit protection is activated, thus resulting in, when recovering to the rated value, sudden reduction of reactive power in the system and sudden drop of voltage in the system;
(5) The recovery characteristics of load further aggravate the drop of system voltage;
(6) The characteristics of the last stage are similar to the characteristics of transient voltage instability, including the stalling of induction motor and the activation of relay protection device, etc.

2.3 Mechanism of Voltage Stability

2.3.1 Voltage Stability of Simple Pure Resistor Element Circuits

Taking the simple pure resistor element circuit shown in Fig. 2.5 as an example, if $R_s = R_L$, then

Fig. 2.5 Circuit diagram of pure resistor

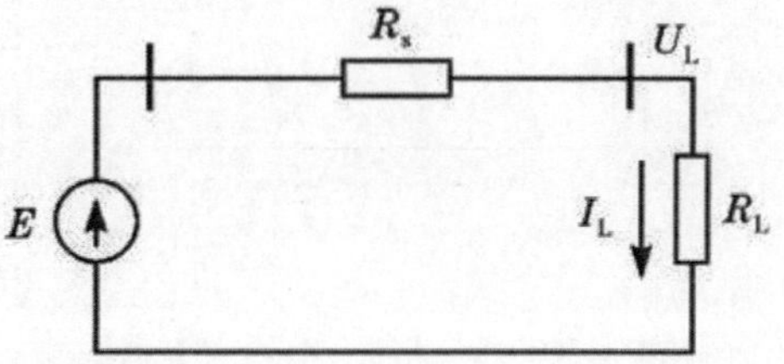

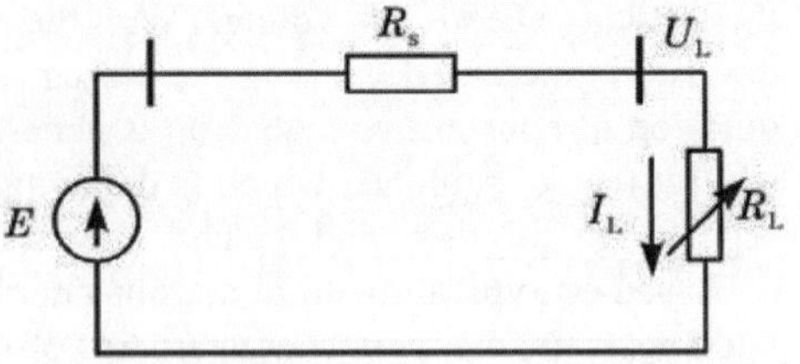

Fig. 2.6 Schematic diagram of pure resistor element circuit with load variation

$$\begin{cases} P_{Lmax} = E^2/(4R_s) \\ U_L = E/2 \end{cases} \tag{2.1}$$

Assuming $E = 10$ V and $R_s = 1\Omega$, it can be seen from Eq. (2.1) that

$$\begin{cases} P_{Lmax} = \frac{10^2}{4\times 1} = 25(w) \\ U_{Lcr} = 10/2 = 5(V) \end{cases} \tag{2.2}$$

Assuming that the pure resistor load is variable, for which the circuit is shown in Fig. 2.6.

The following equation can be obtained:

$$\begin{cases} I_L = \frac{E}{R_s+R_L} \\ U_L = \frac{E}{R_s+R_L} \times R_L \\ P_L = I_L U_L \end{cases} \tag{2.3}$$

If R_L takes different values, the results as shown in Table 2.2 can be obtained.

The curve of power, voltage and current with the variation of pure resistor load R_L is shown in Fig. 2.7.

Thus, when $R_L = R_s = 1\Omega$, the operating point of the system exactly corresponds to the peak point of the power-voltage curve, which indicates that P_L reaches the maximum value. The curve is shown in Fig. 2.8.

Table 2.2 Variation of system parameters with load R_L variation

R_L/Ω	I_L/A	U_L/V	P_L/W
49	0.2	9.8	1.96
24	0.4	9.6	3.84
9	1.0	9.0	9.0
4	2.0	8.0	16.0
1	5.0	5.0	25.0
0.25	8.0	2.0	16.0
0.11	9.0	1.0	9.0
0	10	0	0

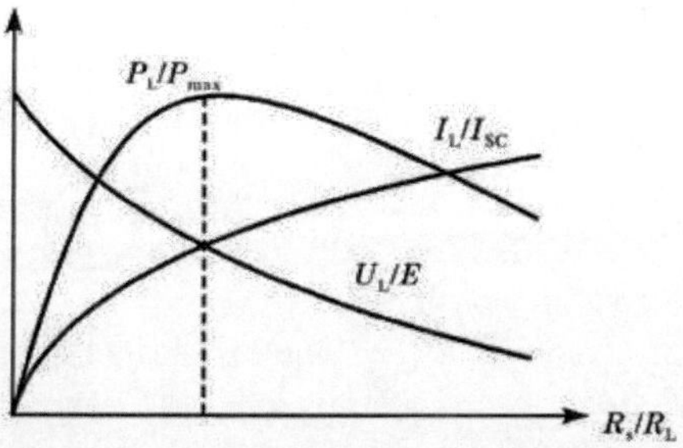

Fig. 2.7 Variation curves of system parameters

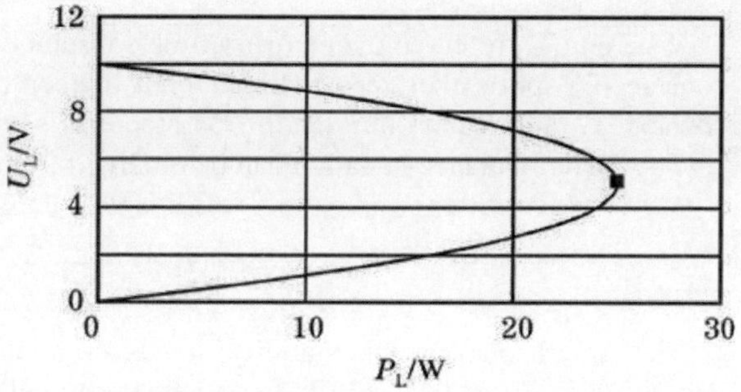

Fig. 2.8 Power-voltage curve

It can be seen that due to the presence of impedance in the power supply, the load voltage will drop when the load attempts to obtain more power by increasing the current. Once the impact of voltage drop on load power is more significant than that of the current increase, the load cannot obtain more power. If the power required by the load exceeds the maximum power that the system can provide and the load attempts to obtain more power by continuously increasing the current, voltage collapse will occur. It is generally believed that in AC systems, there is a close relationship between voltage control and reactive power. Despite the lack of reactive power, this simple DC system example also has the main characteristics of voltage instability in AC systems, which indicates that reactive power is not the only cause of voltage instability.

2.3.2 *Mathematical Description of Voltage Stability of Simple Pure Resistor Element Circuits*

For the simple DC circuit shown in Fig. 2.9, assuming that the final steady state power of the load remains unchanged under the action of some automatic regulating device, then the final steady state power can be expressed by first-order conductance model of Eq. (2.4). Subscript 0 corresponds to the initial time value.

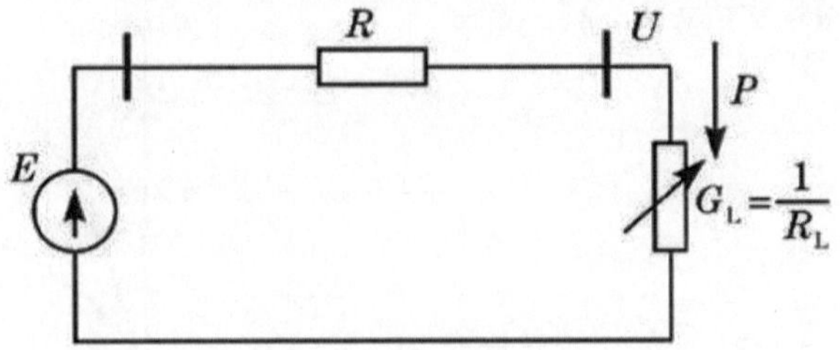

Fig. 2.9 Simple DC circuit diagram

$$T_G \dot{G}_L = P_0 - U^2 G_L \tag{2.4}$$

Wherein: T_G refers to the recovery time constant of conductance. When the time constant T_G approaches zero, the load tends to keep constant power; when the time constant T_G approaches infinity, the load tends to keep constant resistance.

The current balance equation can be obtained from Fig. 2.9:

$$0 = UG_L - \frac{E - U}{R} \tag{2.5}$$

The small disturbance stability in differential algebraic equations (DAE) composed of Eqs. (2.4) and (2.5) can be determined by applying the characteristic analysis method. Through linearizing the Eqs. (2.4) and (2.5) near the balance point, it can be obtained that

$$\begin{aligned} T_G \Delta \dot{G}_L &= -2U_0 G_{L0} \Delta U - U_0^2 \Delta G_L \\ 0 &= G_{L0} \Delta U + U_0 \Delta G_L + \frac{1}{R} \Delta U \end{aligned} \tag{2.6}$$

The characteristic equation can be obtained from Eq. (2.6):

$$T_G \Delta \dot{G}_L = \frac{U_0^2 (RG_{L0} - 1)}{RG_{L0} + 1} \Delta G_L \tag{2.7}$$

From formula (2.7), it can be seen that:

When $RG_{L0} - 1 < 0$ or $R_{L0} > R$, the system remains stable;

When $RG_{L0} - 1 > 0$ or $R_{L0} < R$, the system is not stable;

It can be seen from the *PV* curve shown in Fig. 2.10 that when $RG_{L0} - 1 < 0$ or $R_{L0} > R$, the system is kept in high voltage and low current state, which corresponds to the upper half of the *PV* curve and is stable; when $RG_{L0} - 1 > 0$ or $R_{L0} < R$, the system is kept in low voltage and high current state, which corresponds to the lower half of *PV* curve and is unstable.

It can be seen from Figs. 2.10 and 2.11 that at the balance point of the upper half of the *PV* curve, the disturbance that causes decrease of G_L (increase of R_L) will result in decrease of the power absorbed by the load. Regulation will make G_L increase (R_L decrease) and recover to the initial value. Similarly, the disturbance that

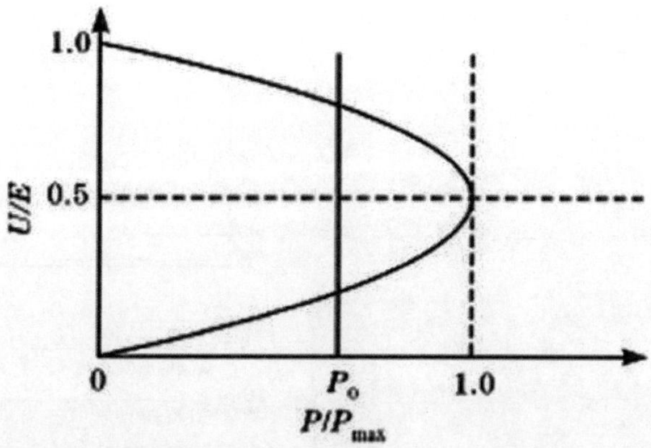

Fig. 2.10 Power-voltage characteristics of the system

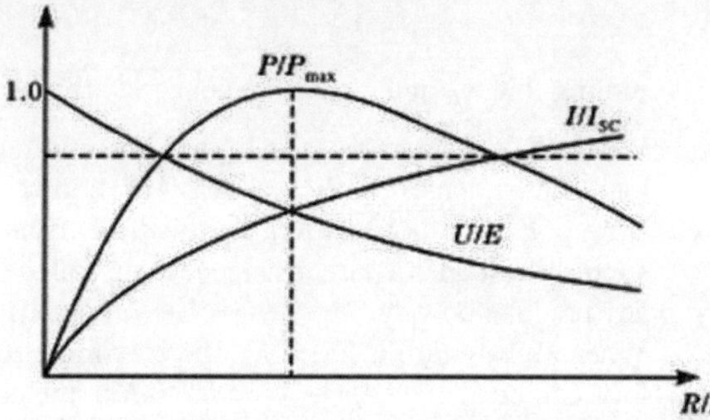

Fig. 2.11 Function relationship of load voltage, current and power to load demand

causes an increase of G_L (decrease of R_L) will result in an increase of the power absorbed by the load. Regulation will make G_L decrease (R_L increase) and recover to the initial value. Similarly, at the balance point of the lower half of the *PV* curve, the disturbance that causes a decrease of G_L (increase of R_L) which will result in an increase of the power absorbed by the load. Regulation will make G_L further decrease (R_L increase) and recover to the balance point of the higher half of the *PV* curve. The disturbance that causes an increase of G_L (decrease of R_L) will result in a decrease of the power absorbed by the load. Regulation will make G_L further increase (R_L decrease) and recover to zero along the *PV* curve. This is a voltage collapse process.

2.3.3 Voltage Stability of AC Circuits

Voltage control is closely connected to reactive power in the HV AC power system dominated by reactance. A simple AC circuit is given as an example, and the load is expressed in constant impedance, as shown in Fig. 2.12.

According to Fig. 2.12, the following equation can be obtained:

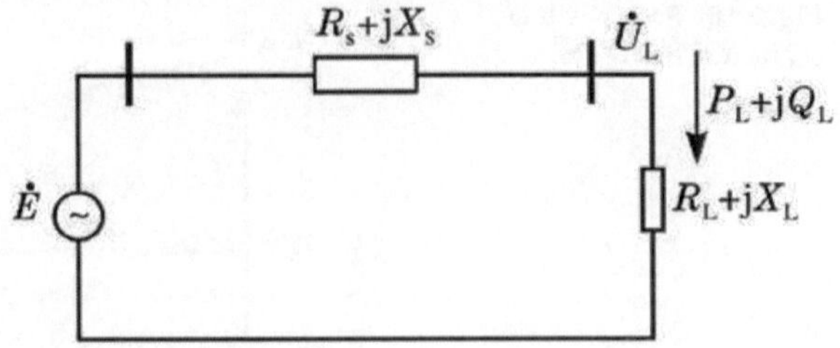

Fig. 2.12 Simple AC circuit diagram

$$\begin{cases} I_L = \frac{E}{\sqrt{(R_s+R_L)^2+(X_s+X_L)^2}} \\ P_L = \frac{R_L E^2}{(R_s+R_L)^2+(X_s+X_L)^2} \\ Q_L = \frac{X_L E^2}{(R_s+R_L)^2+(X_s+X_L)^2} \\ U_L = \frac{\sqrt{R_L^2+X_L^2}}{\sqrt{(R_s+R_L)^2+(X_s+X_L)^2}} E \end{cases} \quad (2.8)$$

Assuming that R_L and X_L can be changed arbitrarily:

(1) When $R_L = R_s$ and $X_L = -X_s$, the maximum active power consumed by the load can be achieved, $P_{Lmax} = E2/(4R_s)$; the corresponding critical voltage is $U_{Lcr} = E\sqrt{R_s^2 + X_s^2}/(2R_s)$. At this time, from the perspective of the voltage source, the load is a pure resistance load, and the power source will not generate any reactive power.

(2) When $R_L = -R_s$ and $X_L = X_s$, the maximum reactive power consumed by the load can be achieved, $Q_{Lmax} = E2/(4R_s)$; the corresponding critical voltage is $U_{Lcr} = E\sqrt{R_s^2 + X_s^2}/(2X_s)$. At this time, from the perspective of the voltage source, the load is a pure reactance load, and the power source will not generate any reactive power.

Reactive power makes the problem complicated in the AC power system. The transmission of active power is the main function of a power system, but reactive power is also an indispensable part of actual power systems. The change of reactive power in the receiving-end system will affect the maximum receiving active power and may also cause voltage instability. But reactive power is definitely not the only source of the voltage problem.

From $R_L = R_s$, $X_L = -X_s$ and $P_{Lmax} = E^2/(4R_s)$, for the AC system, the maximum active power can be achieved only when the capacitive reactance on the load side completely compensates the inductive reactance of the system. For the HV AC transmission system, $X_s \gg R_s$, under normal conditions, reactive power compensation of the load cannot fully compensate X_s. From the critical voltage $U_{Lcr} = E\sqrt{R_s^2 + X_s^2}/(2R_s)$ corresponding to the maximum active power absorbed by the load when R_L and X_s can be changed arbitrarily, if $X_s > \sqrt{3}R_s$, then $U_{Lcr} > E$.

Due to the limitation of operating voltage during normal operation of an AC transmission system, it is impossible to inject a large amount of reactive power compensation to improve the maximum receiving power. In case of system failure or load increase, the injection of reactive power compensation can effectively improve voltage level and receiving capacity. Therefore, the system needs dynamic reactive power reserve.

In conclusion, based on the physical essence and mechanism of voltage stability of power system, this book newly defines the voltage stability and voltage collapse of a power system, as shown below:

Voltage stability refers to the capacity of all buses to maintain stable voltage in a given initial operating state after the electric power system is disturbed. It depends on the system capacity to maintain/restore the balance between load demand and the power supply to the load. Its essence is that when the power provided by the system to the load increases with the increase of current, the system is in a voltage stable state; otherwise, the system is in a voltage unstable state. When the system is in a wide range of voltage instability, if the load is still continuously trying to obtain more power (active or reactive) by increasing the current, then voltage collapse may occur.

2.4 Factors Affecting Voltage Stability

2.4.1 Transient Voltage Stability

Voltage instability is the result of the combined action of system characteristics and load characteristics. When the voltage support of system operating point is strong, the load characteristics have little influence. However, when large disturbance or continuous failure occurs, the system will be in a relatively weak state, and the load characteristics will play a decisive role in determining whether the voltage is unstable or even collapsed. The duration of transient voltage instability is generally 0–10 s, which is also the time category concerned by transient rotor angle stability. The difference between transient voltage instability and transient rotor angle instability is sometimes not very obvious, and they may occur simultaneously under some specific conditions of the system. Sometimes it is difficult to determine whether the voltage collapse causes the rotor angle instability or the rotor angle instability causes the voltage collapse only from the phenomenon.

In a power system operating near the voltage stability limit, the increase of active or reactive power load, the loss of generator or shunt compensation, a drop of sending-end voltage, or the failure of transmission line may cause transient voltage instability of the system. Transient voltage instability is a system problem caused by inadequate system transmission capacity to meet the load demand, rather than the instability of a single element. Transient voltage stability is mainly related to transmission limit of transmission network, dynamic characteristics of load and voltage support of receiving-end system.

1. Transmission capacity limitation of transmission network

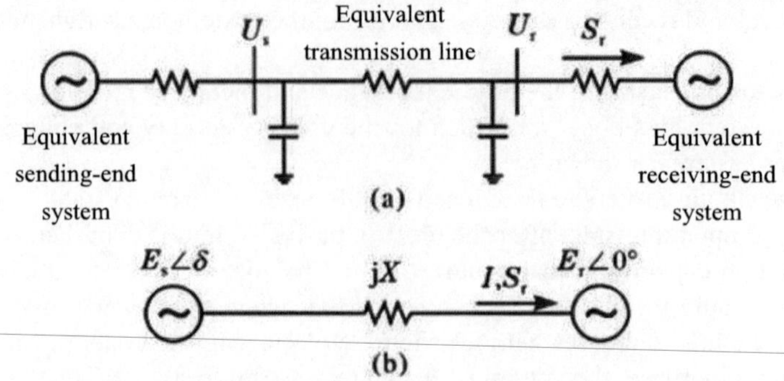

Fig. 2.13 Simple model for calculating active and reactive power transmission

The transmission characteristics of the transmission network are analyzed on the two-machine system model shown in Fig. 2.13.

For transmission lines, the active and reactive power transmitted mainly depends on the amplitude and phase angle of the sending-end and receiving-end voltage. The power expression of the receiving end is as follows:

$$P_r = \frac{E_s E_r}{X} \sin\delta = P_{\max} \sin\delta \tag{2.9}$$

$$Q_r = \frac{E_s E_r \cos\delta - E_r^2}{X} \tag{2.10}$$

(1) Active power transmission problem

From Eq. (2.9), the rotor angle corresponding to the maximum transmission power of the transmission line under ideal conditions is equal to 90°, and Fig. 2.14 is the

Fig. 2.14 Rotor angle curve of simple model

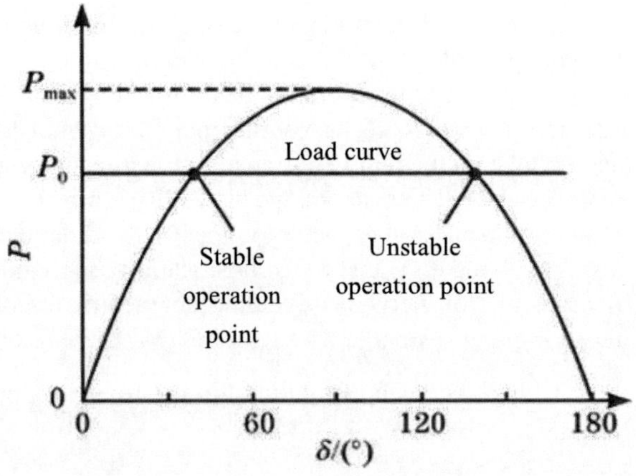

corresponding rotor angle curve. If the line loss or shunt resistive load at the receiving end is considered, the rotor angle δ corresponding to the maximum transmission power will change.

In Fig. 2.14, when the mechanical power of the generator at the sending end is constant, there are 2 intersections between the mechanical power curve and the electromagnetic power curve, of which the left intersection is the stable operating point of the system and the right intersection is the unstable operating point. To illustrate this conclusion, if the receiving-end system is large enough, it can be regarded as an infinite bus with constant voltage phase angle and speed. If the mechanical power of the generator at the sending end increases slightly, the speed of the generator will increase, and the rotor angle will increase accordingly. For the operating point on the right in Fig. 2.14, the increase of the rotor angle will further reduce the electromagnetic power, resulting in the continuous increase of the speed of the generator and the increase of the rotor angle. For the operating point on the left, the increase of rotor angle will increase the electromagnetic power, and finally match the increased mechanical power. The transmission of active power mainly depends on the rotor angle.

(2) Reactive power transmission problem

From the rotor angle curve, when the level of active power transmitted by the line is high, the corresponding rotor angle is large, and the sending and receiving ends need a large amount of reactive power. Assuming that the potentials of the generators at the sending and receiving ends of the system are equal, i.e., $E_s = E_r$, Fig. 2.15 shows the variation characteristics of reactive power corresponding to the rotor angle curve, here $Q_s = -Q_r$. The rotor angle corresponding to the limit operating point of steady state rotor angle stability is equal to 90°, and the reactive power required by the sending and receiving ends is equal to P_{max}.

When the phase angle difference between the two ends of the line is small, cos δ $\approx$ 1, so the following approximate expression is given:

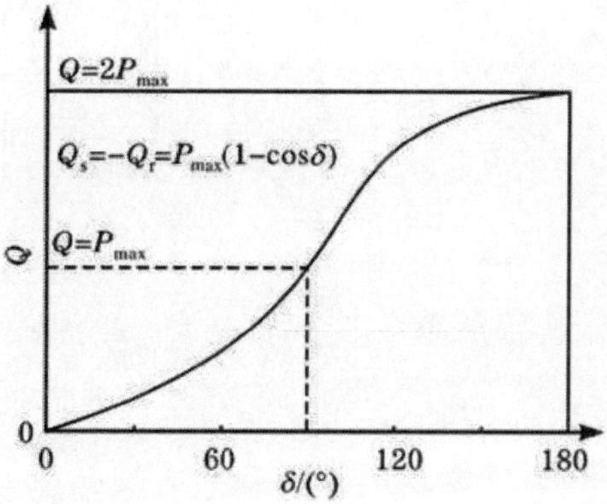

Fig. 2.15 Reactive power-rotor angle curve of simple model

$$Q_{\mathrm{r}} = \frac{E_{\mathrm{r}}(E_{\mathrm{s}} - E_{\mathrm{r}})}{X} \tag{2.11}$$

$$Q_{\mathrm{s}} = \frac{E_{\mathrm{s}}(E_{\mathrm{s}} - E_{\mathrm{r}})}{X} \tag{2.12}$$

From Eqs. (2.11) and (2.12), the transmission of reactive power mainly depends on the amplitude of voltage, and the transmission is from the end with high voltage to the end with low voltage.

It should be noted that when the power system is under heavy load condition, the transmission power and rotor angle are large, and the system will no longer have the above physical characteristics. At this time, the transmission line becomes an element that consumes reactive power, and the reactive power loss of the line is equal to the sum of the reactive power injected by the sending and receiving ends. This is very important for studying the voltage stability problem, because the voltage problem of power system usually occurs under heavy load condition.

(3) Difficulty of reactive power transmission

Difficulty of reactive power transmission: If the voltage phase angle difference between the two ends of the line is large, it is impossible to transmit reactive power through the line even if the voltage amplitude difference between the two ends is large, while long lines or heavy load may cause the voltage phase angle difference between the two ends to increase. According to the operation requirements of the power system, the voltage amplitude at each point should be kept within the range of (1 ± 5%) p.u., which further increases the difficulty of reactive power transmission. Therefore, compared to active power transmission, reactive power cannot be transmitted over a long distance. In addition, in order to reduce the active and reactive power losses of the transmission line, the long-distance transmission of reactive power should also be reduced. The active and reactive power losses on the impedance of the transmission line are I^2R and I^2X respectively, wherein, I^2 is

$$I^2 = \dot{I} \cdot \overset{*}{I} = \left[\frac{P - \mathrm{j}Q}{\overset{*}{U}}\right]\left[\frac{P + \mathrm{j}Q}{\dot{U}}\right] = \frac{P^2 + Q^2}{U^2} \tag{2.13}$$

so,

$$P_{\mathrm{loss}} = I^2 R = \frac{P^2 + Q^2}{U^2} R \tag{2.14}$$

$$Q_{\mathrm{loss}} = I^2 X = \frac{P^2 + Q^2}{U^2} X \tag{2.15}$$

From Eqs. (2.14) and (2.15), it can be seen that in order to minimize the transmission power loss of the line, it is necessary to minimize the reactive power transmitted by the line and maintain a high voltage level under the condition that the active power transmitted by the line is constant, so as to reduce the reactive power loss and help to improve the voltage stability of the power system. It can be seen that reactive power transmission is inefficient under the condition of large active power transmission. A large voltage amplitude drop is required at both ends of the line. In addition, reactive power transmission will increase the active and reactive power losses of the line. Wherever possible, reactive power supply points should be close to reactive power load points, and reactive power should be balanced in layers and zones as far as possible.

(4) Transmission capacity limitation of transmission network

The premise of stable operation of power systems is that there must be a stable balance point, and the most important type of transient voltage instability scenario corresponds to the situation that the balance point no longer exists due to changes in system parameters. In actual operation, transient voltage instability often occurs when the load demand is higher than the maximum transmission power of the system. For example, a large disturbance (such as disconnection) occurs on a heavy-load transmission line, which may cause the network characteristics to change sharply. After disturbance, the maximum power transmitted by the transmission line is reduced compared to that before disturbance, and the power demand of the load cannot be met. In this case, the system will suffer transient voltage instability.

In general, this kind of problem can be studied by steady state rotor angle stability analysis method, which is essentially based on the power flow analysis, and the limit transmission capacity of power grid is taken as the critical point of voltage collapse. In most demonstrations or industrial practice, the *PV* curve—the curve of relation between voltage and active power—is commonly used to determine the voltage stability of the system, as shown in Fig. 2.16. *PV* curve analysis indicates the degree to which the power level of the transmission section leads to the voltage collapse of the whole system by establishing the relation curve between *P* (power transmitted by the transmission section) and *U* (voltage of key nodes).

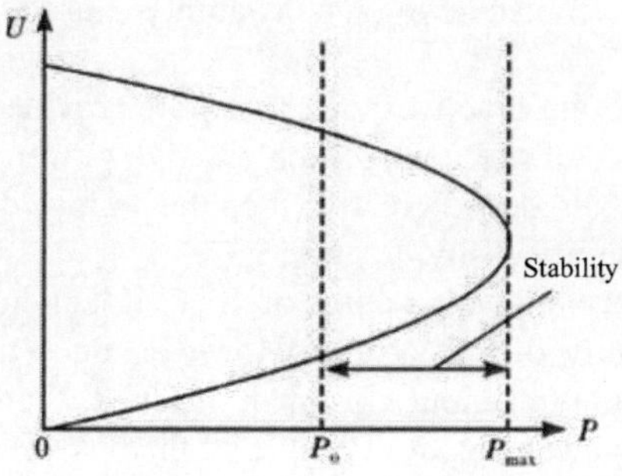

Fig. 2.16 *PV* curve of transmission capacity limitation of transmission network

2. Dynamic characteristics of the load

In general, the load characteristics in transient processes are more sensitive to voltage changes than those in steady state. The dynamic behavior of all kinds of load components and control devices is to recover the load power, at least to a certain extent. The load that affects the transient voltage stability is mainly the induction motor.

As the bus voltage of the load drops, the reactive power absorbed by the load from the system may worsen the regional reactive power balance of the system, forming a positive feedback mechanism of voltage drop. For example, if the voltage drop of a load node is serious (during the slow removal of short-circuit fault), the reactive power absorbed by the induction motor will rise, which may cause voltage collapse, unless a protection device or AC contactor trips the induction motor. After the fault is removed, if some induction motors cannot accelerate any more, these motors will tend to stall and other adjacent motors will also stall. If the instability of induction motor only leads to a voltage drop of individual nodes, the stability of the induction motor cannot be regarded as transient voltage stability, and it shall be called as the instability of induction motor. If the instability of induction motor eventually leads to serious voltage drop/collapse of the system, it can be considered as the transient voltage instability of the system. In transient voltage stability simulation study, induction motor must be expressed as a dynamic load model.

3. Insufficient voltage support of receiving-end system

(1) Transient voltage instability caused by the generator

When a voltage instability event occurs, the power system is generally under heavy load state, and a large amount of power is transmitted from a remote power supply, accompanied by sudden large disturbance. Loss of excitation or tripping of the generator in the receiving-end system may lead to a large amount of reactive power shortage in the receiving-end system, making the remote generator have to provide the required reactive power. Since it is inefficient for remote generators to provide reactive power to load areas, the generation and transmission systems may no longer be able to support reactive power demand and reactive power loss of the system, and the system voltage drops rapidly, resulting in transient voltage instability.

(2) Transient voltage instability caused by shunt capacitor and SVC

Since the reactive power support provided by the shunt capacitor varies with the square of the voltage, the reactive power support provided by the shunt capacitor will be reduced greatly when the voltage drops, which will lead to transient voltage instability easily.

Dynamic regulation of SVC is conducive to improving the transient voltage stability of the system. However, once the maximum output susceptance or the maximum output current is reached, there is a square relation between the reactive power output of SVC and the bus voltage, and its effect is equivalent to that of the shunt capacitor. At this time, SVC loses the ability of reactive power regulation, its ability to provide reactive power support is reduced greatly, and it is also easy to cause transient voltage instability.

(3) Transient voltage instability caused by HVDC

HVDC has unique advantages in long-distance and large-capacity transmission. However, since the DC converter has to absorb a large amount of reactive power (50–60% of the DC active power), the transient voltage stability of AC/DC systems after large disturbances face a severe test, especially the transient voltage stability of weak AC systems connected with DC systems.

When a serious fault occurs in HVDC systems, such as DC bipolar blocking, it will lead to a large transfer of power flow. Once the voltage drops, the reactive power demand of the induction motor load in the receiving-end system may increase greatly. Meanwhile, the reactive power compensation provided by the shunt capacitor will start to decrease, and the whole network voltage may further deteriorate, leading to transient voltage instability. The rapid recovery of DC power after removal of AC system fault is helpful to alleviate the imbalance of active power of the AC system, but the rapid power recovery may cause subsequent commutation failure, leading to transient voltage instability of the AC system. This feature is more obvious for multi-infeed AC/DC power systems.

2.4.2 Medium and Long-Term Voltage Stability

1. Overexcitation limit of the generator

The dynamic state of generator excitation system and its limiting link are important factors affecting the voltage stability of the system. Therefore, the influence of generator regulating systems should be considered accurately when the voltage stability limit of the system is calculated. The over excitation limit (OEL) of the generator is determined by the over-heating load capacity of the generator, and the over-current time of the rotor is related to the peak value of the excitation current.

Excitation regulator is the main means to regulate voltage in the power system, but the regulating range is limited to a certain extent. Once the excitation regulation of a generator in the system reaches the maximum limit value, due to the thermal capacity limitation of the excitation winding, when the over excitation capacity of the unit reaches the set operating time, the over excitation limiter will reduce the excitation current to the rated value, resulting in a sudden reduction of reactive power in the system and a sudden drop of voltage in the system. Since other generators are generally in a state near the limit at this time, the sudden excitation reduction of the generator may cause a chain reaction of other generators, thus causing voltage instability of the system.

2. Operation characteristics of OLTC

The regulating action of OLTC is considered to be one of the main mechanisms leading to voltage instability. When the system voltage drops for a period of time, OLTC restores the voltage on the load side and the power at the same time, which is not conducive to voltage stability. Meanwhile, the constant adjustment of OLTC also makes the voltage on the primary side drop continuously, which may eventually lead to voltage collapse.

For a load with constant power or power close to constant power, the regulating action of OLTC has less influence on the voltage on the primary side. For a load with constant impedance or constant current, the regulating action of OLTC at low voltage makes the load power on the secondary side recover and has adverse influence on the voltage on the primary side. In a complex system, operation of OLTC may cause a drop of voltage of the nodes connected with OLTC, which is not conducive to maintaining the voltage level of the system, but the root cause is that the active or reactive power of the system is out of balance, and the regulation characteristic of OLTC is only one factor that causes the system to lose power balance.

3. Power recovery characteristics of the load

The power recovery characteristics and low voltage instability characteristics of the load determine the voltage collapse of the system and the process of voltage collapse: the power recovery characteristics of the load may push the system with low operating voltage to the edge of collapse, especially the low voltage instability characteristics of the load may lead to a large area of voltage collapse.

References

1. Malkovich (1956) Power system and its operation. Translated by Zhang Zhongjun. Electric Power Industry Press, Beijing
2. USA response to questionnaire sponsored by CIGRE study committee 32 on control of the dynamic performance of future power system. CIGRE report (1976)
3. Duan X, He Y (1995) Review of voltage stability in electric power system. Power Syst Technol 19(4):20–24
4. Liu Y, Chen C, Liang L et al (2013) Review of dynamical analysis art of voltage stability in power systems. Proc CSU-EPSA 15(1):105–108
5. Zhong W (2008) Mechanism research of transient voltage stability in receiving-end grid. PhD thesis, China Electric Power Research Institute, Beijing
6. Zhang X (2007) Investigation of power system voltage stability contingency ranking. Master's thesis, China Electric Power Research Institute, Beijing
7. Lin W (2008) Voltage stability in multi-infeed HVDC transmission systems. Master's thesis, China Electric Power Research Institute, Beijing
8. Wang M, Wu J, Meng D (1991) Large power system technology, 2nd edn. China Electric Power Press, Beijing
9. Hain Y, Schweitzer I (1997) Analysis of the power blackout of June 8, 1995 in the Israel electric corporation. IEEE Trans Power Syst 12(4):1752–1757
10. Cheng H (1998) Research of static-state voltage stability in electric power systems. Doctoral thesis, Shanghai Jiao Tong University, Shanghai

11. U.S.—Canada Power System Outage Task Force (2013) Interim report: causes of the August 14th blackout in the United States and Canada. U.S.—Canada Power System Outage Task Force
12. He D (1996) A preliminary understanding of WSCC disturbance and separation on July 2, 1996 in the United States. Power System Technology 20(9):35–39
13. Lu W, Shu Y, Shi L (1996) WSCC disturbance and separation on August 10, 1996 in the United States. Power Syst Technol 20(9):40–42
14. Zhou X, Zheng J, Shen G et al (2003) Draw lessons from large scope blackout of interconnected North America power grid. Power Syst Technol 27(9):T1
15. Hu X (2003) Rethinking and enlightenment of large scope blackout in interconnected North America power grid. Power Syst Technol 27(9):T2–T6
16. Tang B (2003) Blackout in South of London and its lessons. Power Syst Technol 27(11):1–6
17. J Zhang (1998) Speed up the construction of urban distribution network to reduce power supply interruption. Power Syst Technol 22(8):31–33
18. Qu J, Guo J (2004) Statistics and analysis of faults in main domestic power systems from 1996 to 2000. Power Syst Technol 28(21):60–62
19. Han Z, Cao Y (2004) Power system security and its prevention. Power Syst Technol 28(9):1–6
20. Kundur P (1994) Power system stability and control. EPRI power system engineering series. McGraw Hill, New York
21. Harmand Y, Trotignon M, Lesigne JF et al (1990) Analysis of a voltage collapse-incident and proposal for a time-based hierarchical containment scheme. CIGRE report
22. Kurita A, Sakurai T (1988) The power system failure on July 23, 1987 in Tokyo. In: Proceedings of the 27th conference on decision and control, Austin Texas
23. CIGRE Task Force 38-01-03 of Committee 38 (1986) Planning against voltage collapse. CIGRE report
24. van Cutsem T, Vournas C (1998) Voltage stability of electric power systems. Kluwer Academic Publishers, Norwell
25. Zhong W, Tang Y (2008) Transient voltage stability analysis considering induction motor load. In: 10th youth academic conference of Chinese Society for Electrical Engineering, Jilin
26. Zhong W, Tang Y (2008) Preliminary study on mechanism of transient voltage stability in receiving-end grid. Proc Chinese Soc Electri Eng (Supplement):1–7
27. Tang Y, Zhong W, Sun H et al (2010) Study on mechanism of power system voltage stability. Power Syst Technol 34(4):24–29

Chapter 3
Characteristics and Models of Power System Elements

3.1 Synchronous Generator and Its Control System

The electrical energy of the power system is mainly provided by synchronous generators. Voltage stability of the power system is closely related to synchronous generators. It is very important to understand and master the characteristics of synchronous generators for the study of voltage stability of the power system.

3.1.1 Synchronous Generator

1. Phasor diagram and power expression of synchronous generators

The rotating speed of the rotor of the synchronous generator is the same as that of the rotating magnetic field, which means "synchronization". If rotating speed, armature current frequency, and number of pole pairs are defined as n, f, and p respectively, then

$$n = \frac{60f}{p}(\mathrm{r/min}) \tag{3.1}$$

When the standard current frequency of China's power system is 50 Hz and the synchronous generator is designed as a pair of poles, its rotating speed is 3,000 r/min.

When the prime mover adjusts the synchronous generator to the rated rotating speed, DC excitation current is introduced into the rotor winding, so magnetic flux will be generated in the air gap of the generator, which cuts the stator winding with the movement of the rated rotating speed of the synchronous generator. If the generator has no load, there is no current in the stator winding, the air gap magnetic field is the synchronous rotating main magnetic field generated by the excitation magnetomotive force, and the terminal voltage of the generator is equal to the induced

© Science Press 2021

Y. Tang, *Voltage Stability Analysis of Power System*, Power Systems,
https://doi.org/10.1007/978-981-16-1071-4_3

electromotive force in the winding, which is called no-load electromotive force-, $\dot{E}_0$. If the generator is loaded symmetrically, the stator winding will flow through the load current, and the armature winding will generate the armature magnetomotive force and the corresponding armature magnetic field. If only the fundamental wave is considered, it will rotate in the same direction and at the same rotating speed as the rotor, so that the air gap magnetic field and the induced electromotive force in the winding is changed, and this phenomenon is called the armature reaction. The armature reaction electromotive force induced by the armature magnetic field is expressed with $\dot{E}_a$.

The steady state analysis methods of the non-salient pole synchronous generator and the salient pole synchronous generator are respectively introduced below.

(1) Non-salient pole synchronous generator

When the magnetic saturation is not considered, the superposition principle can be adopted, and the induced electromotive forces of rotor magnetic field and armature magnetic field in stator winding are considered as $\dot{E}_0$ and $\dot{E}_a$ respectively. Therefore, the voltage equation of the stator circuit can be expressed as

$$\dot{E} = \dot{E}_a + \dot{E}_0 = \dot{U} + \dot{I}(r_a + \mathrm{j}x_\sigma) \tag{3.2}$$

wherein, $\dot{E}$ is the synthetic electromotive force; $\dot{U}$ is the terminal voltage of the stator winding; $\dot{I}$ is the stator current; r_a is the stator winding resistance; x_σ is the leakage reactance of stator winding. The equivalent circuit is shown in Fig. 3.1.

Since the armature reaction electromotive force $\dot{E}_a$ can be regarded as a reactance drop generated by phase current, this reactance is called armature reaction reactance, i.e., $\dot{E}_a = -\mathrm{j}x_a\dot{I}$. If x_a and x_σ are combined into an x_s, it is called synchronous reactance and can be directly measured. The Formula (3.2) can be written into:

$$\dot{E}_0 = \dot{U} + \dot{I}\left[r_a + \mathrm{j}(x_a + x_\sigma)\right] = \dot{U} + \dot{I}(r_a + \mathrm{j}x_s) \tag{3.3}$$

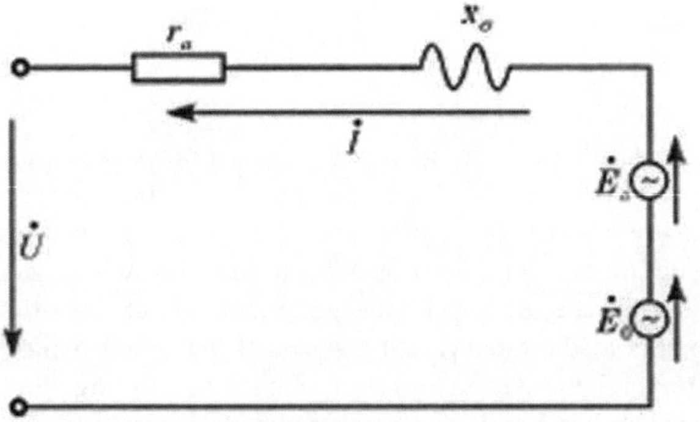

Fig. 3.1 Equivalent circuit of non-salient pole synchronous generators

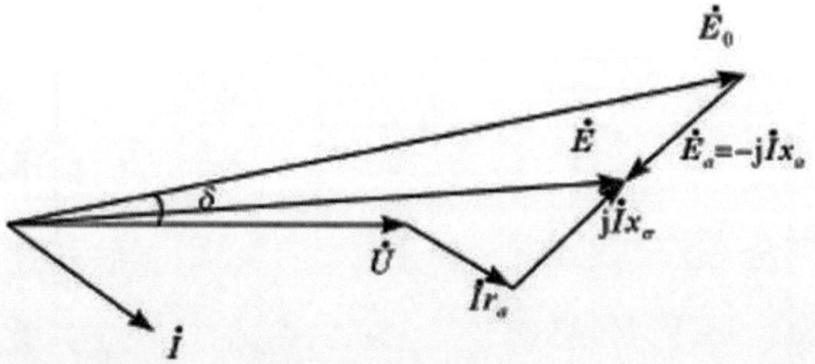

Fig. 3.2 Phasor diagram of non-salient pole synchronous generators

The phasor diagram of non-salient pole synchronous generators is shown in Fig. 3.2.

The expressions of the active power and reactive power of the non-salient pole synchronous generator are as follows respectively.

$$P = \frac{E_0 U}{x_s} \sin\delta \tag{3.4}$$

$$Q = \frac{E_0 U}{x_s} \cos\delta - \frac{U^2}{x_s} \tag{3.5}$$

(2) Salient pole synchronous generator

Because the air gap of salient pole synchronous generator is not uniform and the reluctance is not the same in each air gap, the magnetic conductance under the pole face is large but the magnetic conductance between two poles is small, which is very different, therefore, the two-reaction method is usually used in general analysis. This method is based on the superposition principle. The magnetomotive force $\dot{F}_a$ of the armature fundamental wave is decomposed into the armature reaction magnetomotive forces $\dot{F}_{ad}$ and $\dot{F}_{aq}$ acting on the straight axis and the quadrature axis respectively. Therefore, the straight-axis armature reaction magnetomotive force and the quadrature-axis armature reaction magnetomotive force are respectively established in the stator winding, and their fundamental components are respectively expressed as $\dot{E}_{ad}$ and $\dot{E}_{aq}$. The equivalent circuit diagram is shown in Fig. 3.3.

The equation for the stator winding voltage balance for the salient pole synchronous generator can be obtained, i.e.,

$$\dot{E} = \dot{E}_{ad} + \dot{E}_{aq} + \dot{E}_0 = \dot{U} + \dot{I}(r_a + jx_\sigma) \tag{3.6}$$

For the salient pole synchronous generator, similar to the non-salient pole synchronous generator, the straight-axis and quadrature-axis armature reaction magnetomotive forces $\dot{E}_{ad}$ and $\dot{E}_{aq}$ can be regarded as a reactance voltage drop

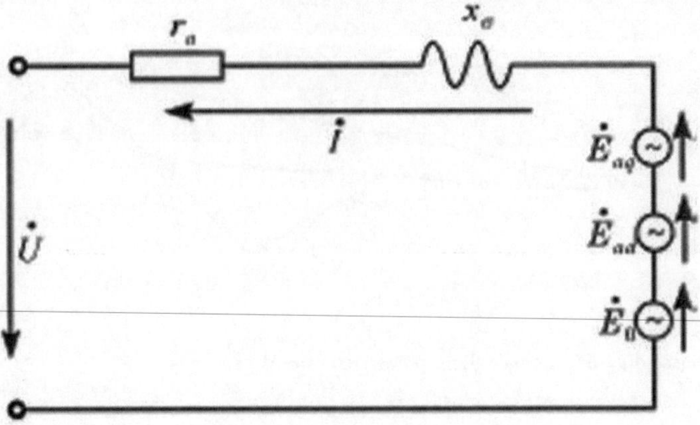

Fig. 3.3 Equivalent circuit of salient pole synchronous generators

generated by the straight-axis and quadrature-axis currents, i.e.,

$$\dot{E}_{ad} = -\mathrm{j}\dot{I}_d x_{ad} \tag{3.7}$$

$$\dot{E}_{aq} = -\mathrm{j}\dot{I}_q x_{aq} \tag{3.8}$$

wherein, $\dot{I}_d$ and $\dot{I}_q$ are straight-axis and quadrature-axis currents respectively; x_{ad} and x_{aq} are straight-axis and quadrature-axis armature reaction reactances respectively.

If the leakage reactance x_σ is combined with the armature reaction reactance, the straight-axis synchronous reactance x_d and the quadrature-axis synchronous reactance x_q can be defined respectively:

$$x_d = x_\sigma + x_{ad} \tag{3.9}$$

$$x_q = x_\sigma + x_{aq} \tag{3.10}$$

Equation (3.6) can be rewritten as

$$\begin{aligned}\dot{E}_0 &= \dot{U} + \dot{I}(r_a + jx_\sigma) - \dot{E}_{ad} - \dot{E}_{aq} \\ &= \dot{U} + \dot{I}(r_a + jx_\sigma) + \mathrm{j}\dot{I}_d x_{ad} + \mathrm{j}\dot{I}_q x_{aq} \\ &= \dot{U} + \dot{I}r_a + \mathrm{j}\dot{I}_d(x_\sigma + x_{ad}) + \mathrm{j}\dot{I}_q(x_\sigma + x_{aq}) \\ &= \dot{U} + \dot{I}r_a + \mathrm{j}\dot{I}_d x_d + \mathrm{j}\dot{I}_q x_q\end{aligned} \tag{3.11}$$

For the making of the phasor diagram of salient pole synchronous generators, the straight-axis component $\dot{I}_d$ and the quadrature-axis component $\dot{I}_q$ of the armature current need to be found respectively. Equation (3.11) can be rewritten as

$$\dot{E}_0 - j\dot{I}_d(x_d - x_q) = \dot{U} + \dot{I}r_a + j\dot{I}x_q \tag{3.12}$$

As the phasor $\dot{I}_d$ is perpendicular to $\dot{E}_0$, $j\dot{I}_d(x_d - x_q)$ and $\dot{E}_0$ are on the same line. Therefore, the position of $\dot{E}_0$ can be determined by the right end term of Eq. (3.12), so as to determine the positions of load currents $\dot{I}_d$ and $\dot{I}_q$. Phasor diagram is shown in Fig. 3.4.

The expressions of the active power and reactive power of the salient pole synchronous generator are as follows respectively.

$$P = \frac{E_q U}{x_d}\sin\delta + \frac{U^2}{2}\left(\frac{1}{x_q} - \frac{1}{x_d}\right)\sin 2\delta \tag{3.13}$$

$$Q = \frac{E_q U}{x_d}\cos\delta - U^2\left(\frac{\sin^2\delta}{x_q} + \frac{\cos^2\delta}{x_d}\right) \tag{3.14}$$

2. Operating capacity limit diagram of synchronous generator

The rated reactive power Q_{GN} output by the generator under rated conditions is

$$Q_{GN} = S_{GN}\sin\varphi_N \tag{3.15}$$

wherein, S_{GN} is the rated apparent power of the generator; φ_N is the rated power factor angle of the generator.

When the generator is operating at non-rated power factor, the reactive power put out by the generator can be smoothly adjusted by changing the excitation current entering the rotor circuit, which can not only put out the reactive power, but also absorb reactive power. Its ability to put out the reactive power is determined by the *PQ* limit of the generator operation. Figure 3.5 is the operating capacity limit diagram of non-salient pole synchronous generator. In the figure, the arc *AC* refers to the rated current of the rotor with B as the center and AB as the radius; the arc

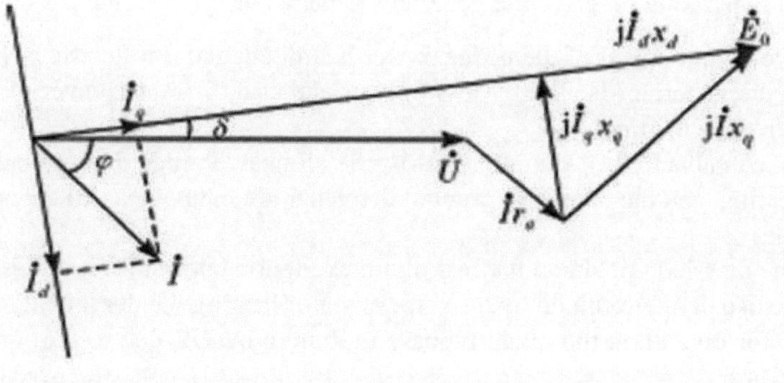

Fig. 3.4 Phasor diagram of salient pole synchronous generators

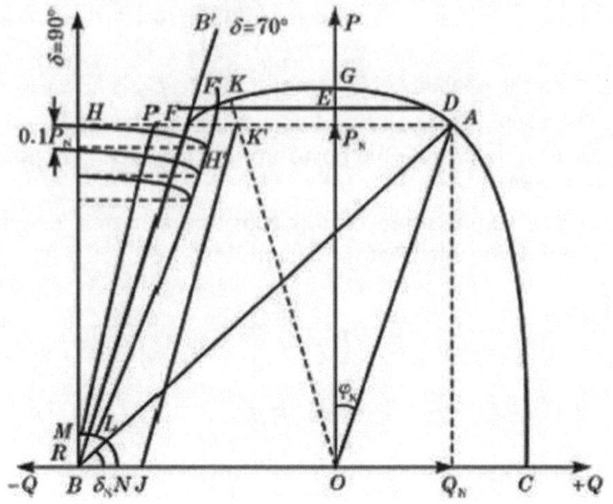

Fig. 3.5 Operating capacity limit diagram of non-salient pole synchronous generator

ADGK refers to the rated current of the stator with *O* point as the center and *OA* as the radius. The intersection point "**A**" of these two arcs is the point that the generator stator and rotor currents reach the rated value at the same time. The projections of *OA* on the vertical and horizontal coordinates are rated active power and rated reactive power respectively. *OB* represents the maximum reactive power that can be absorbed when the generator runs in parallel with infinite power grid and $P = 0$. When $\cos\varphi$ decreases, the generator operating point cannot exceed the arc *AC* due to the limitation of rotor current. When $\cos\varphi = 0$ at the point *C*, the generator will put out the maximum reactive power. When $\cos\varphi$ increases, the armature reaction of the generator decreases and the required excitation current also decreases, so the excitation current does not become a limiting factor. However, due to the limitation of stator current, the operating point of the generator cannot exceed the arc *ADGK*. The active power of the generator cannot exceed its rated power. Therefore, $\cos\varphi$ will continue to increase after passing the point *D*, which is limited by the power limit line (*DEF* line) of the prime mover. Therefore, under steady state conditions, the allowable operating range of the generator depends on the following 5 conditions:

(1) The output power of the prime mover is limited, and the electric power output of the generator is slightly less than or equal to the rated power of the prime mover (*DEF* line).

(2) Rated capacity of generator, that is, the allowable range determined by stator heating, and the operating point of the generator cannot exceed the arc (*ADGK* line).

(3) The magnetic field and the maximum excitation current limit of the generator are usually determined by the rotor or stator heating. Under the lagging power factor operation, the leading phase is limited by *K′J* due to the limitation of *AC* arc.

(4) Static stability limit under the leading phase operation. *BH* is the static stability limit line when the contact reactance between the generator and the system $X_s = 0$ and the generator operating rotor angle is 90°; the arc *RP* is the static stability limit line when $Xs \neq 0$ and the terminal voltage $U = 1$; the straight line *BB′* is the static stability limit when the rotor angle is 70°, i.e., the static stability reserve is 6%; generally, 10% of the static stability reserve is taken as the active power limit (*F′L* line) of the generator under the leading phase operation.

(5) The curve *MLN* is the allowable minimum excitation current limit line, which is usually 10% of the rated excitation current.

Within the *PQ* limit range determined by these conditions, under a certain stator voltage and current, when the power factor decreases, the active output of the generator decreases and the reactive output increases; however, when the power factor increases, the opposite is true. The *PQ* curve of the synchronous generator shown in Fig. 3.5 shows the relationship between the active power and the reactive power during generator operation. The maximum reactive power allowed under the condition of constant active power can be obtained from the figure below.

3. General *PQ* curve of synchronous generators

When the synchronous reactance of the generator is unknown and only its rated power factor is known, the allowable reactive power for different active powers can be determined by checking the general *PQ* curve, as shown in Fig. 3.6. The solid line in the figure indicates a generator with a rated power factor of 0.8, the dotted line indicates a generator with a rated power factor of 0.9, and other power factor values can be obtained using the interpolation method.

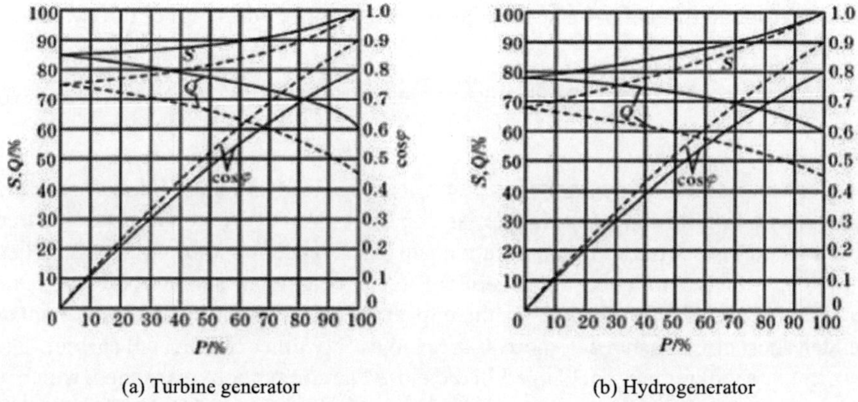

Fig. 3.6 General *PQ* curve of synchronous generators

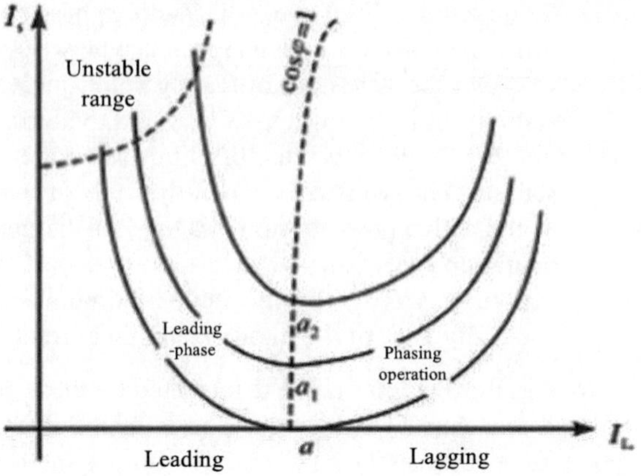

Fig. 3.7 V-shaped curve of synchronous generators

4. V-shaped curve of synchronous generators

The v-shaped curve of the synchronous generators is the relationship curve between the stator current I_s and the excitation current I_L, as shown in Fig. 3.7. Each V-shaped curve in the figure represents an active power output situation, and the connecting line (dotted line) at the lowest point of the curve represents the operation situation of $\cos\varphi = 1$, under which only the active power is put out, no reactive power is outputted, and the stator current value is minimum. The lowest curve indicates the excitation condition that the active output is zero. Point "*a*" indicates the excitation condition of no reactive power output, i.e., no-load excitation. The right side of the dotted line is the over excitation state, which outputs the capacitive reactive power. The excitation current in the unstable area is very small and the excitation electromotive force E_0 is also very small in the figure. When the generator reaches the critical power angle of 90°, the rotor angle cannot be kept stable in case of further decrease of the excitation current.

5. Dynamic characteristics of synchronous generators
(1) Sudden short circuit

Sudden short circuit of generators is a transient process from normal load operating state to short circuit state. In this process, the short circuit current will contain a free component. These free components attenuate based on certain time constants. When the attenuation is completed, the generator enters a steady state short circuit state. Due to the existence of free components, the impact short-circuit current at the moment of sudden short circuit can reach more than ten to twenty times of the rated current. The flux linkage will remain unchanged in the closed circuit without resistance, which is the constant flux linkage theorem in the superconductor closed circuit. Although the resistance always exists in actual closed circuits of the generators, the flux linkage will gradually change due to the influence of the resistance; however, as the flux linkage cannot change abruptly, the flux linkage can still be considered to be unchanged at the moment of sudden short circuit. Parameters of the generator in case of sudden short circuit are described below.

(1) Direct-axis subtransient reactance x''_d

When the rotor is equipped with damper windings, since the damper windings are closed loops and the flux linkage cannot change abruptly, the armature flux will be excluded from the damper windings at the moment of short circuit. That is, the armature flux will pass through the air gap, the leakage magnetic circuit next to the damper winding and the leakage magnetic circuit next to the exciting winding in sequence; the larger the reluctance is, the smaller the corresponding reactance is x''_d. x''_d is called the straight-axis ultra-transient reactance or the straight-axis subtransient reactance. At the moment of short circuit, the short circuit current in the stator winding will be limited by x''_d. Since the induced current in the damper winding attenuates quickly, the stator current will be limited by x'_d after the first few cycles, and finally the stator current will be limited by x_d when reaching the steady state.

Suppose Λ_{ad} is the magnetic conductance of the air gap, Λ_{ld} is the magnetic conductance of the leakage magnetic circuit nest to the damper winding, Λ_f is the magnetic conductance of the leakage magnetic circuit next to the exciting winding, and Λ_σ is the magnetic conductance of the armature leakage. Then the total magnetic conductance Λ''_d of the magnetic circuit of the armature flux is

$$\Lambda''_d = \Lambda_\sigma + \frac{1}{\frac{1}{\Lambda_{ad}} + \frac{1}{\Lambda_f} + \frac{1}{\Lambda_{ld}}} \tag{3.16}$$

Since reactance is proportional to magnetic conductance, x″d can be expressed as follows:

$$x''_d = x_\sigma + \frac{1}{\frac{1}{x_{ad}} + \frac{1}{x_f} + \frac{1}{x_{ld}}} \tag{3.17}$$

wherein, x_{ad} is the straight-axis armature reaction reactance; x_f is the leakage reactance of the exciting winding; x_{ld} is the leakage reactance of the damper winding on straight axis.

The equivalent circuit diagram corresponding to x''_d is shown in Fig. 3.8.

(2) Direct-axis transient reactance x'_d

In case of a short circuit, the flux chained with the rotor winding cannot change abruptly according to the principle that the flux linkage cannot change abruptly. Therefore, the armature reaction flux generated by the short circuit current cannot be chained with the rotor winding through the rotor core, but is forced into the leakage magnetic circuit outside the rotor winding. The reluctance of the path through which the magnetic flux $\dot{\phi}'_{\mathrm{ad}}$ generated by the stator short circuit current passes

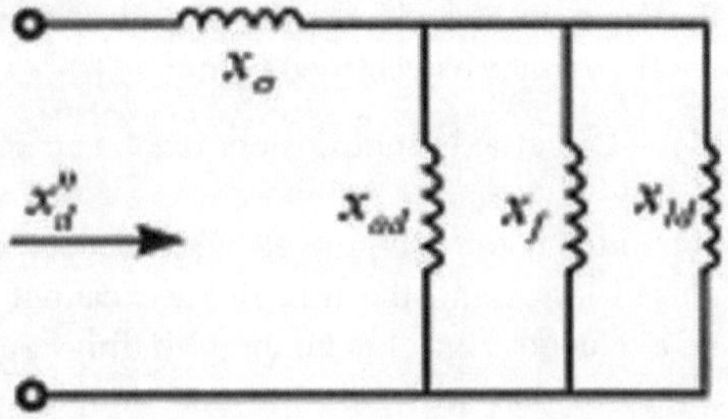

Fig. 3.8 Equivalent circuit diagram for x''_d

becomes larger, which limits the reactance of the armature current and makes it become smaller, resulting in a larger short circuit current at the moment of sudden short circuit. This reactance limiting the armature current is called the straight-axis transient reactance or the straight-axis transient reactance, and is expressed by x'_d. Due to the damping effect of the resistance on the rotor winding, the larger impact current gradually decreases in case of sudden short circuit, finally resulting in the short circuit current being limited by x_d.

If there is no damper winding on the rotor or the induced current attenuation in the damper winding is completed, the armature reaction flux can pass through the damper winding, and its total magnetic conductance Λ'_d is as follows

$$\Lambda'_d = \Lambda_\sigma + \frac{1}{\frac{1}{\Lambda_{ad}} + \frac{1}{\Lambda_f}} \tag{3.18}$$

Similarly, the direct-axis transient reactance (x'_d) is

$$x'_d = x_\sigma + \frac{1}{\frac{1}{x_{ad}} + \frac{1}{x_f}} \tag{3.19}$$

The equivalent circuit diagram corresponding to x'_d is shown in Fig. 3.9.

(3) Quadrature-axis subtransient reactance x''_q.

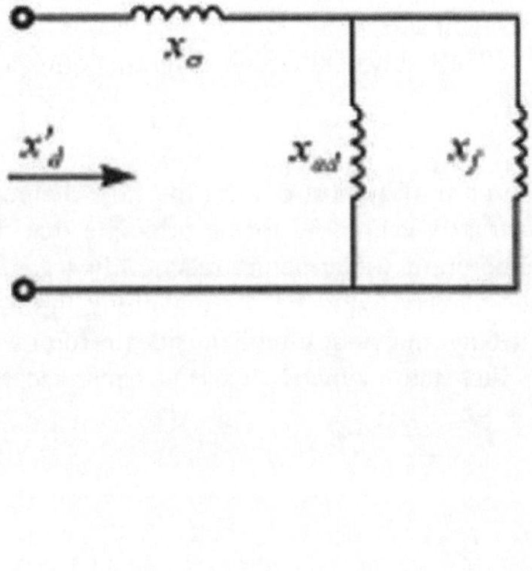

Fig. 3.9 Equivalent circuit diagram for x'_d

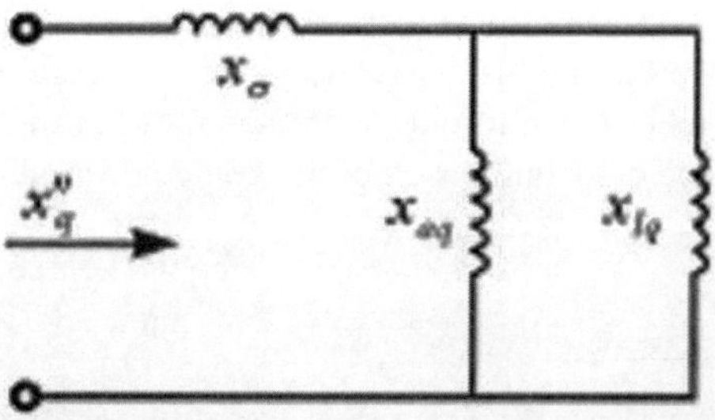

Fig. 3.10 Equivalent circuit diagram for x''_q

In the case of a damper, since the damper winding is asymmetric, its damping effect on the quadrature axis is different from that on the straight axis. The corresponding quadrature-axis subtransient reactance x''_q is as follows.

$$x''_q = x_\sigma + \frac{1}{\frac{1}{x_{aq}} + \frac{1}{x_{lq}}} \tag{3.20}$$

wherein, x_{aq} refers to the quadrature-axis armature reaction reactance; x_{lq} refers to the leakage reactance of damper winding at the quadrature axis.

The equivalent circuit diagram corresponding to x''_q is shown in Fig. 3.10.

(4) Quadrature-axis transient reactance x''_q.

If the generator is not short-circuited at the outgoing terminal, but is subject to the load impedance short circuit, the armature magnetic field generated by the short circuit current has not only the straight-axis component, but also the quadrature-axis component. Because the magnetic circuits along the quadrature and straight axes have different reluctances, the corresponding reactance values are also different. Since the synchronous generator has no exciting winding at the quadrature axis, generally, the quadrature-axis transient reactance is equal to the quadrature-axis synchronous reactance, that is

$$x'_q = x_q \tag{3.21}$$

(2) Generator oscillation

When the generator is running, there is an elastic effect between its resultant magnetic field and rotor magnetic field. When the load increases, the rotor angle δ (referring to the angle between the no-load electromotive force E_0 and U) will increase, but, due to the elastic effect, the new steady-state δ cannot be reached immediately, which will cause oscillation. If the amplitude of the oscillation gradually decreases, the rotor will still run stably at synchronous speed when a new displacement angle is finally obtained. If the amplitude of the oscillation gradually increases, the δ will continuously increase and the synchronous generator will fall out of synchronism with the power grid.

The motion equation of the generator is a core equation for generator oscillation. Assuming that T_a refers to the accelerating torque, and T_m and T_e both refer to the mechanical torque, with the unit of N m, the unbalanced torque of the generator is the essential reason of the generator oscillation:

$$J\frac{\mathrm{d}\omega_\mathrm{m}}{\mathrm{d}t} = T_\mathrm{a} = T_\mathrm{m} - T_\mathrm{e} \tag{3.22}$$

wherein, J, ω_m and t respectively refer to the total moment of inertia of the generator and the turbine, the angular velocity of the rotor, and the time, respectively with the unit of kg m^2, rad/s, and s.

H is defined as the value obtained by the kinetic energy at rated speed divided by the rated power reference value S_{base}, that is

$$H = \frac{\text{Kinetic energy at rated speed(MW s)}}{\text{Rated power reference value(MVA)}} = \frac{\frac{1}{2}J\omega_0^2}{S_\mathrm{base}} \tag{3.23}$$

If the inertia time constant H is used to describe the motion equation of the generator, then

$$\frac{2H}{\omega_0^2}S_\mathrm{base}\frac{d\omega_m}{\mathrm{d}t} = T_\mathrm{m} - T_\mathrm{e} \tag{3.24}$$

Because T_{base} is equal to S_{base}/ω_0, rearrangement results in

$$2H\frac{\mathrm{d}}{\mathrm{d}t}\left(\frac{\omega_\mathrm{m}}{\omega_0}\right) = \frac{T_\mathrm{m} - T_\mathrm{e}}{S_\mathrm{base}/\omega_0} \tag{3.25}$$

The motion equation in the per-unit form can be obtained.

$$2H\frac{\mathrm{d}\omega^*}{\mathrm{d}t} = T_\mathrm{m}^* - T_\mathrm{e}^* \tag{3.26}$$

6. Mathematical model of synchronous generators

(1) Six-order model

In the transient stability simulation analysis of the power system, the six-order generator model is generally used as the detailed model of the generator.

$$\begin{cases} T_j\frac{d\omega}{dt} = T_m - T_e - D(\omega - \omega_0) \\ \frac{d\delta}{dt} = \omega - \omega_0 \\ T'_{d0}\frac{dE'_q}{dt} = E_{fd} - E'_q - \frac{x_d - x'_d}{x'_d - x''_d}(E'_q - E''_q) \\ T''_{d0}\frac{dE''_q}{dt} = E'_q - E''_q - (x'_d - x''_d)I_d \\ T'_{q0}\frac{dE'_d}{dt} = -E'_d - \frac{x_q - x'_q}{x'_q - x''_q}(E'_d - E''_d) \\ T''_{q0}\frac{dE''_d}{dt} = E'_d - E''_d + (x'_q - x''_q)I_q \\ U_d = X''_q I_q + E''_d - r_a I_d \\ U_q = -X''_d I_d + E''_q - r_a I_q \end{cases} \tag{3.27}$$

wherein, T_j, T_m, T_e and E_{fd} respectively refer to the inertia constant, the mechanical power, the electromagnetic power and the excitation electromotive force of the generator; E'_d and E'_q as well as E''_d and E''_q respectively refer to the transient and subtransient electromotive forces of the generator on the d axis and q axis; T'_{d0}, T'_{q0}, T''_{d0} and T''_{q0} respectively refer to the transient and subtransient time constants of the open circuits on the d axis and q axis; X'_d, X'_q, X''_d and X''_q respectively refer to the transient and subtransient reactance on the d axis and q axis; r_a refers to the generator stator resistance; U_d and U_q refer to the projections of the generator terminal voltage on the d axis and q axis; I_d and I_q refer to the projections of the generator output current on the d axis and q axis.

(2) Five-order model

If the electromagnetic transient process of the stator is neglected, but the electromagnetic transients of the f, D and Q windings at rotor side and the electromechanical transient process of the rotor motion are considered, a practical five-order model is obtained:

$$\begin{cases} T_j\frac{d\omega}{dt} = T_m - T_e - D(\omega - \omega_0) \\ \frac{d\delta}{dt} = \omega - \omega_0 \\ T'_{d0}\frac{dE'_q}{dt} = E_{fd} - E'_q - \frac{x_d - x'_d}{x'_d - x''_d}(E'_q - E''_q) \\ T''_{d0}\frac{dE''_q}{dt} = E'_q - E''_q - (x'_d - x''_d)I_d \\ T''_{q0}\frac{dE''_d}{dt} = -E''_d + (x'_q - x''_q)I_q \\ U_d = X''_q I_q + E''_d - r_a I_d \\ U_q = -X''_d I_d + E''_q - r_a I_q \end{cases} \tag{3.28}$$

(3) Three-order model

If the transients of d and q windings of the stator and the effect of the damper windings (D and Q windings) are neglected, a practical three-order model is obtained:

$$\begin{cases} T_j\frac{d\omega}{dt} = T_m - T_e - D(\omega - \omega_0) \\ \frac{d\delta}{dt} = \omega - \omega_0 \\ T'_{d0}\frac{dE'_q}{dt} = E_{fd} - E'_q - (x_d - x'_d)I_d \\ U_d = X_q I_q - r_a I_d \\ U_q = E'_q - X'_d I_d - r_a I_q \end{cases} \tag{3.29}$$

In the practical power system dynamic analysis, when the excitation system dynamic is considered, the simplest model is the three-order model. Because of its simplicity and availability to calculate the excitation system dynamic, it is often used in the power system dynamic analysis with low accuracy requirements, but for which the excitation system dynamic shall still be considered.

(4) Classical two-order model
(1) E' constant model.

This model is widely used in power system analysis. It uses the equivalent circuit composed of the transient electromotive force $\dot{E}'$ after the transient reactance X'_d to indicate the synchronous generator, only considering the dynamic process of the generator rotor. As shown in the following formula:

$$\begin{cases} T_j\frac{d\omega}{dt} = T_m - T_e - D(\omega - \omega_0) \\ \frac{d\delta}{dt} = \omega - \omega_0 \\ \dot{U} = \dot{E}' - (r_a + jX'_d)\dot{I} \end{cases} \tag{3.30}$$

(2) E'_q constant model.

In the three-order model, if E'_q is assumed to be constant, the E'_q constant model can be obtained:

$$\begin{cases} T_j\frac{d\omega}{dt} = T_m - T_e - D(\omega - \omega_0) \\ \frac{d\delta}{dt} = \omega - \omega_0 \\ U_d = X_q I_q - r_a I_d \\ U_q = E'_q - X'_d I_d - r_a I_q \end{cases} \tag{3.31}$$

3.1.2 Excitation Control System of Synchronous Generators

1. Function of excitation systems

Excitation regulator is the main means of regulating voltage in a power system, and as follows, its main functions are to:

(1) Provide sufficient, reliable and continuously adjustable DC current for generator excitation winding.
(2) Automatically maintain the generator terminal voltage. This function is enabled by the voltage deviation feedback control of the excitation system, and the output value and the reference input value are compared to reduce the difference between them, so as to maintain the generator terminal voltage.
(3) Limit and protect excitation current or voltage. In case of a large disturbance in power system, the excitation control can energize the generator to provide instantaneous or short-term reactive power support, which plays an important role in voltage stability in the transient process of the power system. The overload capacity of the generator is limited by the following factors: insulation damage caused by excitation voltage being too high, rotor overheating caused by excitation current too high, stator terminal overheating caused by operation in the state of under-excitation, and heating caused by flux linkage being too high, etc.

Excitation system is an important factor affecting the voltage stability of a power system. Therefore, when the voltage stability of power system is studied, it is necessary to accurately consider the influence of the generator excitation regulation system, and different excitation regulation systems have different influence on the voltage stability of power system. The fast and strong excitation regulation system can quickly restore the voltage of the nodes adjacent to the generator terminal, but it has little effect on the load nodes further from the generator terminal, so it is still necessary to install a reactive compensation device to effectively regulate the voltage of the remote load nodes; if the excitation regulation system is of the weak and slow regulation type, when the system regulation capacity reserve is insufficient, the voltage collapse of the remote load nodes may expand, thereby affecting the voltage of the nodes adjacent to the generator terminal, so that the above adjacent nodes have the risk of continuous low voltage. Therefore, under normal operation, the certain safety margin should be provided for the generator excitation voltage.

In consideration of the voltage stability of a power system, the shunt capacitor bank can be installed at the generator terminal, so that the generator can operate near the power factor 1.0 under normal conditions, thus increasing the spinning reactive power reserve of the generator. Once needed, the reserve capacities can be put into operation quickly to prevent voltage instability. Therefore, the reactive power reserve scale of generator is also a measure of voltage stability level of the power system.

In addition, the over excitation and low excitation limit characteristics of the excitation control system are important characteristics affecting the voltage stability of power system. The detailed analysis is shown in the following contents.

2. Model of excitation control systems

There are many kinds of excitation systems. Only two kinds of excitation regulation systems are introduced here. The block diagrams and parameters of models are briefly introduced as follows.

(1) Conventional or quick response excitation system in separate excitation

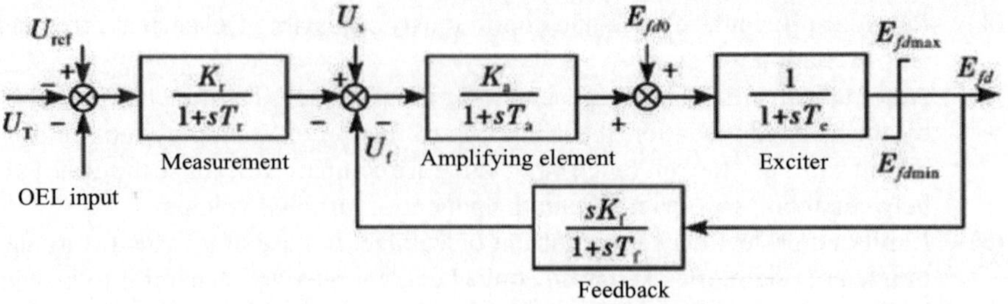

Fig. 3.11 Block diagram of conventional or quick response excitation system in separate excitation

This system is usually an excitation regulation system with exciter, and its model is shown in Fig. 3.11.

In Fig. 3.11, U_T is the generator terminal voltage; U_{ref} is the voltage setting of the excitation regulator; U_s is the input signal of power system stabilizer (PSS); U_f is the input signal of feedback; E_{fd0} is the steady status excitation voltage; E_{fd} is the actual output excitation voltage; $E_{fd\max}$ and $E_{fd\min}$ are the upper and lower limits of excitation output voltage; K_r, K_a and K_f are the magnification times of measurement, magnification and feedback respectively; T_r, T_a, and T_f are the time constants of the above three links respectively; T_e is the exciter time constant.

(2) Model of self-shunt excitation regulation system

The difference between self-shunt excitation regulation system and separate-excitation system is that the excitation power of the former is from the generator terminal, and Fig. 3.12 shows the block diagram of the transfer function of the self-shunt excitation regulation system.

In Fig. 3.12, U_{T} is the generator terminal voltage; U_{T0} is the set voltage of the excitation regulator; U_s is the input signal of PSS; K_r *and* K_a are the magnification times of measurement and magnification respectively; K_2 is the type parameter of transformation, when it is equal to 0, it is the proportional integral; when it is equal to 1, it is the transposition; $T_1, T_2, T_3, T_4, T_{\mathrm{r}}$ and T_{a} are the time constants of the above three links respectively; E_{fd0} is the steady status excitation output voltage; $\tilde{E}_{fd\max}$ and $\tilde{E}_{fd\min}$ are the variable upper and lower limits of excitation output voltage.

3. Over excitation and low excitation limits

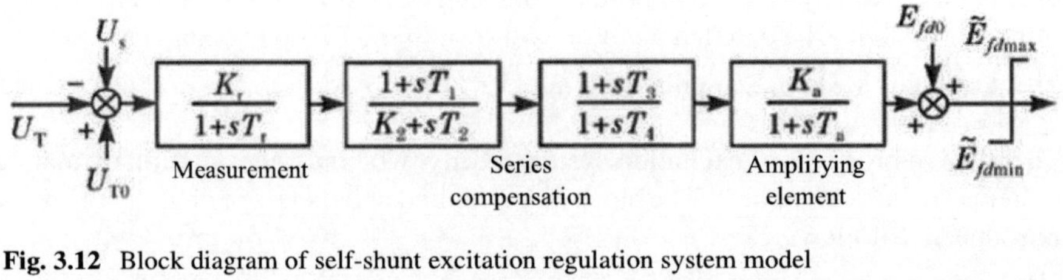

Fig. 3.12 Block diagram of self-shunt excitation regulation system model

(1) Over excitation limit

The over excitation limit (OEL) of the generator is determined by the over-heating load capacity of the generator, and the over-current time of the rotor is related to the peak value of the excitation current. Once the excitation regulation of a generator in the system reaches the maximum limit value, due to the thermal capacity limitation of the excitation winding, when the over excitation capacity of the unit reaches the set operating time, the over excitation limiter will reduce the excitation current to the rated value, resulting in a reduction of reactive power of the generator and a drop of voltage in the system. Since the system voltage decreases and other generators will also be in a state near over-excitation, the sudden excitation reduction of the generator may cause a chain reaction of other generators, thus causing voltage instability of the system.

This book introduces two kinds of over excitation models, one is a practical model (OELCEPRI) [1] proposed by excitation experts of China Electric Power Research Institute (CEPRI) based on long-term engineering practice; the other is a model (OELIEEE) [2] proposed by the IEEE excitation limit working group. The descriptions of the two are as follows.

(1) OELCEPRI model

The block diagram of this model is shown in Fig. 3.13. There are two types of electrical quantities input to the OELCEPRI model:

① Generator rotor current—for brush excitation system only.
② Excitation current of AC exciter—necessary for brushless excitation systems and also can be used for the brush excitation system.

Meaning of variables in Fig. 3.13: I_{fd} is the generator magnetic field current; U_{LR} is the square value of the long-term allowable current of the generator magnetic field; B is the calorific value; C is the over excitation recovery coefficient; K_{L2} is the gain of over excitation limit circuit; T_{L1} is the lead time constant of the over excitation limit circuit; T_{L2} is the lag time constant of the over excitation limit circuit; and U_{L} is the output value of the over excitation limit.

The output of the model is used to limit the output voltage of the regulator so as to limit the rotor current of the generator. The rotor current or voltage cannot be directly limited in an actual device, but can be directly limited in a program simulation.

The principle that the model judging over-excitation is the inverse-time characteristics of equivalent heating, and the limit curve is determined by the two input points (over excitation time and over excitation current).

The process and principle of over excitation determination are as follows:

The over excitation process may be approximately constant over excitation process. For example, the over excitation caused by a regulator fault may be such a process. After the end of the over excitation process, the generator excitation current I_{fd} is a fixed value.

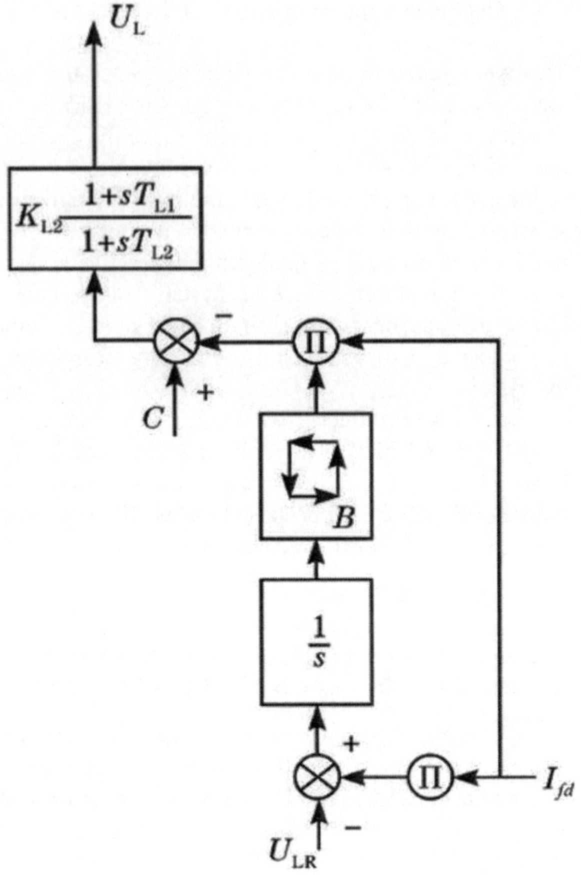

Fig. 3.13 Over excitation limit model (OELCEPRI)

The over excitation may also be a slow process occurring gradually and increasing the over excitation value. For example, if the system voltage drops gradually, the generator excitation current I_{fd} changes from lower than $I_{fd\infty}$ to equal to $I_{fd\infty}$, and then higher than $I_{fd\infty}$ ($I_{fd\infty}$ is the allowable excitation current for long-term operation). The over excitation values are also changing.

Therefore, it is generally not necessary to calculate the over excitation time to determine whether the over excitation is allowed, but to cumulatively calculate whether the calorific value $B = \int\left(I_{fd}^2 - I_{fd\infty}^2\right)\mathrm{d}t$ reaches the allowable value B_0, to determine whether to limit, that is, when $B = B_0$, the limiter acts.

After the limiter has acted, the generator rotor current shall be kept below $I_{fd\infty}$, so as to release the excess heat generated during the over excitation process. Generally, $I_{fd\infty}$ is (0.90–0.95).

When the limiter cannot effectively limit I_{fd}, the over excitation protection action shall shut down the generator, to ensure the safety of units. Therefore, when $B > B_0$, timing shall be carried out. After T seconds, I_{fd} shall be less than $I_{fd\infty}$, and B shall be less than B_0, and it shall be continuously reduced, otherwise, the generator shutdown shall be carried out.

There are four statuses:

① When $B = 0$, $I_{fd} \le I_{fd\infty}$, there was no over excitation or overheating accumulation in the past, and there is currently no over excitation either.

② When $B = 0$, $I_{fd} > I_{fd\infty}$, there was no over excitation or overheating accumulation in the past, and there is over excitation currently, so the overheating accumulation calculation should be carried out $B = \int\left(I_{fd}^2 - I_{fd\infty}^2\right)\mathrm{d}t$.

③ When $B > 0$, $I_{fd} > I_{fd\infty}$, there was over excitation and overheating accumulation in the past, and there is over excitation currently, so the overheating accumulation calculation should be carried out $B = \int\left(I_{fd}^2 - I_{fd\infty}^2\right)\mathrm{d}t$.

④ When $B > 0, I_{fd} \le I_{fd\infty}$, in the past, there was over excitation occurred, with over heating accumulation. At present, there is no over excitation (after the action of limiter or the recovery of system voltage level), the overheating accumulation calculation $B = \int\left(I_{fd}^2 - I_{fd\infty}^2\right)\mathrm{d}t$sshouldalsobecarriedout. The accumulation in the opposite direction is a release process, which is calculated until $B \le 0$. When $B \le 0$, let $B = 0$.

The identification and operation process of over excitation are as follows:

① When over excitation occurred for the first time, the heat accumulation calculation is carried out.

② When the heat accumulation exceeds the given value, i.e., when B is $\ge B_0$, the limit signal is sent out, and the output voltage of the regulator is limited to the level corresponding to $(0.90\text{–}0.95)I_{fd\infty}$. At the same time, timing is carried out, $T = T + \Delta t$.

③ The limiter acts normally. After a certain period of time (1–2 s), I_{fd} will be less than $I_{fd\infty}$. At this time, there is still $B > 0$, and it is necessary to calculate the reverse (release) heat accumulation until $B \le 0$.

④ If the limiter does not act normally, and after $T > T_0$, $B < B_0$ and $I_{fd} < I_{fd\infty}$ are still impossible, then generator shutdown is carried out.

⑤ When the limiter acts normally, although there may be $I_{fd} < I_{fd\infty}$, if the over excitation is caused by low system voltage, it may lead to stator overcurrent or generator to be out-of-step, and it may also cause generator shutdown.

(2) OELIEEE model

The block diagram of model is shown in Fig. 3.14.

The input of the model is one of the excitation current I_{fd}, excitation voltage E_{fd} or exciter excitation voltage V_{FE}. If the above input is larger than the set value I_{TFPU}, the delay unit of the model is started. After a certain period of time, if the value of the variable reference point I_{ref} is less than the input value of the model ($I_{err} < 0$), the output part of the model is started to limit the output of excitation, so as to limit the over excitation.

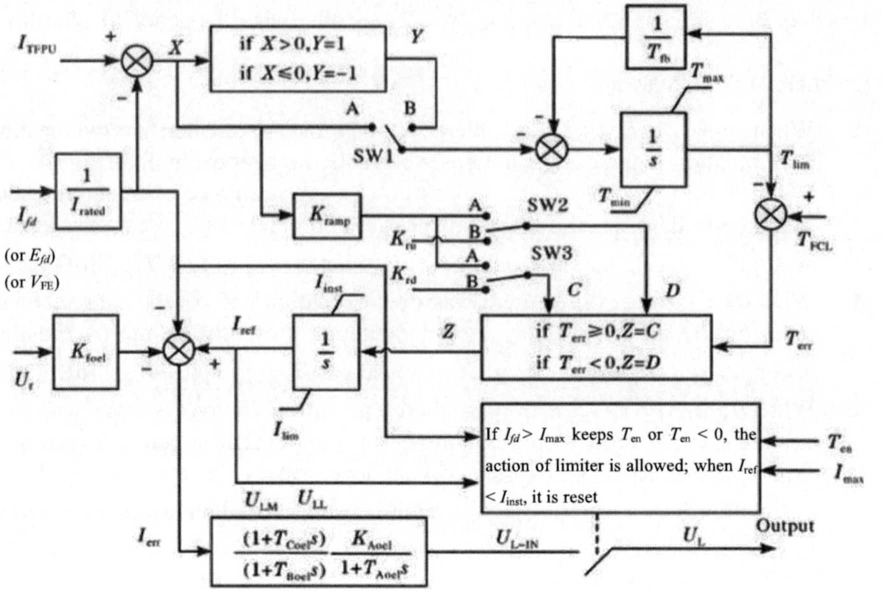

Fig. 3.14 Over excitation limit model (OELIEEE)

The delay unit consists of two comparison links, two integration links and three switches. The integral unit simulates the delay function. Different switch positions can represent the inverse time limit and the fixed time limit.

If the input is larger than the set value I_{TFPU}, the timer 1 (the first integration) starts to integrate. If the position of switch 1 (SW1) is in position "A" in Fig. 3.14, the integration speed is proportional to the input deviation (inverse time limit function); if the position of SW 1 is in position "B" in Fig. 3.14, the integration is at a fixed rate (fixed time limit function).

If the output T_{lim} of timer 1 is larger than the set value T_{FCL}, another timer (the second integration) is started. Similarly, the fixed time limit and inverse time limit functions are realized respectively when SW 2 and SW 3 are in different positions.

If the variable reference point I_{ref} is less than the input excitation current I_{fd}, then the I_{err} is less than 0, and the model output is a negative value, and the over excitation limit is carried out. Under the rated operating status, the output of the model is a large positive number.

The typical values of upper and lower amplitude limit I_{inst} and I_{lim} in the block diagram can be 1.6 and 1.05. When the input of excitation current I_{fd} (or other two inputs, the same below) is at or below the rated value, the value of reference point I_{ref} of the limiter is equal to I_{inst}, which means that when the excitation current I_{fd} reaches 1.6 times of its rated value, the output limiting excitation can be conducted instantaneously without through delay unit.

When in the system there is a short circuit and other faults, in the initial stage of the transient process, in order to maintain the generator terminal voltage at a high level, the excitation system generates a large amount of excitation current in an instant. At this time, the over excitation device should not limit the excitation currents in general. The link containing the maximum excitation current parameter I_{max} and the delay time parameter T_{en} in the model describes this phenomenon.

The model can also accept input U_s from the PSS model.

(2) Low excitation limit

Low excitation is generally caused by the rise of system voltage or AVR fault. After a low excitation of the generator has occurred, the generator will absorb a lot of reactive power, if not limited, it will cause the generator to exceed its stable operating limit or stator heating limit and lead to the operation of generator shutdown.

The principle of the low excitation limiting device is usually to determine the current operation point of the generator according to the voltage and current (or active power P and reactive power Q) at the outlet of the generator. If the operation point exceeds the low excitation action limit of the generator, a signal to limit the excitation current is sent to the excitation control system, to increase excitation to the generator. The range of low excitation limit action can be represented by a circle or a straight line on the PQ plane of the generator, that is, a boundary determined by the active power and reactive power of the generator. Once the dividing line is crossed, a low excitation limit signal is sent out.

Figures 3.15 and 3.16 are a low excitation limit model [3] proposed by IEEE excitation limit working group in 1995 and corresponding model characteristics. The input signal of the model is the AC voltage $\dot{U}_T$ and current $\dot{I}_T$ of the generator terminal, and the signal U_s of the PSS. U_{UR} describes the radius of the action circle and U_{UC} describes the operation point of the generator. When the reactive power absorbed by the generator increases or the active power generated increases, U_{UC} increases. When $U_{UC} > U_{UR}$ and the operation point exceeds the boundary of low excitation action, the low excitation limit signal is put out. The function of U_s is to enhance the damping and stabilize the output of the generator.

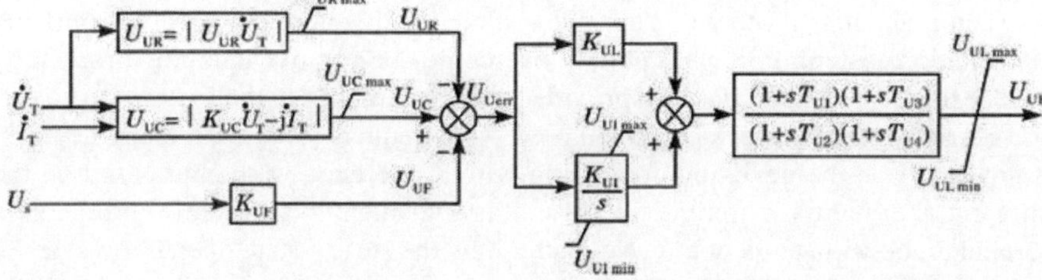

Fig. 3.15 Low excitation limit model

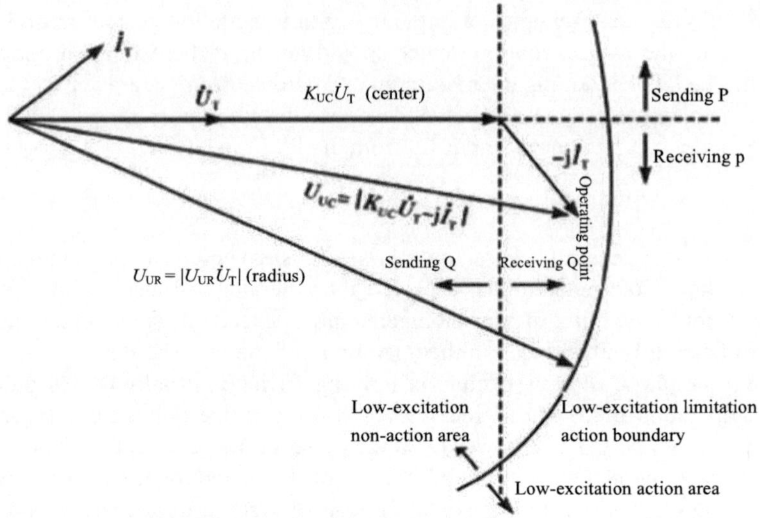

Fig. 3.16 Low excitation limit characteristics

3.1.3 *Power System Stabilizer*

1. Characteristics of power system stabilizers

The power system stabilizer (PSS) developed in 1970s is a breakthrough in science and technology. In order to maintain the generator terminal voltage, the excitation system needs to use larger voltage magnification times. The research shows that this excitation control of high voltage magnification coupled with a quick response excitation system has a significant effect on the voltage control of the generator, but at the same time, it will weaken the damping of the system, even provide negative damping, making the system produce low-frequency oscillation. PSS is also known as the additional control of an excitation system, which can solve the contradiction between maintaining generator terminal voltage and generating negative damping in excitation system. PSS can not only overcome the negative damping brought by quick excitation system, but also provide positive damping to the system.

PSS adopts the feedback control of power and frequency, but the feedback control is not based on the deviation comparing with a reference value, but based on the differential or inertial differential link, so it is equivalent to a dynamic correction in parallel to the voltage feedback. No matter how the steady-state operation values of power and frequency change, they will not affect the change of steady-state voltage value, but only play a role in the dynamic process.

2. Power system stabilizer model

The PSS input signal can take the speed bias or electric power bias or terminal voltage bias as required, or the combination of them. The output U_s acts on the input terminal

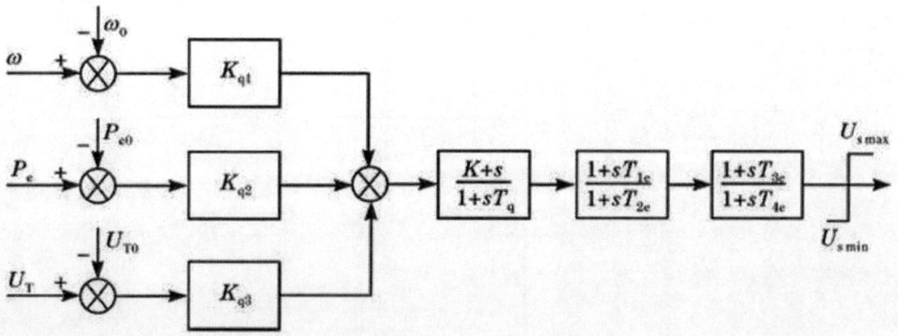

Fig. 3.17 Power system stabilizer model

of the excitation controller to improve the damping characteristics of the unit, that is, to increase the damping effect.

In Fig. 3.17, K_{q1}, K_{q2}, K_{q3} respectively are magnification times of speed deviation, power deviation, and voltage deviation; K is the coefficient of changing link; $U_{s\,min}$ and $U_{s\,max}$ are the upper and lower limits of stabilizer output; T_q is the time constant of blocking operation; T_{1e}, T_{2e}, T_{3e} and T_{4e} are the time constants of phase shift.

3.1.4 Prime Mover Governor

1. Characteristics of governors

The governor is an automatic device for adjusting the generator speed according to the deviation between the actual speed and the given value. Generally, the generator speed regulating system realizes the power frequency regulation by controlling the opening of the steam valve of the steam turbine or the opening of the guide vane of the hydraulic turbine. By changing the parameters and given values of the speed regulating system (generally given speed or given power), the required power frequency regulation characteristics of the generator can be obtained.

In the new stable and balanced status, if the load bearing capacity of the prime mover increases (decreases), but the speed decreases (increases), the speed regulating characteristic of this governor (that is, the relationship between the speed *n*, or the frequency *f* and the output power of the prime mover) is called the characteristic of deviation speed regulation, as shown in the curve ① in Fig. 3.18; if the speed (or frequency) is independent of the output power, it is called the characteristic of no deviation speed regulation, as shown in curve ② in Fig. 3.18.

The characteristic of speed regulation is expressed by the deviation regulation coefficient K_c:

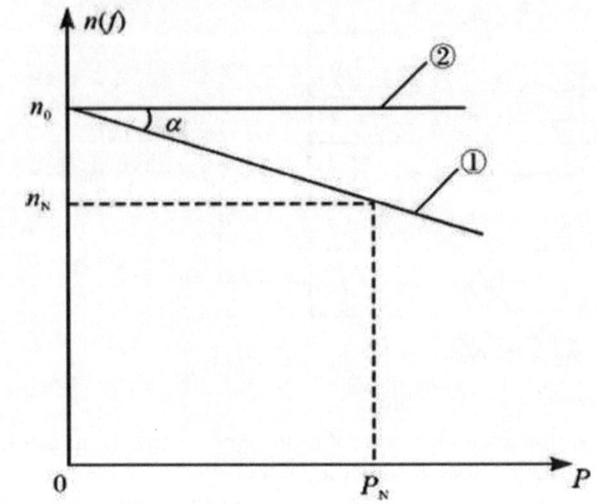

Fig. 3.18 Characteristic of speed regulation

$$K_c = -\frac{\Delta n}{\Delta P} = -\frac{\Delta f}{\Delta P} = \frac{n_0 - n_N}{P_N} = \tan\alpha \tag{3.32}$$

2. Governor model

A general governor model applicable to both water and thermal generator units presented here as shown in Fig. 3.19 can be used in the following three cases.

(1) General turbo generator unit: $T_w = 0$, $k_\beta = 0$, $T_i = 100{,}000$, $\alpha = 1$, $T_{rh} = 100{,}000$.

(2) Intermediate reheat turbo generator unit: $T_w = 0$, $k_\beta = 0$, $\alpha = 1/3$.

(3) General hydraulic turbine generator unit: $\alpha = 1$, $T_0 = T_w/2$.

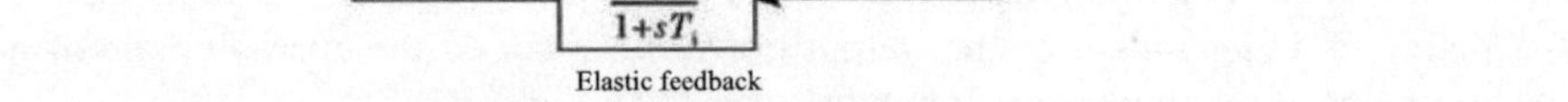

Fig. 3.19 Block diagram of governor transfer function

In which, T_{i} is soft feedback time constant; T_0 is the steam volume time constant; T_{rh} is the intermediate reheated steam volume time constant; α is the proportion of the output power of the HP cylinder.

3.2 Load

3.2.1 Composition of Load

Load is the electric equipment that transforms electric energy into other forms of energy. It can be generally divided into residential load, industrial load, commercial load, agricultural load and auxiliary equipment of power plant, and others.

Residential load is mainly household electrical appliances, including heating, refrigeration, water heater, household electrical appliances, lighting, entertainment and other electric equipment, such as electric heaters, air conditioners, resistance water heaters, electric stoves, refrigerators, washing machines, dishwashers, dryers, incandescent lamps, fluorescent lamps, televisions and others.

The industrial load includes more kinds of electric equipment, but the main ones are motor, arc furnace, rectifier type electrolytic equipment, and others.

The commercial load mainly includes the electric equipment such as incandescent lamps, fluorescent lamps, heat pumps, central air conditioners, window type air conditioners, fans, and motors, of which the central air conditioner is the one with the largest power consumption.

Agricultural load includes rural civil power consumption, production and irrigation power consumption and rural commercial power consumption, and others.

The auxiliary equipment of power plant is mainly the motor.

3.2.2 Static Load Characteristics

The power of loads changes with the change of system operation parameters (mainly voltage and frequency). The curve or the mathematical expression reflecting the change law is called load characteristics. Load characteristics include dynamic characteristics and static characteristics, and static characteristics include polynomial and exponential forms.

Without consideration of the frequency variation, the load static characteristics expressed by polynomial divides the load into three parts: constant impedance (power is proportional to the square of voltage), constant current (power is proportional to voltage) and constant power. The PV relation curves of these three load types are shown in Figs. 3.20, 3.21 and 3.22.

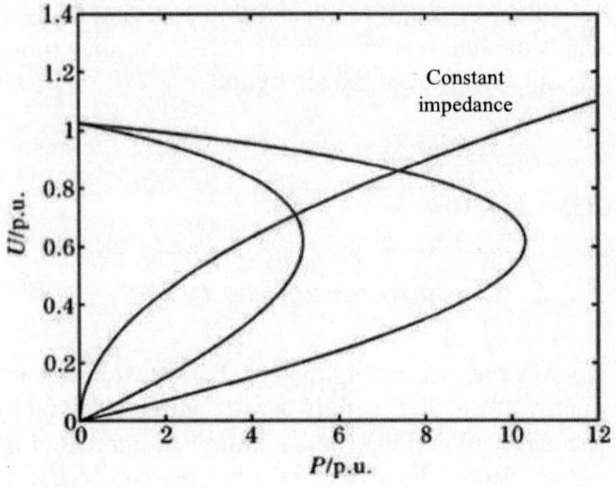

Fig. 3.20 *PV* relation schema of constant impedance load

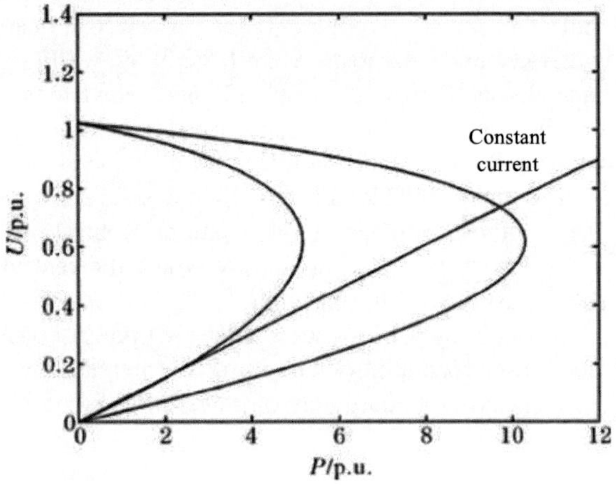

Fig. 3.21 *PV* relation schema of constant impedance current

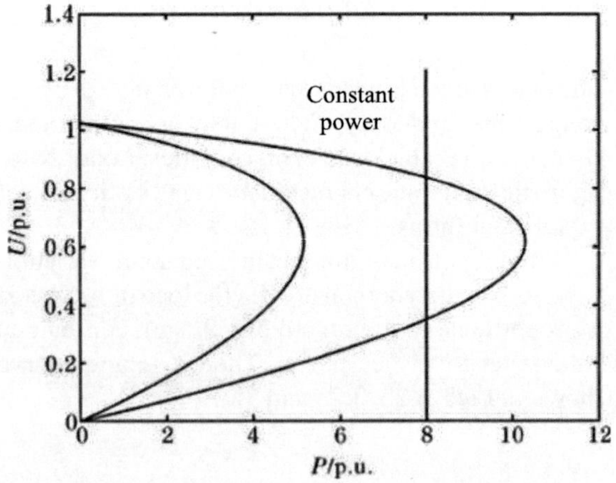

Fig. 3.22 *PV* relation schema of constant impedance power

For the constant impedance load, there is always an intersection between the PV curve and the nose curve of the system, that is, the system always has an operating point. For constant current and constant power load, if the system transmission capacity is reduced due to fault or disturbance, and the nose curve shrinks to the vertical axis, there may be no intersection between the PV curve and nose curve of the load, that is, the system voltage is unstable.

3.2.3 Dynamic Characteristics of Loads

The dynamic process of induction motors can be simulated by the equivalent circuit shown in Fig. 3.23. In the figure, R_s and X_s are the resistance and leakage reactance of the stator winding, R_r and X_r are the equivalent resistance and leakage reactance of the rotor winding, X_m is the excitation reactance, and s is the rotor slip of the induction motor.

Conduct Thevenin equivalent transformation to the stator impedance and excitation reactance in Fig. 3.23. The equivalent circuit is shown in Fig. 3.24.

In Fig. 3.24, the equivalent potential and equivalent impedance are respectively.

$$U_e = \frac{jX_m U_s}{R_s + j(X_s + X_m)} \tag{3.33}$$

$$R_e + jX_e = \frac{jX_m(R_s + jX_s)}{R_s + j(X_s + X_m)} \tag{3.34}$$

Thus, the active power absorbed by the induction motor can be expressed as

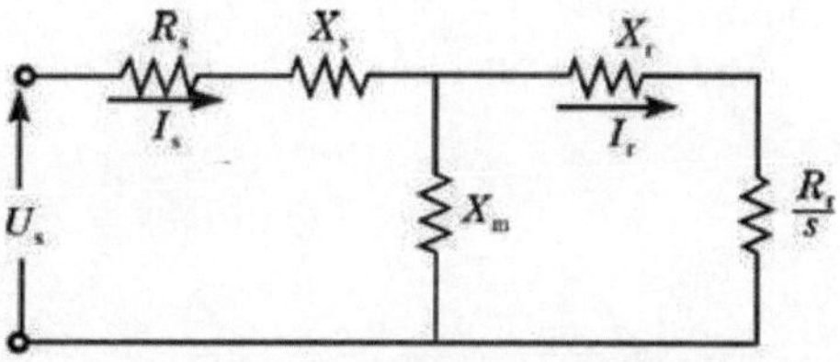

Fig. 3.23 Steady state equivalent circuit of induction motor

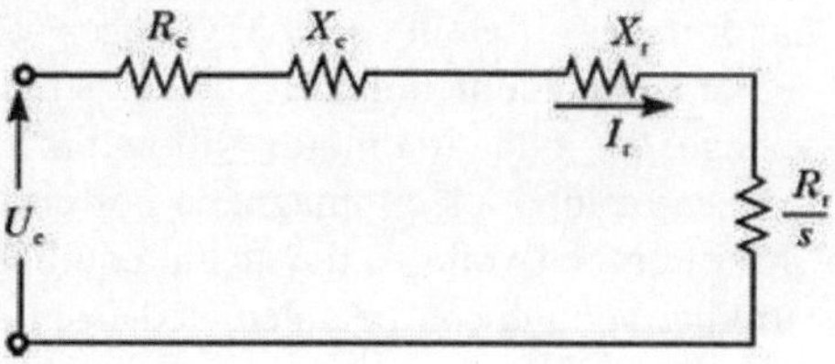

Fig. 3.24 Thevenin equivalent transformation circuit of induction motor

$$P = \frac{U_{\mathrm{e}}^2 (R_{\mathrm{r}}/s)}{(R_{\mathrm{e}} + R_{\mathrm{r}}/s)^2 + (X_{\mathrm{e}} + X_{\mathrm{r}})^2} \tag{3.35}$$

The electromagnetic torque of the corresponding induction motor is

$$T_{\mathrm{e}} = \frac{U_{\mathrm{e}}^2 (R_{\mathrm{r}}/s)}{(R_{\mathrm{e}} + R_{\mathrm{r}}/s)^2 + (X_{\mathrm{e}} + X_{\mathrm{r}})^2} \frac{1}{\omega_s} \tag{3.36}$$

where in, ω_s is the angular velocity of the stator magnetic field. The mechanical load torque of induction motor is

$$T_{\mathrm{m}}(s) = K_{\mathrm{L}}\left[\alpha + (1-\alpha)(1-s)^m\right] \tag{3.37}$$

where in, α is the mechanical load torque coefficient independent of speed; m is the mechanical load torque order related to speed; the load rate coefficient K_{L} can be obtained by the following formula:

$$K_{\mathrm{L}} = \frac{P_{(0)}}{\alpha + (1-\alpha)(1-s_{(0)})^m} \tag{3.38}$$

where in, $P_{(0)}$ is the initial active power of the induction motor, expressed as

$$P_{(0)} = \frac{U_{\mathrm{e}(0)}^2 (R_{\mathrm{r}}/s_{(0)})}{(R_{\mathrm{e}} + R_{\mathrm{r}}/s_{(0)})^2 + (X_{\mathrm{e}} + X_{\mathrm{r}})^2} \tag{3.39}$$

where in, $s_{(0)}$ is the initial rotor slip; $U_{\mathrm{e}(0)}$ is the initial equivalent voltage of induction motor.

The motion equation of load rotor of the induction motor is

$$T_j \frac{\mathrm{d}_s}{\mathrm{d}_t} = T_{\mathrm{m}} - T_{\mathrm{e}} \tag{3.40}$$

where in T_j is the inertia time constant of induction motor.

According to the above formula, the torque slip characteristic curves of induction motor under different terminal voltages can be obtained, as shown in Fig. 3.25.

For an induction motor in stable operation, $T_{\mathrm{e}} = T_{\mathrm{m}}$, that is, for a given terminal voltage, the induction motor will automatically adjust the rotor slip to achieve the balance of input electromagnetic power and output mechanical load power. For a given terminal voltage, the stable equilibrium operation slip corresponding to the terminal voltage can be calculated from the equation $T_{\mathrm{e}} = T_{\mathrm{m}}$. According to the known terminal voltage and slip, the corresponding active power and reactive power of the induction motor can be calculated. With the decrease of terminal voltage, if

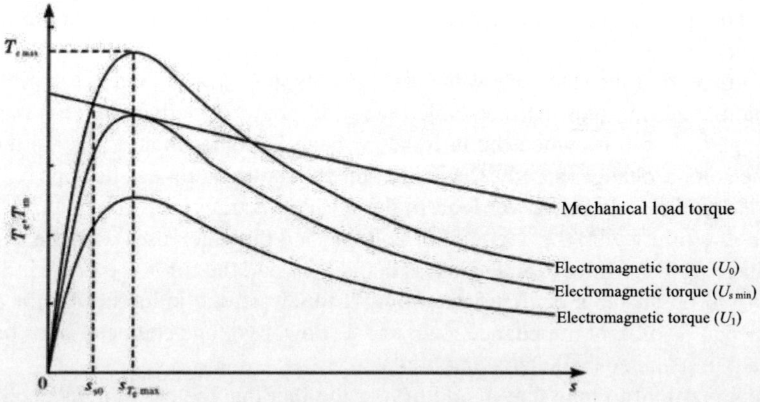

Fig. 3.25 Torque slip characteristic curves of induction motor

the characteristic curve of electromagnetic torque slip in the plane of torque slip is tangent to the characteristic curve of mechanical load torque slip at a certain steady-status voltage of $U_{\text{s min}}$, this voltage is called the critical voltage of a small disturbance stability of the induction motor. When the terminal voltage is less than $U_{\text{s min}}$, the equation $T_{\text{e}} = T_{\text{m}}$ has no solution, that is, there is no operating point, the induction motor enters the locked rotor process, the slip steps 1, and the induction motor stops. According to the relationship between absorbed power and terminal voltage of induction motor determined by equation $T_{\text{e}} = T_{\text{m}}$, the power voltage characteristic curve of induction motor as shown in Fig. 3.26 can be obtained (the

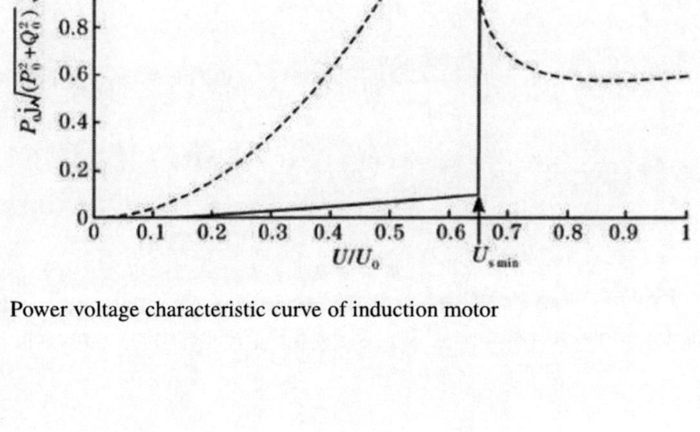

Fig. 3.26 Power voltage characteristic curve of induction motor

power voltage characteristic curves of different types of induction motor are slightly different).

In Fig. 3.26, if the terminal voltage is higher than $U_{s\,min}$, the induction motor has a stable operation point. At the stable operation point, its active power is basically unchanged (therefore, when the terminal voltage is higher than $U_{s\,min}$, in the case that the voltage changes slowly, the load of the induction motor can be approximately regarded as the constant power load in the *PV* curve analysis). The reactive power decreases slightly with the decrease of voltage, and then increases with the decrease of voltage. If the terminal voltage is less than $U_{s\,min}$ and the curve steps, the induction motor will be unstable and stop running. At this time, the induction motor can be regarded as a constant impedance load, and its power voltage characteristics become constant impedance load characteristics.

It is important to note that in addition to the fact that the continuous declination of terminal voltage can make the induction motor go to locked-rotor state, if the low voltage impact during fault or disturbance make the induction motor slip higher than the point of intersection between mechanical load torque and electromagnetic torque curve on the right side of the $s_{T_e max}$ axis as shown in Fig. 3.25, even if the voltage of the fault or disturbance disappeared after restored to the value of the pre-fault, the induction motor will continue to slow down until locked-rotor state. But this situation should belong to the instability of the induction motor itself, not the voltage instability.

3.2.4 Mathematical Description of the Load

1. Static load model

The voltage static characteristic and frequency static characteristic of loads are respectively simulated by polynomial, and quadratic polynomial fitting is usually adopted. The fitting results obtained in this way have sufficient accuracy within a certain voltage range.

The standard static load model recommended by IEEE Task Force in 1995 is as follows [4]:

$$\frac{P}{P_{frac}P_0} = K_{pz}\left(\frac{U}{U_0}\right)^2 + K_{pi}\frac{U}{U_0} + K_{pc} + K_{p1}\left(\frac{U}{U_0}\right)^{n_{pv1}}(1 + n_{pf1}\Delta f) + K_{p2}\left(\frac{U}{U_0}\right)^{n_{pv2}}(1 + n_{pf2}\Delta f) \tag{3.41}$$

$$\frac{Q}{Q_{frac}Q_0} = K_{qz}\left(\frac{U}{U_0}\right)^2 + K_{qi}\frac{U}{U_0} + K_{qc} + K_{q1}\left(\frac{U}{U_0}\right)^{n_{qv1}}(1 + n_{qf1}\Delta f) + K_{q2}\left(\frac{U}{U_0}\right)^{n_{qv2}}(1 + n_{qf2}\Delta f) \tag{3.42}$$

wherein, $K_{pz} = 1 - (K_{pi} + K_{pc} + K_{p1} + K_{p2})$; $K_{qz} = 1 - (K_{qi} + K_{qc} + K_{q1} + K_{q2})$; $Q_0 \neq 0$; P_{frac} and Q_{frac} represent the proportion of active and reactive static load in the total load respectively; K_{pz}, K_{pi} and K_{pc} respectively represent the active power of the constant impedance, constant current and constant power of the total load; K_{qz}, K_{qi} and K_{qc} respectively represent the reactive power of the constant impedance, constant current and constant power of the total load; K_{p1}, K_{q1}, K_{p2} and K_{q2} represent the part related to both voltage and frequency in the total load.

This static model has high adaptability. In practical application, the important terms of the model are often selected to derive some practical models according to the specifics of the problem analyzed.

The static model commonly used in China is usually expressed by the following polynomials:

$$P = P_0(A_p U^2 + B_p U^1 + C_p U^0)(1 + L_{dp}\Delta f) \tag{3.43}$$

$$Q = Q_0(A_q U^2 + B_q U^1 + C_q U^0)(1 + L_{dq}\Delta f) \tag{3.44}$$

wherein, $A_p + B_p + C_p = 1.0$; $A_q + B_q + C_q = 1.0$; $\mathrm{L_{dp}} = \frac{\mathrm{dP}}{\mathrm{df}}\big|_{\mathrm{f=f_0}}$, the value range is 0–3.0, generally 1.2–1.8; $\mathrm{L_{dq}} = \frac{\mathrm{dQ}}{\mathrm{df}}\big|_{\mathrm{f=f_0}}$, the value range is −2.0 to 0, generally −2.0.

The coefficients A, B, and C respectively represent the proportion of the constant impedance (Z), constant current (I), and constant power (P) of the load in the node load, so they are often referred to as the ZIP model.

2. Static load model

In the load dynamic model with the electromechanical transient of the induction motor considered, the stator winding transient is ignored and only the rotor winding transient and rotor motion dynamic are considered. This model accurately reflects the influence of the rotor winding electromagnetic transient on the electromagnetic moment, and has better simulation accuracy than the load dynamic model with only the mechanical transient considered, and has been widely used in power system stability analysis. The equivalent circuit of induction motor load is shown in Fig. 3.27.

The mathematical expression is:

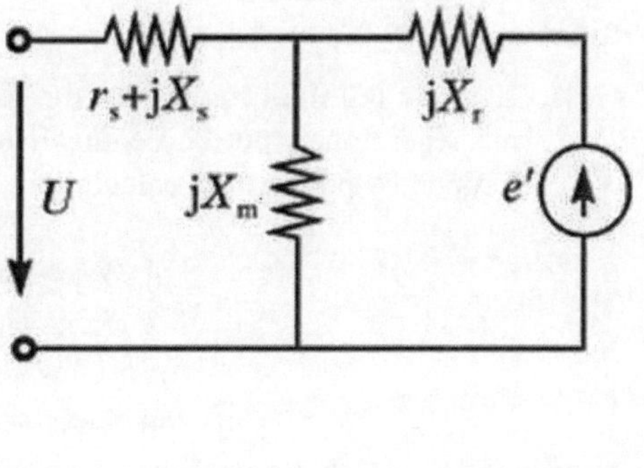

Fig. 3.27 Equivalent circuit of induction motor load

$$\begin{cases} \frac{de'}{dt} = -2\pi f_0 \cdot jse' - \frac{1}{T}\left[e' - j(X - X')I\right] \\ T_j \frac{ds}{dt} = T_m - T_e \\ U = e' + (r_s + jX)I \\ T_m = k\left[a + (1-a)(1-s)^\rho\right] \end{cases} \tag{3.45}$$

wherein, the equivalent reactance of the induction motor $X = X_s + X_m$; the transient reactance of induction motor $X' = X_s + X_r X_m/(X_r + X_m)$, wherein X_s is the leakage reactance of the asynchronous motor, X_m is the excitation reactance, and X_r is the rotor reactance; the stator open circuit rotor loop time constant $T = \frac{X_r + X_m}{\omega_0 r_r}$; T_m is the mechanical moment of the load; T_e is the electromagnetic moment of the motor; a is the proportion of the constant torque part; ρ is a torque parameter related to speed; T is the rotor loop time constant of the motor, which is T_{d0} when the stator is in open circuit; e' is the electric potential in the motor; s is the slip frequency of the asynchronous motor; T_j is the inertia constant.

3. Synthetic load model (SLM)

In the power system simulation calculation, the load is generally connected to the 220-kV or 110-kV bus side of the 220-kV substation, but the motor load, lighting and domestic electrical load cannot be directly supplied by the 220-kV or 110-kV bus, but by the 110-kV/10-kV transmission and power distribution network. Therefore, it is necessary to consider the influence of the distribution power grid in the load model. The equivalent circuit of synthesis load model with the power distribution network considered is shown in Fig. 3.28 [5, 6].

In addition to the simulation of the equivalent static load and motor load, in this load model, the equivalent generator, equivalent distribution power grid, and also capacitance compensation are considered, the input data including induction motor parameters and its proportion in the P_L, the proportion of static load component in P_{static} and Q_{static}, the power factor of the static load η_{pfac}, power distribution network equivalent impedance (per-unit value) based on the initial load capacity of equivalent branches, the active power P_G and reactive power Q_G output by equivalent generator, the unit parameters given by the generator card, and excitation speed regulating system parameters.

The SLM initialization steps in power system transient stability analysis are as follows:

(1) Calculate the initial power of the load according to the load power, equivalent generator output and equivalent impedance loss of distribution network obtained by power flow calculation.

$$\begin{cases} P_L = P_0 + P_G - P_{D0} \\ Q_L = Q_0 + Q_G - Q_{D0} \end{cases} \tag{3.46}$$

wherein, P_{D0} and Q_{D0} are the active power loss and reactive power loss of equivalent impedance of distribution network respectively.

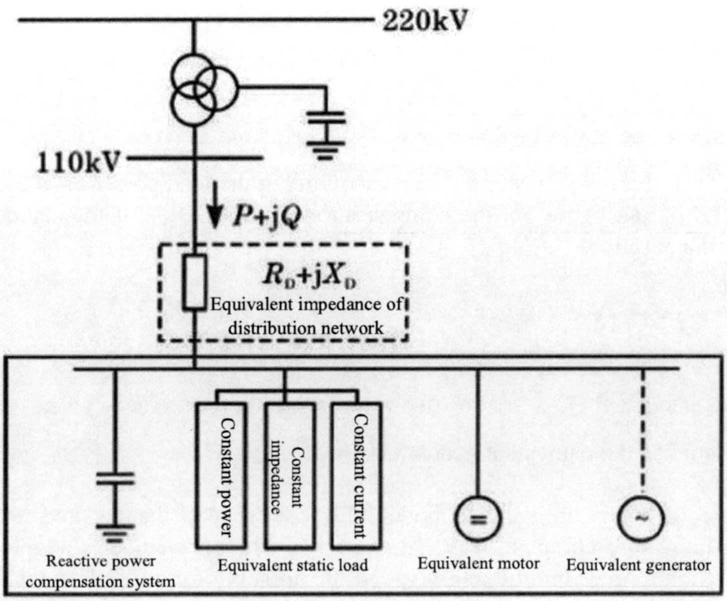

Fig. 3.28 Equivalent circuit of synthesis load model (SML) with power distribution network considered

(2) Initialize the admittance matrix according to the induction motor parameters, static load and generator parameters respectively.

(3) Calculate the bus voltage of the load

$$\dot{U}_{L0} = \dot{U}_{s0} - \frac{P_0 - jQ_0}{\overset{*}{U}_{s0}}(R_D + jX_D) \tag{3.47}$$

wherein, $\dot{U}_{s0}$ is the voltage of the system bus; $\overset{*}{U}_{s0}$ is the conjugate of the bus voltage of the system. If $|\dot{U}_{s0} - \dot{U}_{L0}| > 5\%$, a warning message is sent.

(4) Calculate the power loss of equivalent impedance of the distribution network

$$P_{D0} + jQ_{D0} = -\frac{P_0^2 + Q_0^2}{U_{s0}^2}(R_D + jX_D) \tag{3.48}$$

(5) Calculate Q_{IM0} according to part of the initialization parameters of the induction motor.

(6) Calculate the static active power and static reactive power of the load

$$\begin{cases} P_{\text{staic}} = P_{\text{L}} - P_{\text{IM0}} = P_{\text{Z0}} + P_{\text{I0}} + P_{\text{P0}} \\ Q_{\text{staic}} = P_{\text{staic}} tan\varphi = Q_{\text{Z0}} + Q_{\text{I0}} + Q_{\text{P0}} \end{cases} \tag{3.49}$$

wherein, φ is the power factor angle of static load, *and* $\varphi = \cos^{-1}(\eta_{\text{pfac}})$.

(7) Check the active power balance according to the load power balance formula (3.46), and figure out the compensation capacity Q_{SC0} of the static reactive compensation

$$\begin{cases} 0 = P_0 - P_{\text{D0}} - (P_{\text{IM0}} + P_{\text{Z0}} + P_{\text{I0}} + P_{\text{P0}} - P_{\text{G0}}) \\ -Q_{\text{SC0}} = Q_0 - Q_{\text{D0}} - (Q_{\text{IM0}} + Q_{\text{Z0}} + Q_{\text{I0}} + Q_{\text{P0}} - Q_{\text{G0}}) \end{cases} \tag{3.50}$$

If the calculated Q_{SC0} is a positive value, a warning message is given.

(8) Initialize the equivalent generator

Compared to the traditional load model, SLM can simulate the load and distribution system in a more complete way. The iterative solution method is adopted in the computational procedure of stability, and the amount of computation increases by very little. Characteristics of the SLM are as follows:

(1) The influence of distribution system impedance can be considered for both static load and motor load. The impedance simulation method is adopted for the distribution system to ensure that the model structure is more in line with the relationship between actual distribution system and electrical load. An accurate equivalent impedance of the distribution system can be obtained by using an appropriate equivalent method.
(2) SLM simulates the reactive compensation of the distribution system. Distribution networks and power consumers are equipped with a large number of reactive compensation devices, whose dynamic characteristics have an important impact on the stability of the system, and should be simulated in detail. SLM provides an effective simulation method for the reactive compensation model of distribution system.
(3) The small units of the distribution system can be considered conveniently. In power system simulation analysis (especially disturbance tests and accident analysis), it is sometimes necessary to consider small units connected to the distribution network. SLM contains small units, which can make the simulation of small units more convenient.
(4) Negative constant current and constant power can be avoided in static reactive power loads. The introduction of load power factor ensures that there is no negative constant current and negative constant power in reactive part of static load, which makes the model more in line with the actual situation.

3.3 Reactive Compensation Element

In the process of transmission of active power in the power system, the support of reactive power is needed to establish and maintain magnetic field in electrical equipment and complete the mutual conversion of electromagnetic energy. Reactive power does not act outward, but it provides voltage support to the system. Not only do most network elements consume reactive power, but most consumer loads also consume reactive power.

Reactive power compensation is an effective way to improve both power transmission capacity and voltage stability. The reactive power compensation of power transmission system is used to control voltage, improve power transmission capacity of transmission networks and stability of power system operation. The reactive power compensation of distribution systems is mostly the compensation of load, which is mainly to control reactive power, improve the power factor of load and improve the electric energy quality. The basic requirement of reactive power control is to minimize the reactive power flow between different voltage levels, which is the general principle of reactive power compensation, namely, on-site reactive power compensation by levels and sections.

Reactive compensation can be divided into series compensation and shunt compensation, and also active compensation and passive compensation. This section mainly introduces series capacitor, shunt reactor and capacitor, FACTS device, synchronous condenser and other reactive power compensation equipment.

3.3.1 *Series Capacitor*

In general, series capacitor compensation is associated with the improvement of power transmission capacity and transient stability of long distance power transmission lines.

The series capacitor reduces the inductive reactance of the transmission line and effectively shortens the length of the line to realize the long-distance transmission of power. The reactive power compensation (I^2X_C) it sends out compensates the reactive power consumption (I^2X_L) of the power transmission line. The simple equivalent circuit is shown in Fig. 3.29.

After the addition of series compensator, the net resistance of the circuit is

$$X_{\text{net}} = X_L - X_C = \omega L - \frac{1}{\omega C} \tag{3.51}$$

In this case, the series compensation degree is

$$z_{\text{se}} = \frac{X_L - X_{\text{net}}}{X_L} = \frac{X_C}{X_L} \tag{3.52}$$

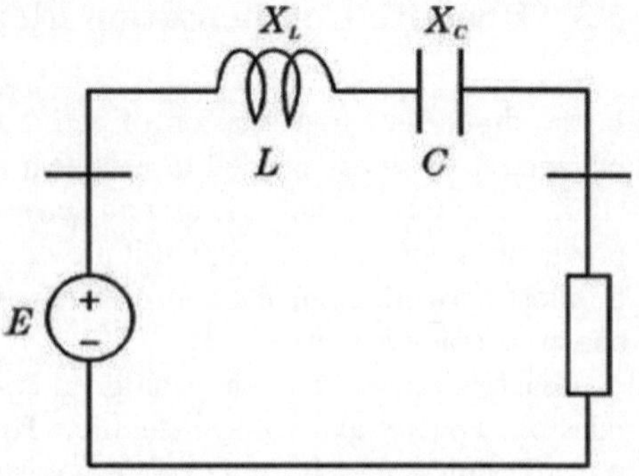

Fig. 3.29 Series capacitor compensation principle

wherein, Z_{se} is the series compensation degree, which usually ranges from 30 to 80%.

The reactive power produced by series capacitors is proportional to the square of the current and independent of the node voltage, so the series capacitors can provide the most reactive power when the system needs the reactive power most. This fast and inherent self-regulation is a very important characteristic of series capacitor compensation, which is very beneficial to voltage stability.

Different from shunt capacitors, series capacitors can reduce the characteristic impedance and electrical length of the circuit, thus effectively improving the voltage regulation characteristics and rotor Angle stability.

Figure 3.30 is a schematic diagram of a typical series capacitor compensation device comprising a capacitor and its protective devices, namely, the metal oxide varistor (MOV), trigger gap, and bypass switches.

The MOV in the fixed series capacitor compensation device (hereinafter referred to as series compensation) protects the capacitor by limiting the voltage at both ends of the capacitor. The function of the MOV can be simulated by using the linearized model. When the line current I_L is higher than 0.98 of capacitor protection level current I_{pr}, MOV is energized and the whole fixed series complement is also replaced by a linearized model, as shown in Fig. 3.31.

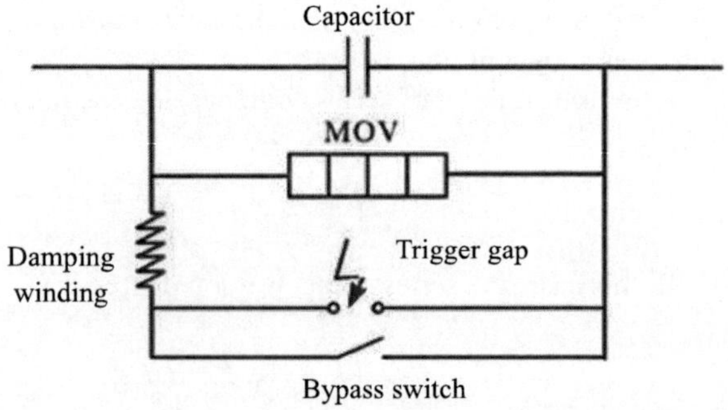

Fig. 3.30 Schematic diagram of typical series capacitor compensation device

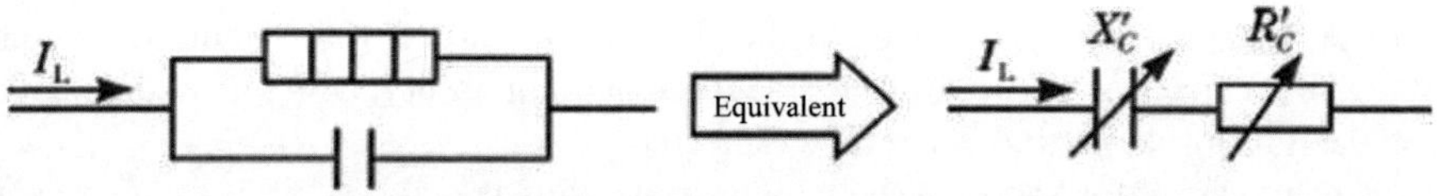

Fig. 3.31 Conducted linearized model of MOV

The equivalent resistance and reactance in the linearized model are functions of the line current, the basic relation is

$$R_C' = X_{C0}(0.0745 + 0.49\mathrm{e}^{-0.243 I_{\mathrm{p.u.}}} - 35.0\mathrm{e}^{-5.0 I_{\mathrm{p.u.}}} - 0.6\mathrm{e}^{-1.4 I_{\mathrm{p.u.}}}) \tag{3.53}$$

$$X_C' = X_{C0}(0.1010 - 0.005749 I_{\mathrm{p.u.}} + 2.088\mathrm{e}^{-0.8566 I_{\mathrm{p.u.}}}) \tag{3.54}$$

wherein, $I_{\mathrm{p.u.}}$ is the per-unit value of the line current relative to the protection level of the capacitor, namely, $I_{\mathrm{p.u.}} = \frac{\mathrm{I_L}}{\mathrm{I_{pr}}}$. Wherein, I_{L} is the line current, I_{pr} is the protection level current of the capacitor; X_{C0} is the capacitive reactance of the capacitor under normal conditions.

After the MOV is powered on, if the line current is less than 0.98 of the capacitor protection level current, the MOV stops the energizing. The MOV can absorb a certain amount of energy after energizing. When the absorbed energy reaches the limit value, the trigger gap triggers the energizing to short circuit the whole series complement device. If the peak current in MOV reaches the set value, in order to avoid the MOV energy from growing too fast, the trigger gap should also trigger the energizing. The trigger gap sends the switch on command to the bypass switch in the process of energizing. When the series complement of bypass is completed, the bypass is put back into operation after a set time.

3.3.2 *Shunt Capacitor and Shunt Reactor*

Shunt capacitors and reactors have been the most widely used important equipment to control the distribution of voltage and reactive power in power systems, and are also an important means of regulating the system voltage.

Since the switching values of shunt capacitors or reactors are discrete and their operating characteristics determine that shunt capacitors or reactors cannot be operated frequently, this reactive power regulation method can only control the voltage of a bus within a certain range. Therefore, the bus voltage that can be controlled by a switchable shunt capacitor or reactor can be set within a range (U_{min}, U_{max}). When the voltage is lower than the lower limit U_{min}, switch on the capacitor or switch off the reactor, while when it is higher than the upper limit U_{max}, switch off the capacitor or switch on the reactor. When the capacity of a switchable shunt capacitor or reactor is exhausted, the voltage of the controlled bus will be out of control, and the switchable shunt capacitor or reactor will lose its regulating ability, which is equivalent to a capacitor or reactor that is fixedly connected.

A shunt capacitor can effectively expand its voltage stability limit by increasing the load power factor of the receiving system, and it can also be used to release the "rotating reactive power reserves" of the generator, allowing nearby generators to operate near the power factor of 1.0, which is equivalent to increasing the dynamic reactive power reserve that the system can quickly respond to and is very beneficial to voltage stability.

However, from the viewpoint of voltage stability and control, shunt capacitors have several inherent limitations:

(1) In a system in which shunt capacitor compensation is widely applied, the voltage regulation capacity becomes worse.
(2) The reactive power generated by the shunt capacitor is proportional to the square of the voltage, and the output of reactive power decreases during the LV of the system, which is a vicious circle. This is completely opposite to the self-regulation performance of series capacitors.
(3) In the transient process of power systems, the switching speed is not fast enough to prevent transient voltage instability. If voltage collapse causes power system disconnection, the stable part of the system may suffer from overvoltage at the moment of disconnection. In the process of voltage attenuation, switching of shunt capacitors will aggravate the overvoltage of the power system.

In general, the above discussion on shunt capacitors is also applicable to switchable shunt reactors. A shunt capacitor is always installed on the bus instead of the line, and the shunt reactor can be connected to either the line or the bus or the third winding of a transformer.

Figure 3.32 is the schematic diagram of a system including a switchable shunt capacitor.

Fig. 3.32 Schematic diagram of a system including a shunt capacitor

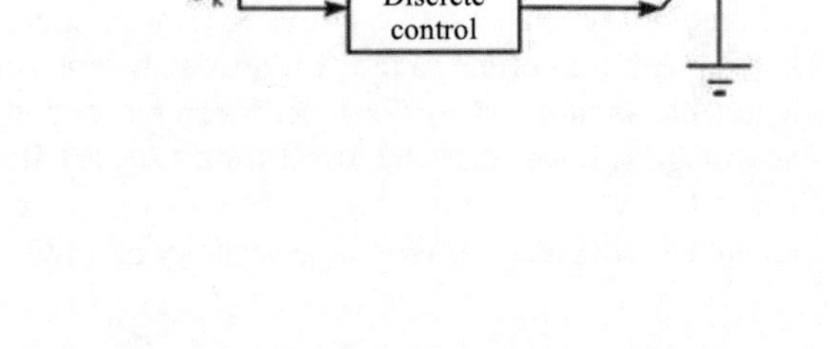

3.3.3 Static Var Compensator

A static var compensator (SVC) is a piece of reactive power compensation equipment. General Electrical and Westinghouse in the United States manufactured the world's first SVC in the late 1970s. SVC is a compensation device that can control reactive power. It overcomes the shortcomings of mechanical switching of shunt capacitors and reactors, and can track the fluctuation of reactive power in the power grid or load, compensate the reactive power in real time, and regulate voltage quickly and accurately.

SVC is developed to solve flicker caused by the electric arc furnace. It can regulate capacitive or inductive current output to maintain or control some specific operating parameters of power systems. SVC is no longer only used in power transmission networks, but also widely used in power distribution systems, such as electric arc furnaces, rolling mills and other industrial loads and smaller loads (as small as 1 Mvar). SVC can reduce voltage drop in case of starting a large motor, and can also be applied to single-phase loads such as electric welding machines and electrified railway power supply systems.

SVC is a device that is not limited by the lead-lag range, mostly has no response delay, and can quickly regulate reactive power. The voltage is regulated according to its slope characteristic. This slope is related to the steady-state gain of the regulator, and usually the regulation range is 1–5%. When the limit is reached, SVC becomes a simple capacitor and loses its voltage control capability.

Under normal conditions, SVC works in the inductive output or no-deviating regulation range so as to obtain fast capacitive reactive power output in case of disturbance. The schematic diagram of a system with a SVC connected is shown in Fig. 3.33a, which is similar to the switchable shunt capacitor or reactor, but SVC can smoothly change the equivalent reactance of its system, and the voltage of a bus in the control system is near the given value.

The actual operating characteristics of SVC are shown in Fig. 3.33b. According to its operating characteristics, the SVC model can be described in the following three operating states.

1. Linear control state

(a) Schematic Diagram of System with a SVC Connected (b) Operating Characteristics of a SVC

Fig. 3.33 SVC model

If SVC is used for no-deviating regulation and the voltage of controlled bus K is kept constant, bus K can be regarded as *PV* node controlled by SVC. Therefore, SVC connected to bus I shall be regulated accordingly and injected with appropriate reactive power.

If SVC is used for deviating regulation, the reactive power injection of SVC can be known from the operating characteristics of SVC as follows

$$Q_{\mathrm{SVC}} = U_{\mathrm{L}} \frac{U_{\mathrm{ref}} - U_{\mathrm{K}}}{X_{\mathrm{sl}}} \tag{3.55}$$

In the formula, X_{sl} is a reactance representing SVC deviating regulation characteristics; U_{ref} is the given reference voltage of the controlled bus.

2. Capacitive limit state

If the equivalent susceptance in a system with a connected SVC reaches the maximum $B_{\max}$ and is still insufficient to maintain the voltage of bus K, SVC will lose the ability to regulate the voltage, and is equivalent to a fixed susceptance shunt to bus I, then

$$Q_{\mathrm{SVC}} = U_{\mathrm{I}}^2 B_{\max} \tag{3.56}$$

3. Inductive limit state

If the equivalent susceptance in a system with a connected SVC connected reaches the minimal $B_{\min}$ and is still insufficient to suppress the voltage of bus K, SVC will lose the ability to regulate the voltage, and is equivalent to a fixed susceptance shunt to bus I, then

$$Q_{\mathrm{SVC}} = U_{\mathrm{I}}^2 B_{\min} \tag{3.57}$$

Under normal conditions, SVC works in the inductive output or no-deviating regulation range so as to obtain fast capacitive voltage rise. In order to obtain capacitive reserves, shunt capacitor banks or shunt reactors near SVC can be switched on and off.

Figure 3.34 is a model block diagram of SVC.

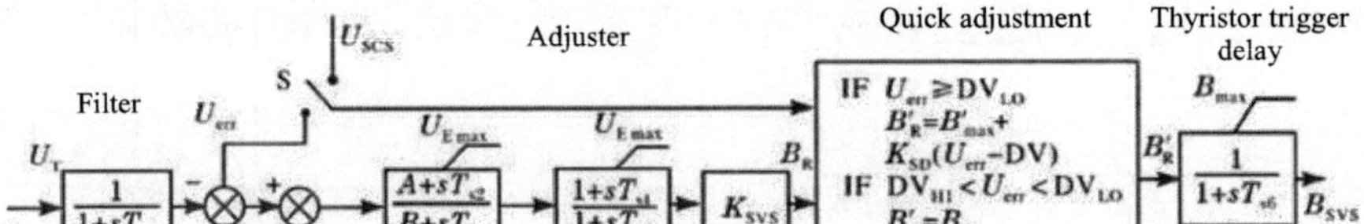

Fig. 3.34 Block diagram of SVC model

3.3.4 Static Compensator

With the emergence of new high-power switching devices and the introduction of phase control technology, pulse width modulation (PWM) technology and four-quadrant converter technology, power electronic converter technology has developed rapidly. A new generation of FACTS device developed on this basis, static compensator (STATCOM), has been a research hotspot due to its many excellent performances such as smooth control of reactive power, regulation of system voltage, correction of power factor and load balancing, and has been gradually applied at home and abroad.

STATCOM is mainly used for power flow control, reactive power compensation and system stability improvement in power transmission networks, and is mainly used to improve power quality and power supply reliability in power distribution networks. Its operating principle is to exchange reactive power with AC system by regulating the amplitude and phase of output voltage. The three-phase AC voltage output by STATCOM is synchronized with the three-phase AC voltage of the connected power grid and connected in shunt to the transmission line. Although its output characteristics are similar to those of a rotating synchronous condenser, it can provide fast response speed and symmetrically leading or lagging reactive current, thus regulating the reactive power of the transmission line, balancing the voltage of the transmission line, and keeping the static or dynamic voltage of the transmission line running within the allowable range of the system. Also, its smooth and continuous control can reduce the voltage fluctuation generated by passive devices to a minimum, which is conducive to improving the stability of the power system.

The basic operating principle of STATCOM is shown in Fig. 3.35. Its principle is: When the output voltage amplitude of STATCOM is less than the equivalent voltage amplitude of the system, STATCOM absorbs leading reactive power, which is equivalent to the function of a reactor; when the output voltage amplitude of STATCOM is equal to the equivalent voltage amplitude of the system, STATCOM

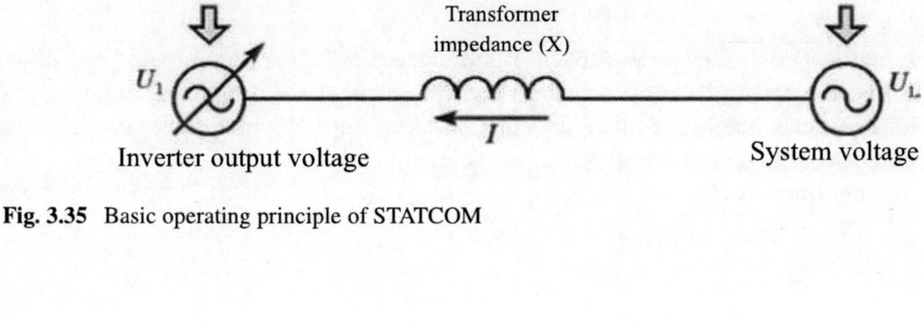

Fig. 3.35 Basic operating principle of STATCOM

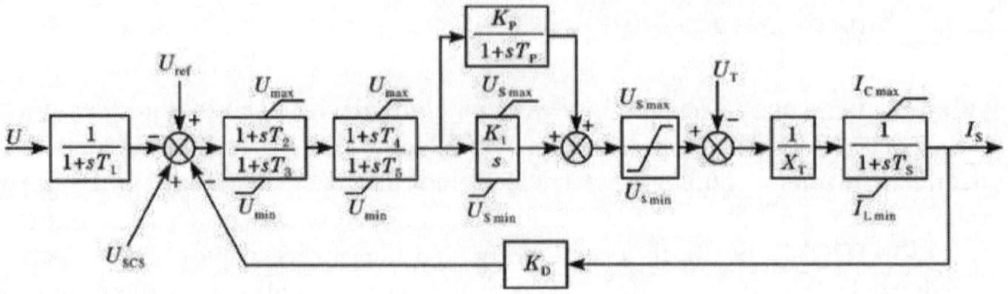

Fig. 3.36 STATCOM model

has no reactive power exchange with the system; when the output voltage amplitude of STATCOM is larger than the equivalent voltage amplitude of the system, the STATCOM outputs the leading reactive power, which is equivalent to the function of a capacitor. In fact, the operation of STATCOM is so complicated that its functions far exceed those of capacitors and reactors.

An outstanding feature of STATCOM is that it can provide dynamic voltage support. Installing dynamic reactive power compensation devices with appropriate capacity at appropriate locations in the power grid is the main means to improve the voltage stability of the system. STATCOM has better reactive power compensation characteristics than SVC at LV. Research shows that a STATCOM is equivalent to about 1.3 times of a SVC with the same capacity at LV, and the STATCOM response speed is about 20 ms, which is higher than SVC response speed.

Since STATCOM uses power electronic converters to generate reactive power, it has the characteristics of fast response speed, no need of load capacitance/reactance, and better transient reactive power compensation characteristics, etc., thus it has the functions of controlling node voltage, realizing instantaneous reactive power compensation, damping system oscillation, enhancing system transient stability, and improving power quality, etc. STATCOM can provide strong support for the voltage near the control point during the transient process after the system accident, thus improving the voltage recovery capability after the power grid accident and greatly improving the transient stability problem caused by the voltage drop after the accident, so it has received extensive attention.

STATCOM model is shown in Fig. 3.36.

3.3.5 *Thyristor Controlled Series Capacitor*

A thyristor controlled series capacitor, abbreviated as TCSC, was proposed by Vithayathil and others in 1986. As a means to quickly regulate the impedance of power grid lines, it can smoothly regulate the compensating impedance of power transmission lines in a large range through series variable impedance in the middle of power transmission lines. It is mainly used for network power flow control, improving power

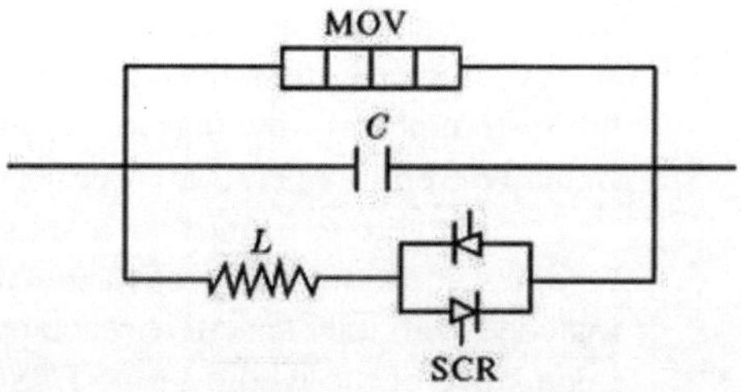

Fig. 3.37 Basic structural diagram of TCSC

grid flow distribution, eliminating circulation, improving system transient stability, damping control of power system, and inhibiting low frequency and sub-synchronous oscillation.

The basic operating principle of TCSC is that the thyristor controlled reactor (TCR) can be partially turned on within half a cycle of power frequency by utilizing the fast controllable capability of the thyristor, thus changing the charging and discharging state of the capacitor, while the line current remains basically constant, thus changing the equivalent reactance of TCSC. According to the different conduction states of TCR, the total equivalent reactance of TCSC can be either capacitive reactance or inductive reactance. Therefore, as long as the conduction angle of the thyristor is accurately regulated, the equivalent reactance of the TCSC can be quickly, continuously and accurately controlled.

The basic structural diagram of TCSC is shown in Fig. 3.37. It is mainly composed of 4 components: capacitor *C*, bypass inductor *L*, two anti-shunt high-power thyristors SCR, metal oxide voltage limiter MOV, and the actual device also includes a bypass circuit breaker, etc. Compared with fixed series compensation, an inductance branch controlled by bidirectional thyristor is added in TCSC, and the equivalent reactance of TCSC can be smoothly regulated by regulating the trigger angle of controllable series compensation.

Under normal circumstances, TCSC has four basic operating modes: thyristor blocked, capacitive vernier operation, inductive vernier operation, and thyristor bypassed.

(1) Thyristor blocked. At this time, the thyristor is locked, which is equivalent to the trigger angle $\alpha = 180°$. TCSC is equivalent to fixed series compensation, and the corresponding capacitive reactance value is called the basic capacitive reactance value. The line compensation of TCSC in this state is called the basic compensation. TCSC must operate in thyristor blocked mode before it is switched on, so this mode is the most basic mode of TCSC operation.
(2) Capacitive vernier operation. When $\alpha_{\min} \le \alpha < 180°$, the capacitive reactance value of TCSC is between its minimum value (basic capacitive reactance value) and maximum value (usually 1.7–3 times the basic capacitive reactance value), which mainly depends on conditions such as line current and short-term overload capacity of TCSC. TCSC is usually operated in capacitive vernier operation. In the transient process, the transient stability of the system can be improved by increasing its capacitive reactance value. In the dynamic

process, its impedance can be controlled to suppress system oscillation. During steady-state operation, the capacitive reactance value can be regulated to make the system power flow reasonably distributed and reduce the power loss.

(3) Inductive vernier operation. In case $90° \leq \alpha \leq \alpha_{max}$, TCSC presents as an inductive regulatable reactance. If a set of TCSC device is composed of multiple modules, the inductive reactance regulating mode of different modules is matched with the capacitive reactance regulating mode, so that the equivalent impedance of the whole set of TCSC device can be continuously regulatable in a larger range.

(4) Thyristor bypassed. At this time, the trigger angle $\alpha = 90°$, the thyristor is fully conducted, and the TCSC is used as a small inductive reactance. During the short circuit fault period of the system, TCSC operates in thyristor bypassed mode and uses its small inductive reactance to increase the line impedance, thus reducing the fault current.

The control block diagram of TCSC adopts linear control combined with Bang-Bang control, as shown in Fig. 3.38.

The linear control part in Fig. 3.38 adopts a control mode similar to PSS or PID control, as shown in Fig. 3.39.

The linear control part can adopt a combination of power control and voltage control, or only power control or voltage control, and each control module can use PID control or PSS-like control.

The amplitude limiting of TCSC mainly considers the voltage and current tolerance of capacitor. The reactance determined by the capacitor voltage and current tolerance is

$$X_{\mathrm{TCSC,max}}^{\mathrm{T}} = X_{\mathrm{TCSC,rated}} \left(\frac{I_{\mathrm{T}}}{I_{\mathrm{N}}} \right)^{\beta} K_{\mathrm{T}} \tag{3.58}$$

In addition, the control block diagram of TCSC is also provided with two open-loop controls, i.e., transient control and bypass control. When the three phases are disconnected after the short circuit of the line, the transient stability control acts to regulate the capacitive reactance of the TCSC of the shunt line to the maximum, and after compensation for a period of time, it is switched to the normal regulating mode; the compensation time needs to be preset.

MOV is included in TCSC. Its function is to protect the capacitor by limiting the voltage across the capacitor. The linearized model is used to simulate the function of MOV, which is consistent with the function and model of MOV in fixed capacitor compensation.

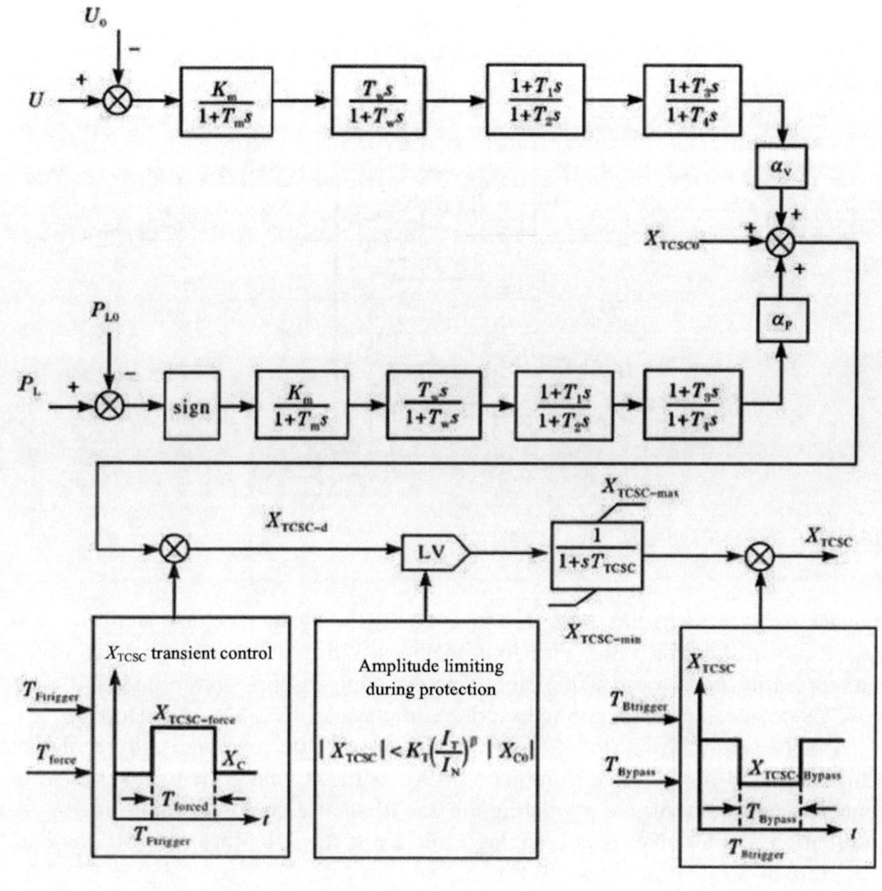

Fig. 3.38 Control block diagram of TCSC

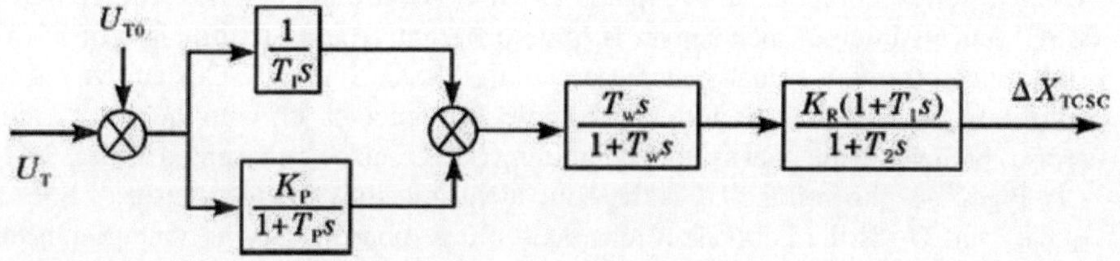

Fig. 3.39 Block diagram of linear control part of TCSC

3.3.6 Static Synchronous Series Compensator

STATCOM for shunt reactive power compensation has long attracted people's attention, and some products have been put into commercial operation. It realizes reactive

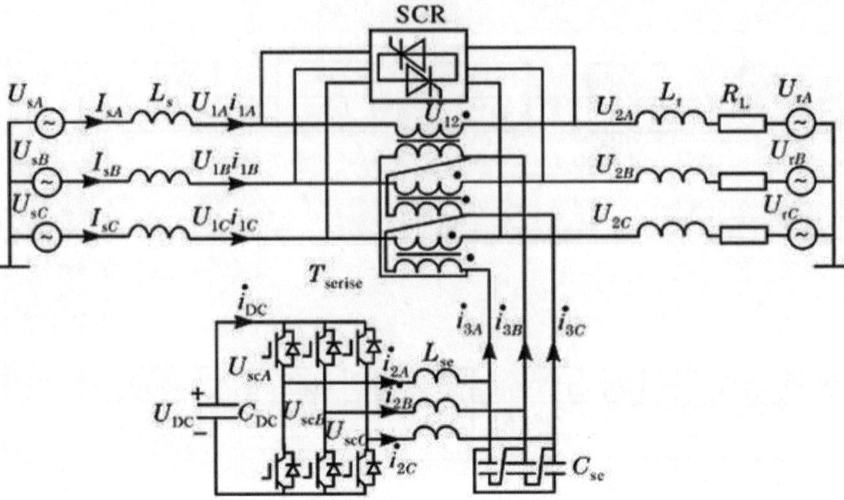

Fig. 3.40 Basic structural diagram of SSSC

power exchange with the system and controls the system voltage with the VSC based on full-control power electronic devices. Using the same principle, VSC is connected in series to provide reactive power compensation in the line, and a new FACTS device, static synchronous series compensator (SSSC), is developed.

SSSC realizes the control of line voltage by connecting a voltage with regulatable amplitude and phase angle difference of 90° with the line current in series in the line, and plays the roles of regulating the line transmission power, inhibiting system oscillation and the like, thus realizing effective system stability control. The basic structure of SSSC is as shown in Fig. 3.40.

SSSC is equivalent to having a variable resistor in series in the circuit. From the external characteristics, it is similar to series capacitors (fixed capacitor, thyristor switched series compensation capacitor TSSC, controllable series compensator TCSC), but its internal mechanism is quite different. The basic principle of SSSC is introduced through a dual system containing SSSC. Since SSSC is equivalent to connecting a voltage source with phase angle difference of 90° with the line current in series in the line, the dual system containing SSSC can be represented as Fig. 3.41.

In Fig. 3.41, the voltage $\dot{U}_q$ is the equivalent injecting voltage phasor of SSSC, $\dot{U}_1$, $\dot{U}_2$, and $\dot{U}_L$ is the equivalent terminal voltage phasor and line voltage phasor at both ends of the system respectively, and I is the line current phasor. The SSSC equivalent injecting voltage phasor in Fig. 3.41 can be expressed as $\dot{U}_q = K\dot{I}e^{-j90^\circ}$, wherein K is a constant. The phasor diagram of the compensation effect of this equivalent injecting voltage can be shown in Fig. 3.42. Due to the use of VSC based on fully-controlled power electronic device, the magnitude of SSSC injecting voltage can be independent of the line current and system impedance, and the controllable voltage $\dot{U}_q$ in series in the line and the line reactance voltage drop phase are opposite (capacitive compensation) or the same (inductive compensation) through control in

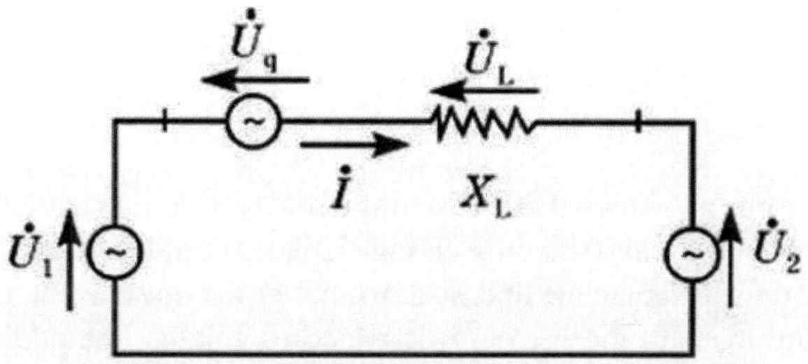

Fig. 3.41 Equivalent circuit of dual system with SSSC

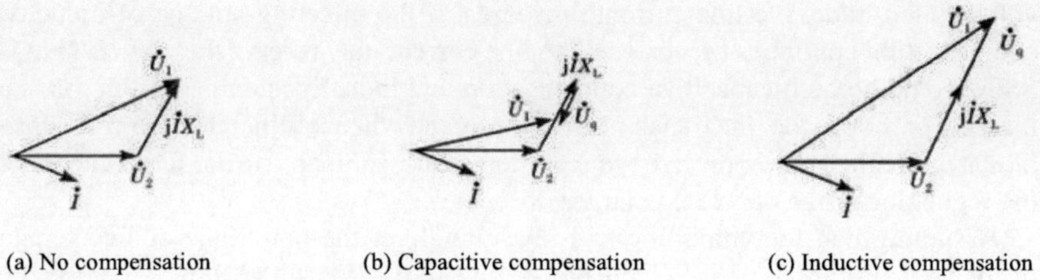

Fig. 3.42 SSC compensation phasor diagram

order to perform a similar effect of series capacitance or series inductance, therefore, its external characteristics are similar to those of TCSC. Especially when SSSC is in capacitive compensation and on the premise of maintaining the same transmission power, the transmission capacity and stability margin of the system are improved by reducing the voltage drop and phase difference of both ends of power transmission line.

SSC regulates line current by controlling the injecting voltage amplitude. Assuming that in Fig. 3.42, $\dot{U}_1 = U_1\angle 0°$, $\dot{U}_2 = U_2\angle{-\delta}$, $\dot{I} = I\angle - \theta$, $\dot{U}_q^{(C)} = U_q\angle(-\theta - \pi/2)$ refers to capacitive compensation, $\dot{U}_q^{(L)} = U_q\angle(-\theta + \pi/2)$ refers to inductive compensation, X_L is line reactance (including coupling transformer leakage reactance), and $U_1 = U_2 = U$, *then*

$$\dot{I} = \frac{\dot{U}_1 - \dot{U}_2 - \dot{U}_q}{jX_L} = \frac{U\sin\delta - jU(1-\cos\delta)}{X_L}\left[1 \pm \frac{U_q}{U\sqrt{2(1+\cos\delta)}}\right] \tag{3.59}$$

The line current amplitude without SSSC is

$$I_{\text{Without SSSC}} = \frac{\sqrt{U_1^2 + U_2^2 - 2U_1U_2\cos\delta}}{X_L} = \frac{U\sqrt{2(1-\cos\delta)}}{X_L} \tag{3.60}$$

The line current amplitude with SSSC is

$$I_{\text{With SSSC}} = \frac{U\sqrt{2(1-\cos\delta)}}{X_L}\left[1 \pm \frac{U_q}{U\sqrt{2(1+\cos\delta)}}\right] = I_{\text{Without SSSC}} \pm \frac{U_q}{X_L} \tag{3.61}$$

The above formula shows that phase angle of the line current is not affected by the injecting voltage U_q and the only change is the amplitude. SSSC is equivalent to adding an injecting current to the line, so as to change the line current. In case of capacitive compensation, with the increase of injecting voltage, the additional injecting current amplitude increases which makes line current increase, the maximum line current depends on SSSC the compensation capacity and the limitations on line transmission capacity; in case of inductive compensation, with the increase of injecting voltage amplitude, the line current decreases, if the injecting voltage of inductive compensation continues to increase, the line current may reserve the power. That is, SSSC could be both capacitive compensation and inductive compensation that can increase or reduce the line transmission power, and when it is in inductive compensation, it can realize power reverse transmission. Thus, compared to the series capacitor, the regulating range of SSSC is larger.

Assuming that the amplitude and phase angle at the both ends of line remain unchanged after input of SSSC, the power flow of the line with SSSC is the function of rotor angle difference δ and SSSC equivalent voltage $\dot{U}_q$, the transmission active power of the line can be expressed as

$$P_1 = \frac{U_1U_2\sin\delta}{X_L}\left(1 \pm \frac{U_q}{\sqrt{U_1^2 + U_2^2 - 2U_1U_2\cos\delta}}\right) = \frac{U^2}{X_L}\sin\delta \pm \frac{U}{X_L}U_q\cos\frac{\delta}{2} \tag{3.62}$$

Therefore, the change of line active power after input of SSSC is

$$\Delta P = \pm\frac{U}{X_L}U_q\cos\frac{\delta}{2} \tag{3.63}$$

As the line resistance is ignored, the change of active power is entirely raised due to SSSC, for different U_q amplitudes (capacitive compensation is positive), the relation curve of line power P and phase angle difference δ between the buses on both sides is shown in Fig. 3.43.

SSSC injecting voltage has an obvious control effect to the active power of the line, and can improve the characteristic curve of rotor angle. Then the maximum value of active power is shifted, so as to improve line power with the same rotor angle difference or keep the same line transmission power with the smaller rotor angle difference. Under the condition of capacitive compensation, the operation of SSSC effectively improves the rotor angle curve, reduces the acceleration area and increases the deceleration area, so as to improve the steady state rotor angle stability margin of the system.

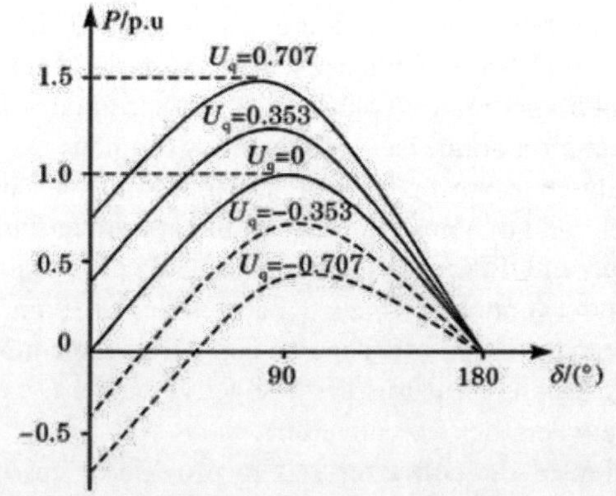

Fig. 3.43 rotor angle curve of line under series voltage of SSSC with different amplitudes

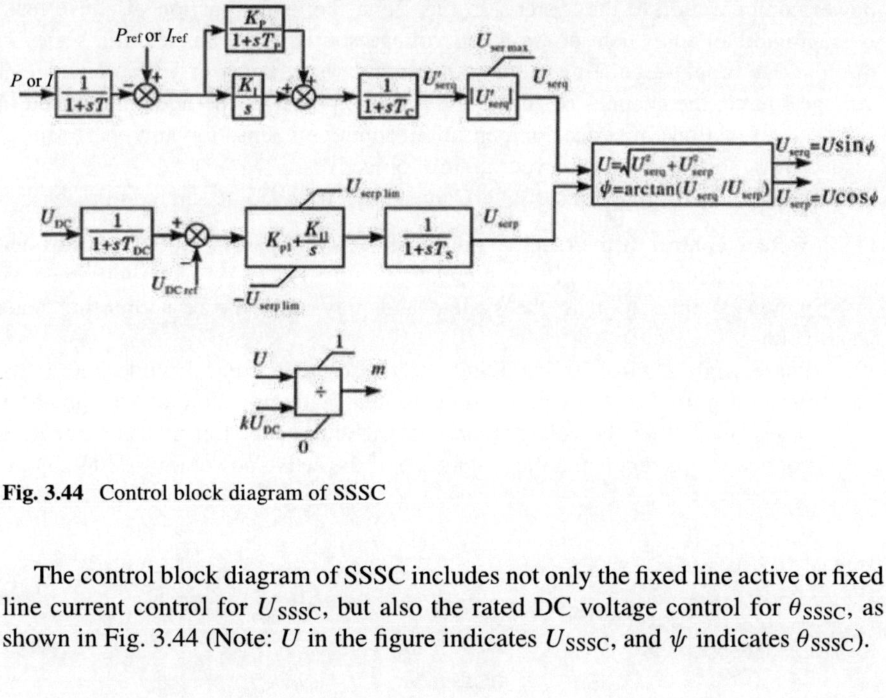

Fig. 3.44 Control block diagram of SSSC

The control block diagram of SSSC includes not only the fixed line active or fixed line current control for U_{SSSC}, but also the rated DC voltage control for θ_{SSSC}, as shown in Fig. 3.44 (Note: U in the figure indicates U_{SSSC}, and ψ indicates θ_{SSSC}).

3.3.7 Unified Power Flow Controller

Unified power flow controllers (UPFC) is the most representative FACTS controller with wide application and strong power flow control function, and is a new-generation

FACTS device first proposed by Mehta and others in 1992. It combines the advantages of STATCOM and SSSC devices, with four basic functions, namely, series compensation, shunt compensation, phase shift, and terminal voltage regulation, and a comprehensive function comprised of these basic functions. This enables UPFC to control the steady-state power flow of transmission lines well, and effectively improve the transient and dynamic stability of the system at the same time.

The basic structure of UPFC is as shown in Fig. 3.45. The main feature of UPFC is that it has two converters connected back to back, which are composed of power electronic devices that can be turned off. The two converters are coupled to a DC connection capacitor. This structure ensures that the active power is randomly transmitted in two directions between the two converters.

The main function of the converter 1 is to provide or absorb the active power required by the converter 2 through the DC connection. The DC power is converted into AC power and coupled to the line through a shunt transformer. In addition, converter 1 can also be regarded as STATCOM to provide independent reactive power compensation to the system. In Fig. 3.45, the main function of converter 2 is to provide an equivalent controllable voltage source U_{pq} connected in series in the line. By timely regulating the amplitude and phase angle of UPFC equivalent voltage source, the changes of active and reactive power of the line is regulated to realize the functions of series compensation, shunt compensation and phase shifter respectively, so as to realize various control objectives.

The basic control functions of UPFC are shown in Fig. 3.46, including:

(1) Voltage control. If the reactive power of the shunt part of UPFC is controlled independently, then UPFC is equivalent to a STATCOM to provide reactive power compensation to the system and plays the role of supporting node voltage.
(2) Phase angle control. If UPFC adopts phase angle control mode, the active power required by the load is compensated by the injecting of voltage at the series side. When the voltage phase at both ends of the transmission line does not need to be regulated, the magnitude of the active power transmitted by the

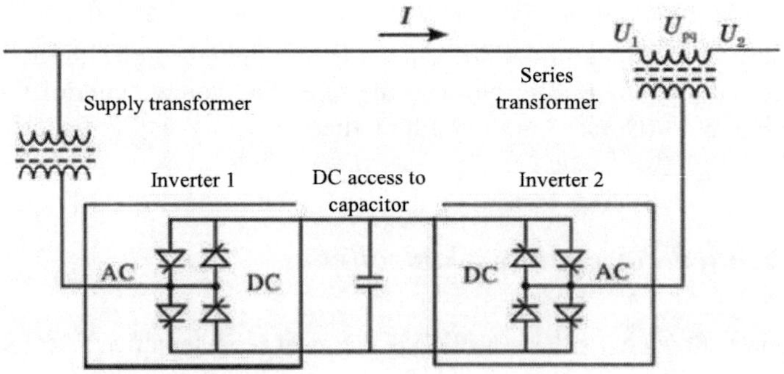

Fig. 3.45 Basic structural diagram of UPFC

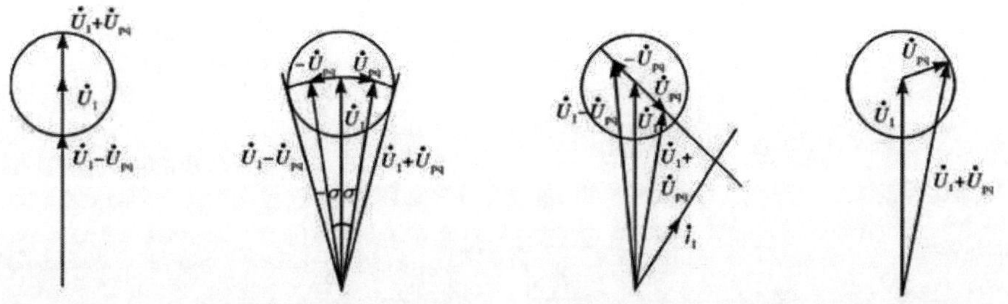

(a) Voltage control (b) Phase angle control (c) Line reactance control. (d) Integrated control

Fig. 3.46 Basic control functions of UPFC

transmission line is continuously regulated, so that the direction and magnitude of power flow in the power system are economical and reasonable.

(3) Line reactance control. If the injecting voltage vector of the series part is perpendicular to the line current, then the UPFC is equivalent to a series compensation device. It not only regulates continuously, but also compensates in both directions (increasing and decreasing the voltage), and it does not cause *LC* oscillation under proper control. It is an advanced and very effective technology for regulating and controlling power grid node voltage, compensating line inductive reactance, enhancing power system transmission power limit, and improving power system stability.

These several functions are combined to give full play to the powerful functions of UPFC. Line reactance control and phase angle control are interrelated functions and can be integrated into a universal controller to control the line active and reactive power to the required level. In addition, the shunt compensation part of UPFC not only performs independent reactive power compensation to control the voltage, but also coordinates with the general controller to jointly control the line power flow. Several control functions of UPFC are switched from one function to another in real time. The flexibility of this function makes UPFC have great potential in solving various problems of the power system.

The equivalent circuit of the dynamic UPFC model on the DC side is shown in Fig. 3.47.

Combined with the basic principle of PWM technology, it can be seen that in Fig. 3.47, the voltage amplitude relationship between the AC side and DC side of UPFC converter and converter 2 can be expressed as follows

$$U_{\text{shunt}} = m_1 k_1 U_{\text{DC}} \tag{3.64}$$

$$U_{\text{series}} = m_2 k_2 U_{\text{DC}} \tag{3.65}$$

In the formula, m_1 and m_2 are modulation ratios, and $0 \le m_1, m_2 \le 1$; k_1 and k_2 are internal parameters of UPFC shunt converter; U_{DC} is DC voltage. The phase δ_1

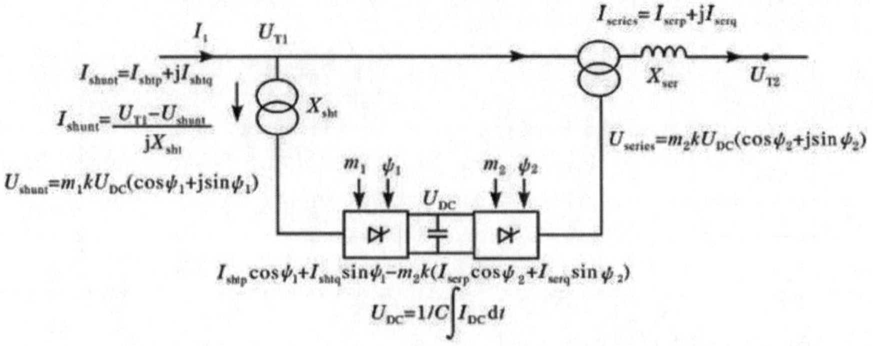

Fig. 3.47 Equivalent circuit of the dynamic UPFC model on the DC side

and phase δ_2 of U_{shunt} and U_{series} are determined by ψ_1 and ψ_2 of converter 1 and converter 2 respectively. With the phase angle θ_{T1} of the terminal voltage U_{T1} of the line installed by UPFC ($\theta_{\text{T1}} = 0$ in Fig. 3.47) as reference, the phase angle relation expression is

$$\delta_1 = \psi_1 - \theta_{\text{T1}} = \psi_1 \tag{3.66}$$

$$\delta_2 = \psi_2 - \theta_{\text{T1}} = \psi_2 \tag{3.67}$$

The DC side voltage of UPFC is supported by DC capacitors, so the dynamic equation of the UPFC DC side is regarded as the dynamic process of capacitor charging and discharging:

$$C\frac{\mathrm{d}U_{\text{DC}}}{\mathrm{d}_t} = I_{\text{DC}},\ \text{i.e.,}\ U_{\text{DC}} = \frac{1}{C}\int I_{\text{DC}}\mathrm{d}t \tag{3.68}$$

According to the principle of power balance (the converter is considered as an ideal converter here, and there is no active power loss), the power balance formulas of the DC side and AC side can be obtained: $U_{\text{DC}}\ I_{\text{DC}} = U_{\text{shunt}}\ I_{\text{shunt}} - U_{\text{series}}\ I_{\text{series}}$, from which the following formula is further obtained with formula (3.64) and formula (3.65)

$$I_{\text{DC}} = m_1k_1\left(I_{\text{shtp}}\cos\psi_1 + I_{\text{shtq}}\sin\psi_1\right) - m_2k_2\left(I_{\text{serp}}\cos\psi_2 + I_{\text{serq}}\sin\psi_2\right) \tag{3.69}$$

The control goal of UPFC is to maintain all of the line active power, reactive power, node voltage U_s and DC side capacitor voltage U_{DC} to be set values. Wherein the power is regulated by the control of the series-side converter 2 with the control quantity of m_2 and ψ_2; and the voltage is regulated by the control of the shunt-side converter 1 with the control quantity of m_1 and ψ_1. The control methods of these control quantities are all regulated by PI. Coupled with some inertial delay links in the actual system, the model block diagram of the UPFC control system is obtained, as shown in Figs. 3.48 and 3.49.

As can be seen from Fig. 3.48, the series part of UPFC control model mainly includes four parts: active power PI regulator, reactive power PI regulator, lag link of series converter time constant and modulation ratio limiter.

As can be seen from Fig. 3.49, the shunt part of UPFC control model mainly includes four parts: node voltage PI regulator, DC side voltage PI regulator, lag link of shunt converter time constant and modulation ratio limiter.

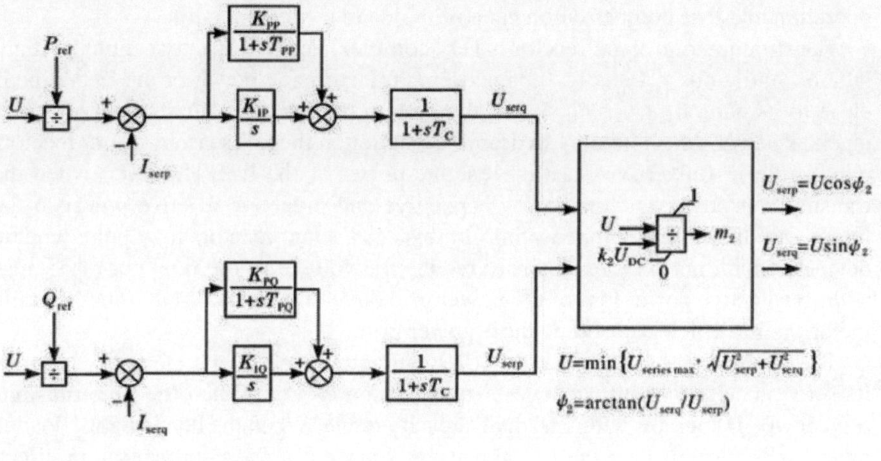

Fig. 3.48 Power flow control

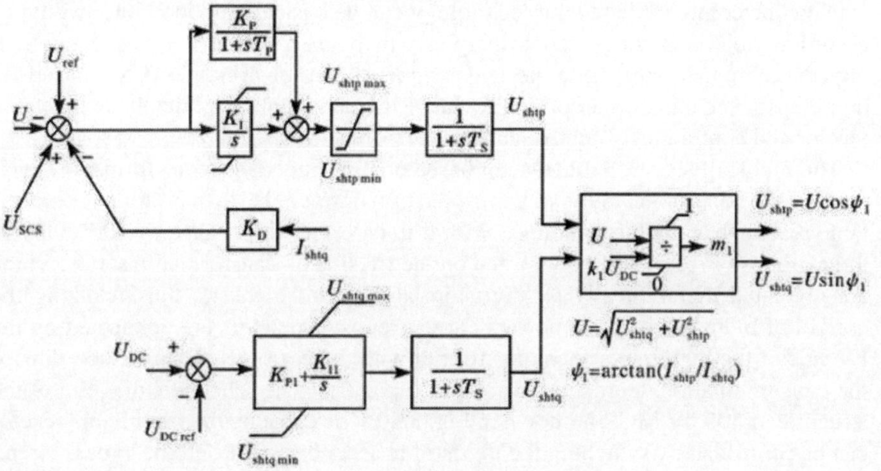

Fig. 3.49 Node voltage control and DC side voltage control

3.3.8 Controllable Shunt Reactor

In EHV long-distance transmission line, due to unacceptable power frequency voltage rise and corresponding transient oscillation overvoltage caused by capacitance effect, equipment such as synchronous condenser or static compensator connected to transformer LV or medium-voltage winding will be cut off simultaneously by the switch on the side of the line, which cannot limit power frequency overvoltage. Therefore, at present EHV shunt reactors are widely used in China. The reactors are directly connected to long-distance transmission lines and cannot stop operation and lose compensation capability due to switch operation.

The structure of a shunt reactor is very complex, and its capacity cannot be regulatable continuously. Usually, the maximum transmission power of the EHV line is close to its natural power P_0. When the system transmits small amounts of power (such as power transmitted by hydropower station in the dry season), shunt reactors play a role in fully compensating reactive power of the line. However, when the transmission power is close to P_0, capacitive and inductive reactive power in the line is essentially self-compensating. In this case, shunt reactors may be redundant devices, which not only excessively reduce the voltage in the line, but also cause additional active power loss of the power grid due to reactive current, thus reducing the economic efficiency of the whole power grid.

The controllable shunt reactor (CSR) can automatically and smoothly regulate its own capacity with the change of transmission power. When the line transmits high power, it operates within a small capacity range. When the line is lightly loaded or unloaded, it will increase its capacity and present a deep compensation effect. In a word, it can limit the rise of power frequency voltage and greatly improve the economic efficiency of the power grid.

Now the controllable reactor technology can be basically divided into two types according to the composition principle, which are respectively based on high impedance transformer principle and magnetic control principle. CSR based on high impedance transformer principle can be divided into hierarchically controllable reactor and continuously controllable reactor.

Hierarchically controllable reactor based on high impedance transformer is developed on the basis of the thyristor controlled transformer (TCT) of SVC. The controllable reactor has double windings, with a transformer impedance of 100%. Then, three groups of reactors are connected on the LV side of transformer in series, which are regulated hierarchically by thyristors and circuit breakers, thus realizing the control of inductive reactive power. Three groups of reactors are connected on the LV side of the high impedance transformer with three groups of taps. According to the tap combination, there are four output capacities. The schematic diagram of the structure of 500 kV MCR is shown in Fig. 3.50. The capacity of controllable reactor can be controlled by switching the thyristor at secondary side and the bypass circuit breaker. Hierarchical grading capacities are 25, 50, 75, and 100%.

The continuous CSR has three windings, including a pair of high impedance windings (the impedance is generally 85%). Phase-controlled thyristor valve is connected

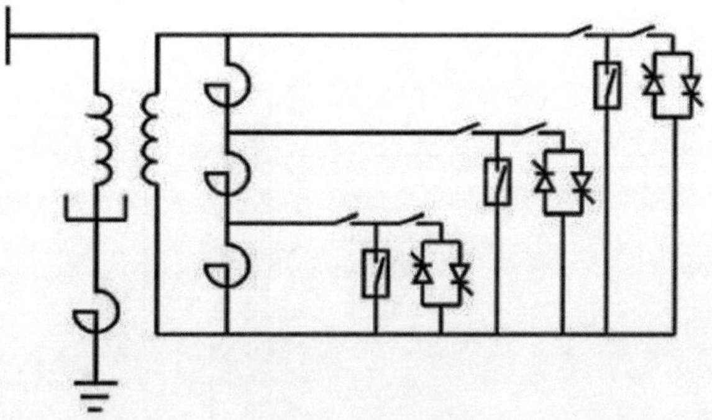

Fig. 3.50 Structure diagram of controllable reactor (Single Phase)

on its LV side. The third winding is designed as a filter for filtering out high harmonic current generated during thyristor valve control. The mode features that the inductive reactive power can be controlled continuously and quickly as required. Its wiring principle is shown in Fig. 3.51.

In terms of the controllable reactor based on the magnetic control principle in the capacity regulating range, only the iron core is saturated, while iron yoke is in unsaturated linear area. The thyristor control system changes the reactor capacity by changing the saturation degree of iron core. Therefore, it is commonly referred to as magnetically controlled reactor (MCR). MCR consists of reactor body and control system. When the DC excitation current is zero, the limb core is not saturated in the whole power frequency period, and the reactor is in the unloaded state with the

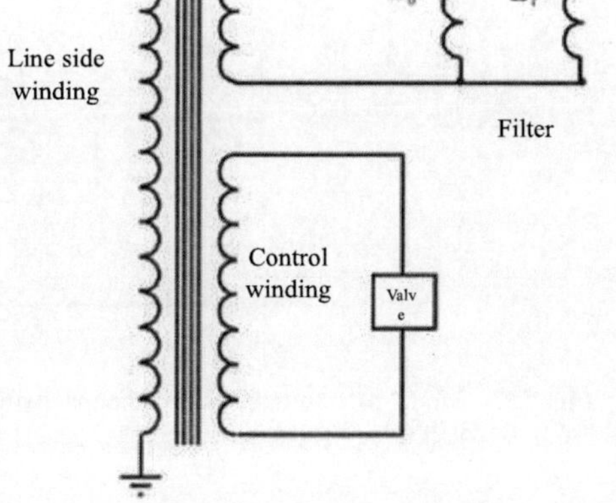

Fig. 3.51 Wiring diagram of continuous CSR (Single Phase)

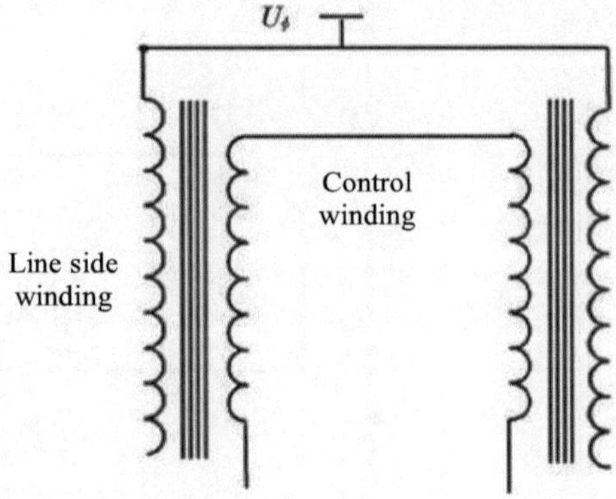

Fig. 3.52 MCR operating schematic diagram

minimum capacity. As the control current increases, the saturation time of limb core increases, and the capacity of reactor also increases. When the limb core is fully saturated in a power frequency period, the capacity of the reactor also reaches the limit Q_m. The operating schematic diagram of a MCR is shown in Fig. 3.52.

The controllable high reactance model in this section is established based on the controllable high reactance data of Jiangling Substation developed by China Electric Power Research Institute, which is a magnetically controllable high reactance model. In the operation, the controllable high reactance reactor mainly adopts automatic constant capacity or automatic constant voltage control. Automatic constant capacity mode can perform feedback control for the capacity of controllable high reactance reactor and maintain the capacity as a set value by automatically regulating the magnitude of excitation current; automatic constant voltage control mode can perform feedback control for the voltage at mounting point of controlled high reactance reactor and maintain the voltage as a set value by automatically regulating the capacity of controllable high reactance reactor. The system analysis models of both control modes are shown in Figs. 3.53 and 3.54.

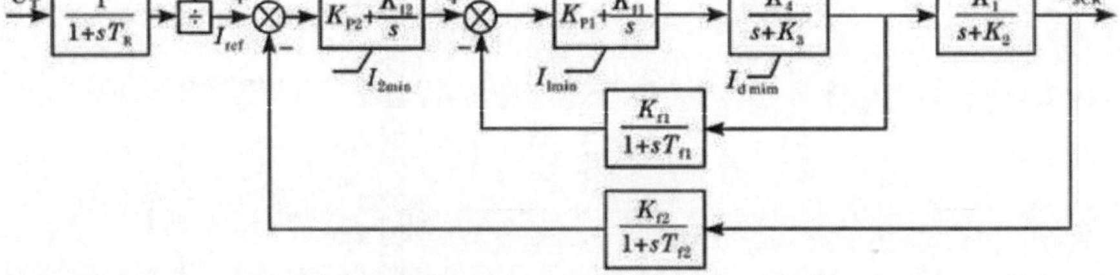

Fig. 3.53 Model of automatic constant capacity control system

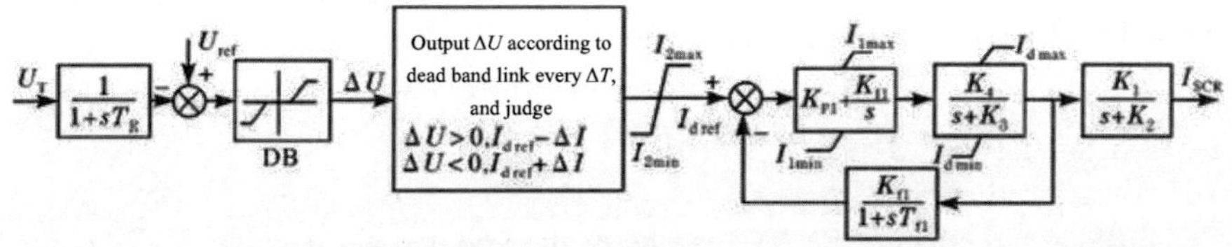

Fig. 3.54 Model of automatic constant voltage control system

3.3.9 Synchronous Condenser

The main advantage of synchronous condenser is that it can continuously regulate the value of reactive power. However, considering that the synchronous condenser is a rotating machine which has large active power loss, a loss in full load operation of about 1.5%, a loss in no-load operation of about 0.5%, complicated operation and maintenance and large investment cost, the synchronous condenser can generally be installed only when large-capacity reactive power compensation is necessary.

In terms of motor principle, the synchronous condenser is a synchronous motor in no-load operation. Its rotors are designed into salient-pole structure just like the synchronous motor. The synchronous condenser also achieves absorbing reactive power or outputting reactive power from the power grid through changing the magnitude of excitation current. During over-excitation operation, the synchronous condenser supplies inductive reactive power to the system, thus acting as a reactive power supply; during under-excitation operation, the synchronous condenser absorbs inductive reactive power from the system, thus acting as a reactive power load. This operation mode is also called “leading phase operation”. A synchronous condenser equipped with automatic excitation device can smoothly regulate the reactive power input or output according to the voltage.

Technically speaking, the application of synchronous condenser in a weak voltage network is somewhat superior to SVC. When the network voltage drops, the synchronous condenser can instantly increase the reactive power output (the same as the reactive power output of the generator), and can prevent the attenuation of potential or flux linkage (armature reaction) in the condenser through excitation control. Both the condenser and the generator have an overload operation capability of tens of seconds. As previously mentioned, SVC will exhibit capacitor characteristics at the limit point. The reactive power output is proportional to the square of voltage. Compared to SVC, the condenser can maintain the rated current at the limit point.

In order to improve the voltage stability of power system, TEPCO has chosen synchronous condenser instead of SVC. One of the reasons is that the critical voltage corresponding to the maximum power point of synchronous condenser is relatively low. As a solution to the voltage stability problem, some generators in the power system have been disengaged from the shaft system and replaced with synchronous condensers for operation. At present, the application of most synchronous condensers is related to HVDC transmission projects.

3.4 OLTC

3.4.1 Characteristics of OLTC

The on-load tap changer (OLTC) is one of the voltage and reactive power control devices widely used in power systems. OLTC can be controlled manually or automatically. At present, most of OLTCs used in the transmission network of China are operated in manual control mode.

The OLTC tap is generally located at the HP side (main network side). When the voltage level at the LV side (load side) is extremely low due to disturbance in the system, OLTC can regulate the tap to recover the voltage level at the LV side and therefore recover the power of loads at the LV side. However, the tap regulation only achieves the regulation for distribution relationship of reactive power in different voltage classes instead of the increase of reactive power. In case of insufficient reactive power support, improvement of the voltage level at load side will inevitably reduce the voltage level at the main network side, thus affecting the voltage at the load side. This feedback process may eventually lead to voltage collapse. In order to avoid and prevent voltage collapse, OLTC in automatic control mode should lock the OLTC tap if the change of tap is not conducive to voltage stability of the system, and then unlock the tap once the voltage is recovered. One simple method is to lock the tap when the voltage of OLTC at power supply side (HV side) drops, and unlock the tap once the voltage is recovered.

The OLTC tap should be regulated step by step in the monitoring of the tap position and the variation of voltage and current. It will take several seconds for the tap changer to complete a gear change. However, in order to avoid frequent operation of tap changer for voltage regulation, delay time should be set for voltage booster and drop. Generally, delay time of 30–60 s should be set for 110 kV voltage regulation, and 60–120 s seconds for 35 kV voltage regulation. According to the range of such delay time, the operation of OLTC mainly affects the medium and long-term voltage stability.

There is a relatively large regulating range for OLTC, generally over 15%. The regulating range for each gear of tap is generally 0.5–1.5%. In China, the regulating range is $\pm 3 \times 2.5\%$ for 110 kV OLTC, and $\pm 4 \times 2\%$, $\pm 8 \times 1.25\%$ and $\pm 8 \times 1.5\%$ for 220 kV OLTC.

3.4.2 Model of OLTC

1. Block diagram of model

The block diagram of the OLTC model is shown in Fig. 3.55.

2. Control and action process

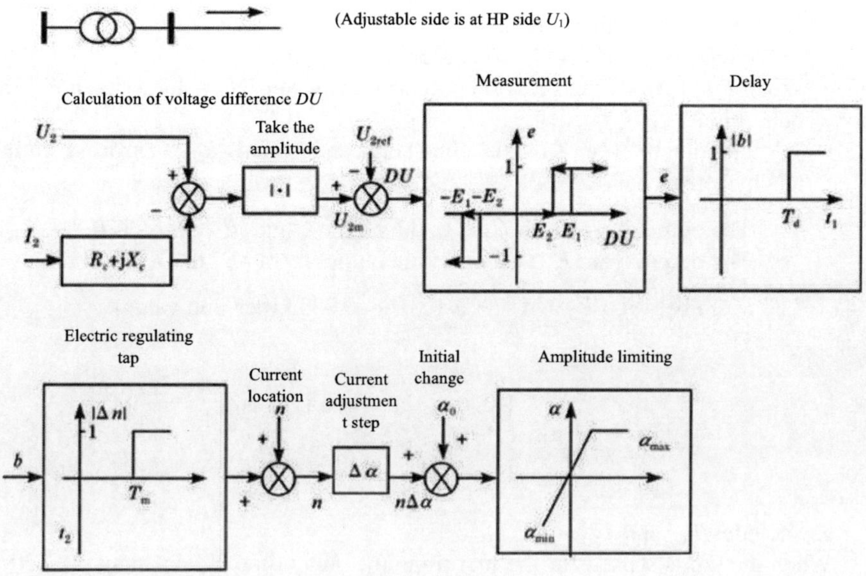

Fig. 3.55 Automatic regulating model for OLTC tap

The OLTC tap is generally connected to the HV side. The positive sequence voltage of loads at the LV side is obtained through measurement and calculation. If the load voltage is lower than the setting valve, the tap actuator should be started after certain delay T_d to change the tap position, thus achieving the voltage regulation at the load side. The action delay T_m is relatively large for an actuator which features a mechanical structure, generally 5–10 s. The delay time T_d is generally 30–120 s taking into account the instantaneous voltage drop when the fault is to be avoided. If, after the first action, the voltage is still lower than the setting value, the delay T_d for the second and subsequent actions can be set to 0 or equal to the first delay. The detailed process is as follows.

(1) Calculation of voltage difference *DU*

Approximate voltage at load side:

$$U_{2m} = \left|\left(U_{2x} + jU_{2y}\right) - \left(I_{2x} + jI_{2y}\right)\left(R_c + jX_c\right)\right| \tag{3.70}$$

$$DU = U_{2m} - U_{2ref} \tag{3.71}$$

All the above-mentioned voltages and currents are positive sequence voltages and currents.

R_c and X_c are compensating impedances. The product of such compensating impedances and the current I_2 flowing through OLTC is the voltage drop across the circuit. The larger the current variation is, the larger the voltage deviation *DU* is.

(2) Action process of measurement
(1) When DU ranges from $-E_2$ to E_2, the output is 0.
(2) When DU is less than $-E_1$, the output is –1; when DU is larger than E_1, the output is 1.
(3) The variation between E_1 and E_2 and between $-E_1$ and $-E_2$ is a process similar to a hysteresis curve.

① The output from E_2 to E_1 is 0; the output from $-E_2$ to $-E_1$ is 0.
② The output from E_1 to E_2 is 1; the output from $-E_1$ to $-E_2$ is –1.

(4) The width between E_2 and $-E_2$ is 0.0166–0.0333 (per-unit value).
(3) Delay

$$b = \begin{cases} 0, t_1 \le T_\mathrm{d}, e = anyvalue \\ 1, t_1 > T_\mathrm{d}, e = 1 \\ -1, t_1 > T_\mathrm{d}, e = -1 \end{cases} \tag{3.72}$$

T_d includes $T_{\mathrm{d}0}$ and $T_{\mathrm{d}1}$.

When the tap is started for the first time, the delay time $T_{\mathrm{d}0}$ is about 30–120 s. $T_{\mathrm{d}0}$ can be constant or has the characteristic of inverse time limit. At this time, T_d is calculated as follows

$$T_\mathrm{d} = T_{\mathrm{d}0}/(|DU/E_1|) \tag{3.73}$$

If, after the first action, the voltage deviation still ranges from $-E_2$ to E_2, the regulating device will act again. $T_{\mathrm{d}1}$ refers to the delay for the second and subsequent actions.

Once the measurement meets the action conditions, it will take T_d seconds to start the tap. Therefore, T_d can reduce the influence of short-term voltage fluctuation and avoid other unnecessary tap regulation.

T_m refers to the mechanical delay of the regulating device, typically 5–10 s.

For every action, the tap is regulated to another grade.

(4) Mechanical regulation

$$\Delta n = \begin{cases} 0, t_2 \le T_\mathrm{m}, b = anyvalue \\ 1, t_2 > T_\mathrm{m}, b = 1 \\ -1, t_2 > T_\mathrm{m}, b = -1 \end{cases} \tag{3.74}$$

$\alpha = \alpha_0 + \mathrm{n}\Delta\alpha, \alpha_{\min} \le \alpha \le \alpha_{\max}$.

The parameter Δn refers to the step pace in each step change of OLTC, which is generally about ± 16 steps.

The maximum value $\alpha_{\max}$ and minimum value $\alpha_{\min}$ of the variation in no-load voltage ratio of OLTC are about $\pm 10\%$ of rated voltage.

Locking and unlocking of OLTC:

If the voltage is lower than U_Block, then Flag_Fault is assigned as 1. If the voltage increases to beyond U_UnBlock (U_UnBlock > U_Block), then Flag_Fault is immediately equal to 0. If the duration that Flag_Fault equals to 1 is equivalent to T_Fault seconds, then the device will be locked after T_Fault_Del seconds. Unlocking can only be done manually.

3.5 HVDC Transmission System

With the rapid development of the power industry in China and the gradual formation of a supergrid, the structure of the power system becomes increasingly complicated. The distinguishing feature of the power system is the grid-connected operation of ultra-high voltage and long-distance high voltage direct current (HVDC) transmission system. Most of the electricity is transmitted to the load center of the receiving-end power grid. In addition, the imbalance of resource and economic development makes it inevitable that multi-circuit DC transmission lines will fall at the same AC receiving end. In China, an AC/DC system with the most fall points and the most complicated structure in the world will be formed. The HVDC transmission capacity will account for an increasing proportion of the system. The AC system will become relatively weak. Therefore, although the multi-infeed AC/DC transmission system has higher transmission capacity and flexibility, the stability of the system faces a severe test.

3.5.1 Reactive Power Characteristics of Converter

In the DC infeed system, no matter for long-distance DC transmission or back-to-back DC transmission, special attention should be paid to voltage stability in the consideration of the power system stability. The cause is that the DC transmission system is required to consume reactive power while providing active power for the receiving-end AC system, i.e., about 40–60% of the DC transmission power. In case of fault to the AC system, the DC system is still required to absorb reactive power from the AC system. For the receiving-end AC system, the DC system presents unfavorable "characteristics of reactive power load".

The reactive power absorbed by the converter can be expressed as

$$Q_{\rm DC} = P\tan\varphi \tag{3.75}$$

$$\tan\varphi = \frac{(\pi/180)\mu - \sin\mu\cos(2\alpha+\mu)}{\sin\mu\sin(2\alpha+\mu)} \tag{3.76}$$

$$\mu = \cos^{-1}\left[U_{\rm d}/U_{\rm dio} - (X_{\rm c}/\sqrt{2})(I_{\rm d}/E_{11})\right] - \alpha \tag{3.77}$$

$$U_d/U_{dio} = \cos\alpha - (X_c/\sqrt{2})I_d/E_{11} \quad (3.78)$$

wherein: U_{dio} refers to the ideal no-load DC voltage of the converter, kV; $U_{dio} = 3\sqrt{2}E_{11}/\pi$; P refers to the power of converter at the DC side, MW; Q_{DC} refers to the reactive power consumption of the converter, Mvar; φ refers to the power factor angle; X_c refers to commutation reactance of each phase, Ω; I_d refers to the DC operating current, kA; α refers to trigger angle of the silicon; E_{11} refers to effective value of no-load voltage for valve-side winding in converter transformer, kV; U_d refers to polar DC voltage, kV. If the converter is operated in an inverter mode, α in the equation is replaced by γ. γ refers to the turn-off angle at the inverter side.

It can be seen that the reactive power of the converter is not only affected by the active power, but also related to many other operating parameters. The most sensitive operating parameters are trigger angle α and turn-off angle γ [6]. The converter can be operated in different control modes. In different control modes, the curves of reactive power absorbed by the converter with the converter power are different. For the same active power operating point, the reactive power absorbed by the converter may vary greatly.

In actual system operation, for normal operation areas, especially areas where the converter power approaches to the rated DC power, various possible control methods should be adopted to minimize the reactive power consumed by the converter. For the inverter, applying a constant turn-off angle control mode is the easiest way to achieve this objective. For rectifiers, constant DC voltage is applied, while reasonable tap control of converter transformer is necessary to ensure that the turn-off angle of inverter will not exceed a certain value during normal operation. Constant current control is the common control mode. In order to ensure that the rectifier will not easily lose its constant current control capability and reduce reactive power consumption as much as possible, the tap control of converter transformer is necessary, thus reducing the trigger angle of rectifier as small as possible.

If the operating power of the converter is far less than the rated power, a certain number of filters should be switched on for filtering, which may result in excessive reactive power of the converter station, then the converter needs to absorb more reactive power. At this time, reactive power of the converter station can be balanced by increase of the trigger angle of the converter. However, long-term operation under this condition will adversely affect the service life of the system equipment. Therefore, if the converter transformer has a large positive voltage regulation range, the high tap should be regulated as much as possible to reduce the valve-side voltage of the converter transformer.

If an AC system connected to the converter transformer is especially weak, which has especially strict requirements for reactive power, the reactive power of the converter station and the AC system should be accurately controlled. At this time, the converter will be used for the accurate control of reactive power. For example, if the reactive power is insufficient, a set of reactive power compensation equipment should be switched on. However, after the reactive power compensation equipment is switched on, the reactive power is excessive. In this case, the converter is forced to absorb more reactive power when a set of reactive power compensation equipment is switched on. Once the active power increases enough (reactive power consumption also increases), the converter will be switched from the forced reactive power control mode to the minimum reactive power mode. With the increase of active power, the unbalanced reactive power of the system increases. Another set of reactive power compensation equipment is switched on in due course.

In order to, in a weak receiving-end AC system, reduce the influence of filter or capacitor switching on AC system voltage and avoid inverter commutation failure or out-of-limit transient voltage variation due to switching, γ-kick function [7] should be introduced to DC reactive power control, i.e., γ-angle step function. Once a filter or a shunt capacitor is switched on, the γ setting value is instantaneously increased to improve the reactive power consumption of the converter and further limit the voltage step of the AC system. The γ setting value after switching on will return to the angle before switching on in a very short period. The increase of γ angle should correspond to the response time of AC system voltage control.

3.5.2 Reactive Power Compensation Characteristics of Converter Stations

1. Reactive power demand and reactive power support capability of AC systems

If the converter station is located in the vicinity of a power plant or a power plant fleet: When the DC system is under heavy load operation, some reactive power supplies of the AC system can be used to achieve installation of less capacitive compensation equipment; when the DC system is under low load operation, some over-compensated reactive power of the converter station can be absorbed relying on the leading phase capability of the generator, so as to achieve installation of less inductive reactive power compensation equipment. In case of sudden shutdown of the DC system, the overvoltage level of load shedding can be reduced, and the equipment cost of the converter station can be correspondingly reduced. Therefore, it is greatly important to reasonably utilize the reactive power regulation capability of AC systems. If the AC system is very strong, except for the reactive power provided by the necessary AC filter, the rest of the reactive power can be provided by the AC system, while the weak system requires a large quantity of reactive power compensation, especially for the weak receiving-end system. The compensation equipment not only provides the reactive power consumed by the converter, but also supplies the reactive power consumed by the load [6].

2. Reactive power compensation equipment in converter station

The reactive power compensation equipment in converter station is divided into three types [8]: (1) Mechanically switched capacitor and reactor. The capacitor is necessary for filtering. The minimum filtering capacitance capacity accounts for about 30% of the converter capacity. (2) SVC. When the converter bus voltage drops significantly, SVC inputs a large amount of capacitance at the moment of fault, which reduces the system strength and deteriorates the HVDC recovery characteristics; and, may result in commutation failure, thus leading to the interruption of DC transmission. Due to those shortcomings, SVC is not widely used [7, 8]. At present, SVC is only applied in the British-side converter station of the Channel Tunnel DC project and the Chatuchak back-to-back DC project in Canada. (3) Condenser. In terms of performance, the condenser is the most ideal reactive power compensation equipment for inverter station. In addition to providing certain reactive power, the condenser can also improve the short-circuit ratio of an inverter station, increase the moment of inertia for the system, improve commutation conditions of converter and reduce overvoltage, which is widely used in early DC transmission. In particular for the receiving end of the transmission project from the remote power station to the power grid in the load center, two typical examples are the receiving-end converter station of the Itaipu DC project and the receiving end of the Nelson River DC project. However, due to the large investment, large land occupation, low operation reliability and heavy maintenance workload of the condenser, only mechanically switched capacitors and reactors are generally used in conventional converter stations. If the switchable passive elements cannot meet the performance requirements of the system and a condenser is necessary, further research is necessary.

The switching control modes of reactive power compensation equipment in converter stations generally include unbalanced reactive power control and AC voltage control. The former keeps the reactive power exchanged between converter station and AC system within a certain range, while the latter keeps the voltage variation of AC bus in the converter station within a certain range. The reactive power control mode is applied in general DC transmission projects. If the AC system connected to the converter station is weak, AC voltage control mode is applied.

The principle of unbalanced reactive power control is to calculate the reactive power shortage of the system in real time in the steady-state process of the DC converter station, so as to meet the demand of the system by switching the reactive power compensation equipment. In case of a reactive power shortage for the system, the filter and other reactive power compensation equipment should be switched on in advance to improve Q_{AC}. Otherwise, if the voltage is too high, it is necessary to, by reducing QAC, switch on the capacitive compensation equipment on late or switch off the capacitive compensation equipment in advance, which is likely to be limited by the minimum number of filter banks or even the absolute minimum number of filter banks [7, 8]. At this time, HV reactor and LV reactor can be switched on, or DC transmission power can be reduced to an acceptable level of the system.

The voltage control of AC system is to control the voltage of AC bus in converter station by using reactive power compensation equipment in converter station. Before or during the unlocking of the converter, filter banks of the minimum number are switched on. With the increase of DC power, the AC bus voltage drops and the reactive power compensation equipment is switched on accordingly. Similarly, with the decrease of DC power or the variation of other operating parameters, the AC bus voltage increases and the reactive power compensation equipment is switched off accordingly. This control may also be limited by filter banks. The handling method is the same as that of unbalanced reactive power.

3.5.3 *Characteristics of Fault DC System*

The operation of DC transmission system is affected by the faults of DC transmission line, converter or AC system. The influence of such faults is reflected by the action of converter control. Converter control plays an extremely important role in making DC transmission system respond satisfactorily to DC or AC system fault. The fault of DC transmission line will not affect the connected AC system as destructively as the fault of AC system. However, converter fault will lead to the stop of entire pole from power transmission. For the transient disturbance of AC system, the power response of DC system is generally much faster than that of AC system. It is necessary to reduce power of the DC system or shut down the DC system until the AC system is recovered enough to restart the DC system and recover the power. Therefore, commutation failure and fault recovery at this time become important problems affecting the operation of DC system.

1. Fault AC bus at rectifier side

When the converter bus voltage at rectifier side drops, the commutation voltage of the rectifier also drops, which will result in drop in DC voltage of the rectifier and also drop in DC current. At this time, the constant current control increases the current by decreasing the "α" angle, so as to recover the current value. When the DC voltage at inverter side drops, the DC current increases under the constant extinction angle control, which increases the commutation angle. In consideration of the constant extinction angle control, it is necessary to increase the trigger angle β. If α angle reaches the minimum trigger angle limit, then the control is constant minimum trigger angle control at rectifier side and constant current control at inverter side. If the voltage continues to decrease, the low voltage dependent current order limiter (VDCOL) may regulate the current and transmission power, or the converter transformer tap control is activated to recover the voltage and current to normal values. If the voltage drops significantly, the DC system can be shut down under the control of VDCOL until fault recovery. Although DC power can be transmitted at very low rectifier voltage, it comes at the cost of increasing reactive power consumption, which is detrimental to AC system.

2. Fault AC bus at inverter side

In case of fault of the AC system, the voltage of AC bus in converter station drops instantaneously. While the time constant of no-load voltage ratio adjustment for converter transformer is large, generally about 10 s. Therefore, there is no time to activate the converter transformer tap control, which increases μ instantaneously. Since the voltage drop is instantaneous, the DC transmission system controller has no time to increase β, that is, β still keeps the original initial value, thus resulting in instantaneous drop of γ to beyond the critical value of commutation failure. Whether the inverter suffers from commutation failure or not relates with the voltage level of converter bus at inverter side before the failure.

During commutation failure, angle α increases greatly due to the action of the constant current controller at rectifier side, and suppresses the peak value of DC current. Angle β also increases instantaneously due to the action of the constant turn-off angle for the inverter, which can make the inverter removed immediately from commutation failure.

In case of fault to the AC system at inverter side, under the constant current control at rectifier side and the constant extinction angle control at inverter side, the reactive power loss of the system increases, which may further reduce the AC voltage and even form a vicious cycle, thus leading to voltage collapse of the system, especially for the case in which a weak AC system is connected to the inverter end. Therefore, in addition to constant current control and constant extinction angle control, constant current control and constant voltage control are also proposed. The constant current control is still applied as the control at rectifier side while the inverter side is regulated in a certain way according to the terminal voltage of DC transmission line. In this way, when the converter bus voltage at inverter side drops, under the constant voltage control, β is decreased, the power factor of the inverter is increased, and the consumed reactive power is reduced, which is conducive to preventing the AC voltage from further drop.

3. Commutation failure

Commutation failure of the inverter is the most common fault of inverters. If, after the commutation between the two bridge arms, the valve that has just withdrawn from conduction fails to recover its blocking capability within a period under the action of the reverse voltage, or the commutation process has not been completed during the reverse voltage period, the valve that has been subject to commutation when the valve voltage is transformed to the forward voltage will arrange commutation to the valve that was originally scheduled to withdraw from conduction, i.e., commutation failure. The causes of commutation failure include: drop of AC voltage, increase of DC current, relative movement of line voltage zero crossing point due to asymmetrical fault of the AC system, excessively small lead trigger angle β or excessively small set lead turn-off angle γ. When γ is less than $\gamma_{\min}$ required by the inverter, commutation failure occurs.

4. DC locking

The DC system will be locked in case of consequential commutation failure for a long time. Once the DC system is locked, the reactive power at this time is excessive due to the minimum filter limit, thus resulting in the reactive power impact to the AC system, and the voltage fluctuation. In addition, the frequency of AC systems at rectifier side increases due to excessive active power, while the frequency of AC systems at inverter side decreases due to insufficient active power. Therefore, in case of locking, the frequency and voltage of the power grid will be affected due to variation in active power and reactive power.

5. Recovery of DC power (unlocking)

After the fault, the voltage is higher than the low threshold of DC VDCOL, and if it rises, DC can recover from commutation failure. When DC unlocking starts to transmit power, the control system put the corresponding AC filter into operation according to the minimum AC filter demand. At this time, the reactive power consumed by the converter is far less than the reactive power provided by the AC filter, which will also have a reactive power impact on the AC system, resulting in voltage fluctuation.

3.5.4 Model of DC Transmission System

1. Mathematical model of a DC transmission line

When the quasi-steady state model is adopted, the steady-state relation equation between voltage, current and trigger angle, commutation angle and extinction angle on both AC and DC sides of the converter station is shown in Formula (3.79). In the equation, α, β, γ and δ respectively indicate the lag trigger angle, lead trigger angle, commutation angle and extinction angle respectively. Subscript "d" indicates DC side, "R" and "I" respectively indicate rectifier and inverter sides, "0" indicates no load, and "(1)" indicates a fundamental component.

The steady-state equation of the rectifier is

$$\begin{cases} U_{\text{dR}} = U_{\text{dR0}}\cos\alpha - R_{\text{cR}} I_{\text{dR}} = U_{\text{dR0}} \frac{\cos\alpha + \cos(\alpha+\gamma_R)}{2} \\ U_{\text{dR0}} = \frac{U_{\text{R}}}{n_{\text{R}}},\ n_{\text{R}} = \frac{\pi}{3\sqrt{2}} k_{\text{R}} \end{cases} \tag{3.79}$$

wherein, U_{dR} is the effective value of AC bus voltage on the rectifier side; k_{R} is the transformation ratio of the rectifier transformer; R_{cR} is equivalent commutation resistance, $R_{\text{cR}} = \frac{3}{\pi} x_{\text{R}}$, x_{R} is commutation impedance.

When the converter losses are not taken into account, the AC side power, fundamental current and power factor are respectively

$$\begin{cases} P_{\text{R}} = U_{\text{dR0}} I_{\text{dR}} = \sqrt{3} U_{\text{R}} I_{\text{R(1)}} \cos\varphi_{\text{R(1)}} \\ Q_{\text{R}} = \sqrt{3} U_{\text{R}} I_{\text{R(1)}} \sin\varphi_{\text{R(1)}} \\ I_{\text{R(1)}} = \frac{\sqrt{6}}{\pi} \frac{I_{\text{dR}}}{k_{\text{R}}} \\ \cos\varphi_{\text{R(1)}} \approx \frac{1}{2}\left[\cos\alpha + \cos(\alpha + \gamma_{\text{R}})\right] \approx \frac{U_{\text{dR}}}{U_{\text{dR0}}} \end{cases} \tag{3.80}$$

Corresponding to the rectifier, the inverter steady-state equation is

$$\begin{cases} U_{\mathrm{dI}} = U_{\mathrm{dI0}} \cos\delta - R_{\mathrm{cI}} I_{\mathrm{dI}} = U_{\mathrm{dI0}} \frac{\cos\delta + \cos(\delta + \gamma_{\mathrm{I}})}{2} \\ U_{\mathrm{dI0}} = \frac{U_{\mathrm{I}}}{n_{\mathrm{I}}}, n_{\mathrm{I}} = \frac{\pi}{3\sqrt{2}} k_{\mathrm{I}} \end{cases} \tag{3.81}$$

$$\begin{cases} P_{\mathrm{I}} = -U_{\mathrm{dI0}} I_{\mathrm{dI}} = -\sqrt{3} U_{\mathrm{I}} I_{\mathrm{I}(1)} \cos\varphi_{\mathrm{I}(1)} \\ Q_{\mathrm{I}} = -\sqrt{3} U_{\mathrm{I}} I_{\mathrm{I}(1)} \sin\varphi_{\mathrm{I}(1)} \\ I_{\mathrm{I}(1)} = \frac{\sqrt{6}}{\pi} I_{\mathrm{dI}} k_{\mathrm{I}} \\ \cos\varphi_{\mathrm{I}(1)} \approx \frac{1}{2}\left[\cos\delta + \cos(\delta + \gamma_{\mathrm{I}})\right] \approx \frac{U_{\mathrm{dI}}}{U_{\mathrm{dI0}}} \end{cases} \tag{3.82}$$

2. Mathematical model of a converter

The working principle of the converter regulation system is to input different regulation signals to the regulator, and its output acts on the phase control circuit to change the phase of the converter trigger pulse through the pulse generator, thus realizing the regulation function. Using different regulation signals will obtain different regulation characteristics and make the whole DC transmission system have different operation characteristics. In order to obtain good operation characteristics, the regulation characteristics of the rectifier should be matched with the regulation characteristics of the transformer. The possible control modes at the rectifier side include constant current I_{dz}, constant extinction angle α_{N}, constant minimum arc burning angle $\alpha_{\min}$, constant power P_{dz}, maximum transformation ratio $T^{\max}$ $_{\mathrm{R}}$ of the transformer at the constant rectifier side and minimum transformation ratio $T^{\min}$ $_{\mathrm{R}}$ of the transformer at the constant rectifier side. The common basic regulation modes include constant current and constant power. Possible control modes at the inverter side include constant minimum extinction angle $\gamma_{\min}$, constant voltage U_{dN}, constant current I_{dz}, constant power P_{dz}, maximum transformation ratio $T^{\max}$ $_{\mathrm{I}}$ of transformer at the constant inverter side and minimum transformation ratio $T^{\min}$ $_{\mathrm{I}}$ of the transformer at the constant inverter side. The common regulation modes include constant voltage and constant extinction angle. The basic regulation modes of the rectifier side and the inverter side will be combined into four regulation modes of the whole DC system. In addition, some control measures are often added under the basic regulation mode. For example, when the constant current regulation mode is adopted at the rectifier side, the minimum extinction angle limit is often added, so that the trigger angle is not less than a minimum safety limit; when the constant voltage regulation mode is adopted at the inverter side, the minimum extinction angle limit is often added. Sometimes, in order to improve the operation performance of the whole AC/DC system by utilizing the fast regulation characteristics of the DC transmission system, some additional input signals (such as the transmission power of some lines of the AC system, the generator speed, and the frequency of the system, etc.) are introduced into the regulator of the converter, which is commonly referred to as modulation control. The following is a brief introduction to their common mathematical models only for basic regulation modes.

(1) Constant current regulation mode

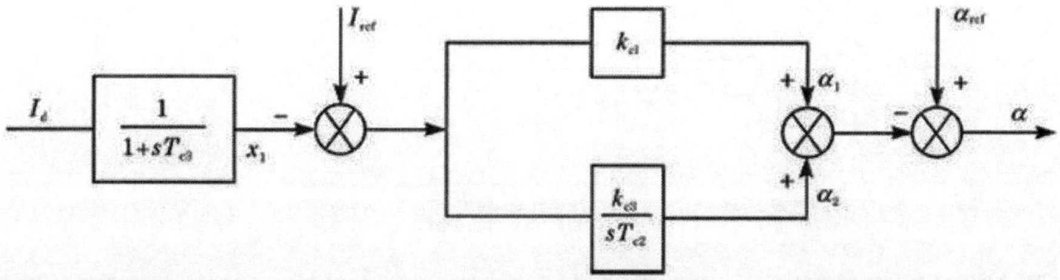

Fig. 3.56 Transfer function block diagram of dc transmission regulation system—constant current regulation mode

The output current I_d of the DC transformer is compared with the reference current I_{ref}. After the error signal obtained is amplified (or regulated by PI composed of integral at the same time), it acts on the phase shift control circuit to change the trigger angle of the converter, thus achieving constant current regulation. The transfer function block diagram is shown in Fig. 3.56.

(2) Constant power regulation mode

The DC power P_d (or active power at AC side) obtained after multiplying the measured DC current and voltage is compared with the given power P_{ref}. The error signal obtained after conversion and amplification acts on the constant current regulator to realize constant power regulation. The transfer function block diagram is shown in Fig. 3.57.

(3) Constant voltage regulation mode

Similar to the constant current regulation mode, by comparing the voltage measurement U_d with the given value U_{ref}, the obtained error signal is amplified and phase-shift controlled to change the lead trigger angle β of the inverter. The transfer function block diagram is shown in Fig. 3.58.

(4) Constant extinction angle regulation mode

By comparing the measured extinction angle δ with the given value δ_{ref}, the error signal is amplified and phase-shift controlled to change the lead trigger angle β of

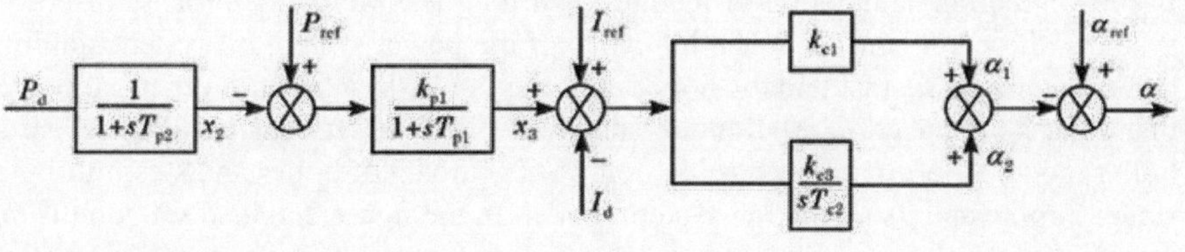

Fig. 3.57 Transfer function block diagram of DC transmission regulation system—constant power regulation Mode

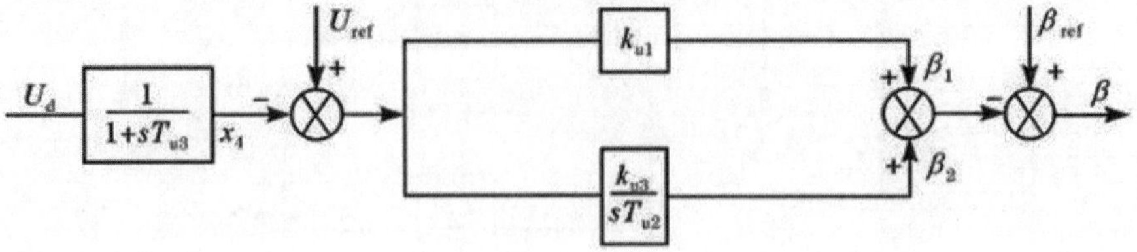

Fig. 3.58 Transfer function block diagram of DC transmission regulation system—constant voltage regulation mode

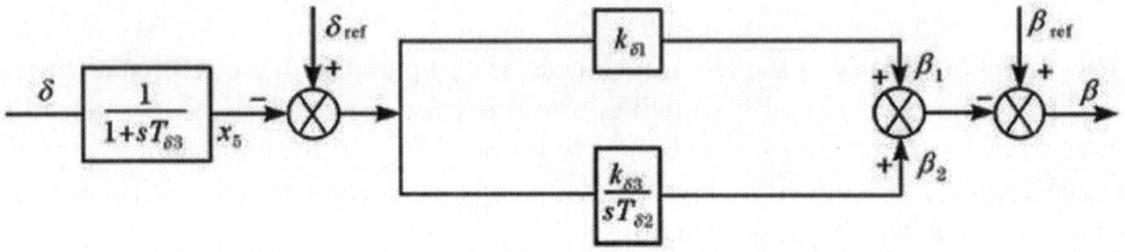

Fig. 3.59 Transfer function block diagram of DC transmission regulation system—constant extinction angle regulation mode

the inverter. The transfer function block diagram is shown in Fig. 3.59. In fact, the extinction angle cannot be directly measured, but is indirectly obtained through the interval between the valve voltage and valve current zero crossing point.

Generally, the basic operation mode during normal operation is as follows: constant extinction angle control and current control at the rectifier end and minimum extinction angle control and voltage control at the inverter end. Other operation modes are transformed from the basic mode due to the change of system conditions, and some quantities exceed the limit.

3.6 Power System of Power Plant

3.6.1 Power System Model of Thermal Power Plant

In the simulation calculation of medium and long-term dynamic process of power systems, the main concern is the influence of the power system on system stability and frequency. The mathematical model of the power system of the thermal power plant and its connection relation are shown in Fig. 3.60. It consists of three parts: boiler model, coordinated control system (CCS) model of boiler and steam turbine, steam turbine and its regulation system model. In the figure, the load set point is the set point of unit power. T_D is the valve position command of steam turbine output by CCS; B_D is the boiler combustion command output by CCS; P_E is electric power input; ω is the unit speed; C_V is the steam valve opening of the steam turbine; P_{MEC}

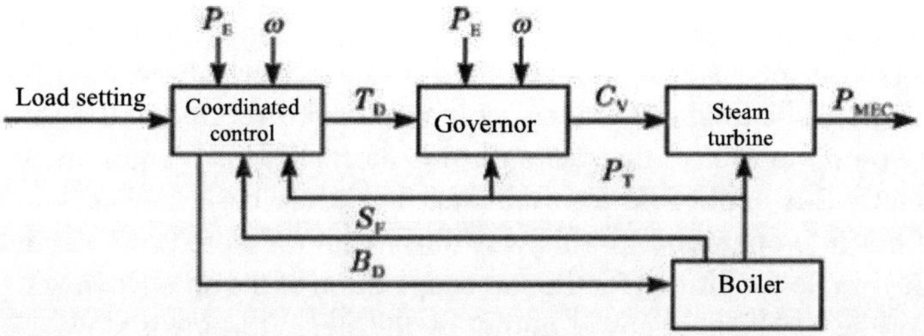

Fig. 3.60 Power system model of thermal power plant

is the mechanical power output of the unit; P_T is the main steam pressure; S_F is steam flow.

In general electromechanical transient calculation, since the calculation process is only in the range of tens of seconds, it can be assumed that the main steam pressure remains unchanged during the transient process, and the steam turbine output power is only controlled by the steam valve opening controlled by the governor. In the medium and long-term dynamic process simulation, the time span is relatively large, and the steam turbine pressure will change in a long time. It is no longer appropriate to assume that the main steam pressure remains unchanged, so the dynamic characteristics of the boiler must be considered.

There are three basic modes of steam turbine/boiler control and regulation in thermal power plants.

(1) Regulation mode in which boiler follows steam turbine

When the load requirement is changed, the steam valve opening of the steam turbine is changed, so the air intake volume of the steam turbine is changed to make the steam turbine adapt to the load requirement. Then the main steam pressure of the steam turbine is also changed, and this pressure signal is sent to the pressure controller, which changes the signals such as the coal-feed and the air supply volume of the boiler through coal control. When the load changes, the rapid change of the output power of the steam turbine depends entirely on the thermal storage energy of the boiler.

This regulation method can change the mechanical power output of the steam turbine at a faster speed, which is beneficial to the stability of the system, but it has a negative impact on the boiler system because of the rapid pressure change. This method is not applicable to unit-made high-power units, because the boiler has limited thermal storage capacity. Using this method alone will cause severe fluctuations in the outlet pressure of the boiler, which will affect the stable operation of the boiler and reduce the primary frequency modulation capacity of the unit.

(2) Control by steam turbine following boiler

The change of the load requirement is sent to the combustion control of the boiler to regulate the coal consumption and air supply volume of the boiler. When the steam quantity of the boiler starts to change, the steam pressure of the boiler also changes. According to the main steam pressure signal of the boiler, the steam turbine changes the steam valve opening, thus changing the air intake volume of the steam turbine, so that the output power of the steam turbine meets the load requirement.

This method is applicable to a unit that bears the base load or a unit that has just been put into operation. In this way, the steam pressure can be stabilized, thus creating favorable conditions for the stable operation of the unit. However, the change in speed of the mechanical power output by the steam turbine is slow.

(3) Coordinated control of boiler and steam turbine

When the load requirement is changed, the coordinated control mode is adopted to send the load regulation signals to the boiler and the steam turbine respectively, and change the signals such as the coal-feed and the air supply volume of the boiler and the air intake volume of the steam turbine. At the same time, according to the change of steam pressure, the steam valve opening of the steam turbine is correspondingly limited, and the boiler regulation function is appropriately strengthened.

The combined regulation mode of the steam turbine and boiler is generally adopted for the high-power intermediate reheating unit.

In the above three modes, the load requirement is mainly composed of power signal obtained by auto generation control (AGC) (or power signal formed by frequency bias) and manually controlled power signals.

1. Dynamic model of a boiler system [9]

As shown in Fig. 3.61, the boiler model can simulate the dynamic characteristics of coal-fired and oil-fired power plants.

In Fig. 3.61, Q is the calorific value; T_W is the lag time constant of water wall; m_W is steam quantity; C_D is the time constant of drum heat storage volume; K is the

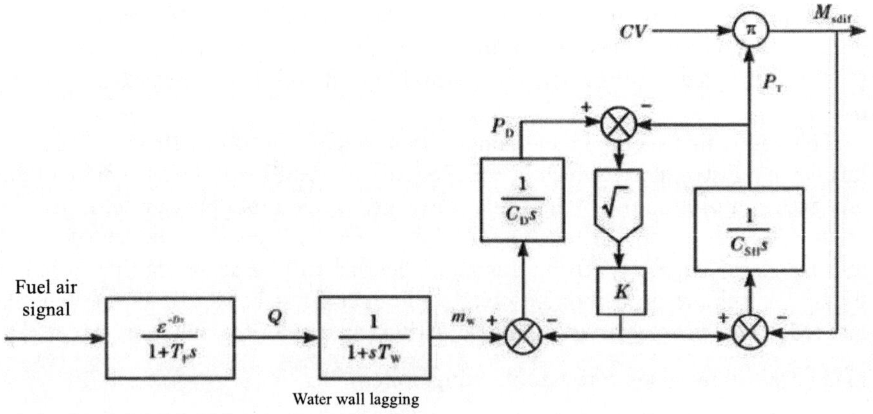

Fig. 3.61 Block diagram of the boiler model

flow coefficient of superheater and themain steam pipeline; C_{SH} is the time constant of superheater volume; P_D is drum pressure; P_T is the main steam pressure; CV is the steam valve opening; M_{sdif} is steam flow.

For the combustion loop in the model, the inertia link $\frac{1}{1+T_{F}s}$ and delay link ε^{-Ds} can be used to represent the dynamic characteristics of coal. The inertia time constant T_F varies greatly for different types of boilers: for intermediate storage pulverized coal systems, if fuel needs to be increased or decreased, only the speed of the coal pulverizer needs to be regulated, and the value is small; for direct-fired pulverized coal systems, the time constant is much larger because the fuel supplied needs to be regulated from the coal feeder, pulverized by the coal feeder, separated by the coarse coal separator, and then transported to the furnace by the pipeline. However, the dynamic characteristics of the fuel are expressed by inertia link $\frac{1}{1+T_{F}s}$, and the delay link is ignored. For coal-fired units, the typical value of T_F is 30 s and the typical value of delay D is 40 s; for oil-fired or gas-fired units, the typical value of T_F is 5 s and the typical value of delay D is 0 s.

When the regulating valve changes, the pressure in front of the regulating steam valve will change, which will lead to the imbalance of the output and input energy of the boiler and then to the change in the drum pressure. The drum pressure is a typical integral process. The value of drum integral time constant C_D depends on the boiler capacity and its type, generally 90–300 s.

Typical values of other parameters in the model are T_W: 5–7 s; C_{SH}: 5–15; K: 3.5.

2. Steam turbine model

The block diagram of this model is shown in Figs. 3.62 and 3.63. The main link of this model is the same as the traditional steam turbine model, except that it is

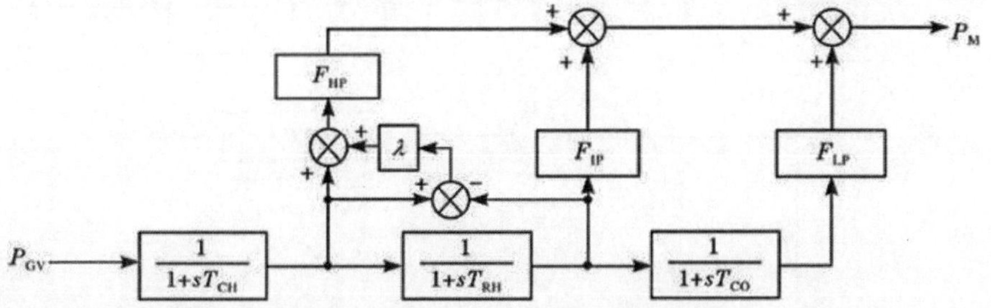

Fig. 3.62 Steam turbine model

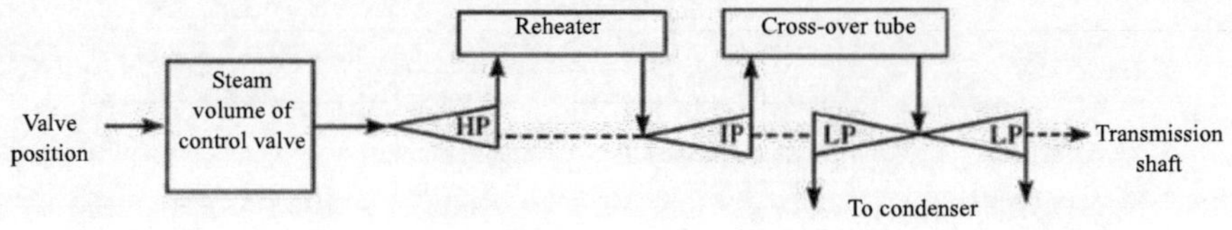

Fig. 3.63 Schematic diagram of physical equipment corresponding to steam turbine model

possible to consider the power change caused by the steam flow difference in the high and intermediate pressure cylinders in terms of this model, which is shown in the block diagram of the model as an additional correction link with λ compared to the traditional model.

In Fig. 3.62, T_{CH} is the steam volume time constant, s; F_{HP} is the power ratio of high pressure cylinder; T_{RH} is the time constant of reheater; F_{IP} is the power ratio of the intermediate pressure cylinder; T_{CO} is the time constant of the cross pipe; F_{LP} is the power ratio of the low pressure cylinder; λ is the natural overregulation coefficient of the high pressure cylinder power; HP is a high pressure cylinder; IP is an intermediate pressure cylinder; LP is a low pressure cylinder.

Typical parameter data:

$F_{HP} = 0.3$, $F_{IP} = 0.4$, $F_{LP} = 0.3$;

$T_{CH} = 0.1$–0.4 s, $T_{RH} = 4$–11 s, $T_{CO} = 0.3$–0.5 s;

$F_{HP} + F_{IP} + F_{LP} = 1$.

3. Coordinated control model of boilers and steam turbines (BCON)

The BCON type coordinated control model can simulate three modes of boiler following steam turbine, steam turbine following boiler and coordinated control.

In Fig. 3.64, *LD* is a fixed power value; K_F is the frequency regulation coefficient; K_{PE} is the power gain; T_{PE} is the power measurement time constant; K_{FF} is feedforward coefficient; K_P and K_G are coordinated control coefficients; P_{ref} is the steam pressure reference value; *LR* is the valve switch command; Fuel is the combustion command.

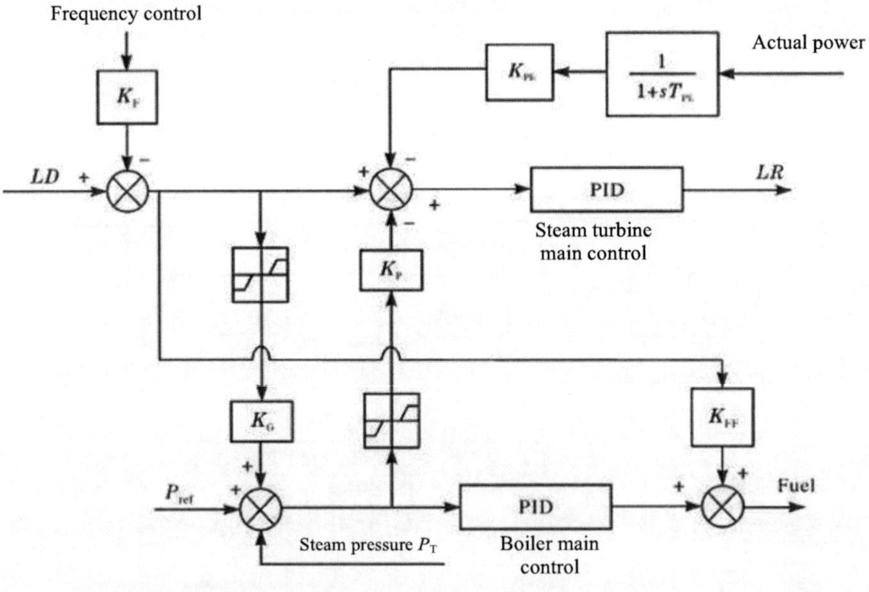

Fig. 3.64 General coordinated control model of boiler and steam turbine

3.6.2 Dynamic Model of Hydropower Plant

This section briefly introduces the water turbine and its control system model proposed by IEEE Committee in 1992 [10].

1. Basic block diagram of a dynamic model of water turbine and its control system

The dynamic model of water turbine and its control system is shown in Fig. 3.65.

2. Dynamic model of water turbine conduits
(1) Model of a single conduit water turbine

Make the following assumptions:

(1) Water is an incompressible liquid;
(2) The turbine is rigid (elasticity is not considered).

The model of the single conduit water turbine is shown in Fig. 3.66.

The calculation formula is as shown below

$$\frac{\mathrm{d}q}{\mathrm{d}t} = \frac{1 - h - h_1}{T_\mathrm{w}}$$

$$T_\mathrm{w} = \left(\frac{L}{A}\right)\frac{q_\mathrm{base}}{h_\mathrm{base}g}$$

$$q = G\sqrt{h}$$

$$P_\mathrm{m} = A_\mathrm{t}h(q - q_\mathrm{nl}) - DG \cdot \omega$$

$$h_1 = f_\mathrm{p}q^2$$

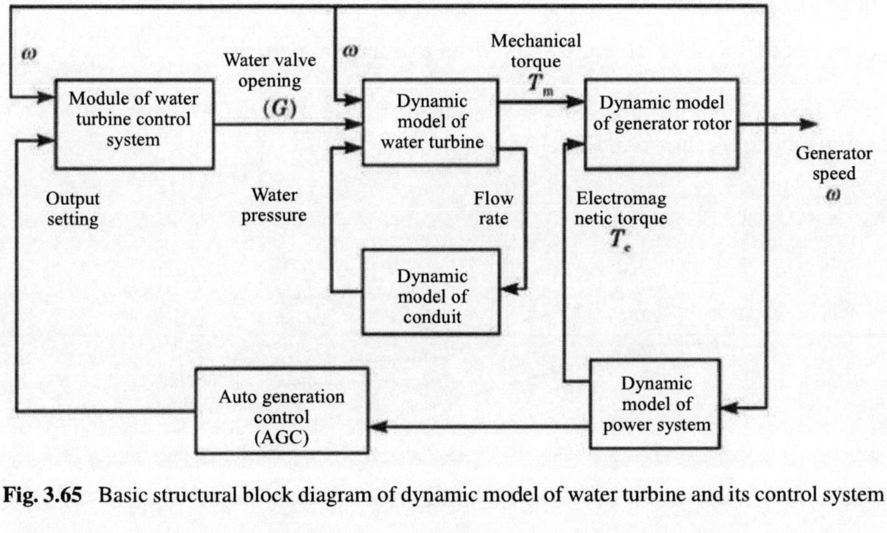

Fig. 3.65 Basic structural block diagram of dynamic model of water turbine and its control system

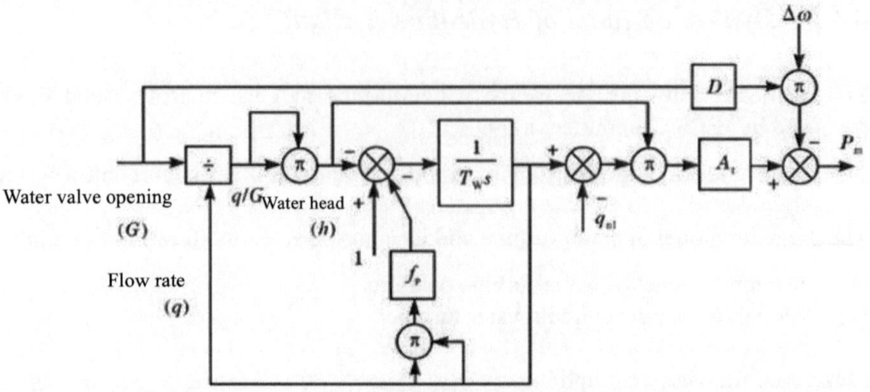

Fig. 3.66 Block diagram of dynamic model of conduit of single conduit water turbine

$$A_{\mathrm{t}} = \frac{\text{Rated power of water turbine (MW)}}{\text{Rated megavolt ampere (MVA) of generator } h_{\mathrm{r}}(q_{\mathrm{r}} - q_{\mathrm{nl}})} \tag{3.83}$$

wherein, q is the per-unit value of conduit water flow, q_{base} is the flow when the water valve is fully opened ($G = 1$), m^3/s; h is the per-unit value of water head at the inlet of water turbine, h_{base} is the static water head of the conduit, m; hl is the per-unit value of water head loss; T_{W} is the water time constant, s; L is the conduit length, m; A is the conduit section, m^2; g is the acceleration of gravity, m/s^2; P_{m} is the per-unit value of water turbine power, and the reference value is the reference megavolt ampere of the generator; q_{nl} is the per-unit value of no-load flow; A_{t} is the proportional coefficient, constant; h_{r} is the water head at the inlet of water turbine at rated flow, per-unit value; q_{r} is the flow at rated load, per-unit value; f_{p} is the water head loss coefficient.

(2) Model of multi-conduit water turbine in the same tunnel.

The structure of multi-conduit water turbine in the same tunnel is shown in Fig. 3.67, and its model is shown in Fig. 3.68.

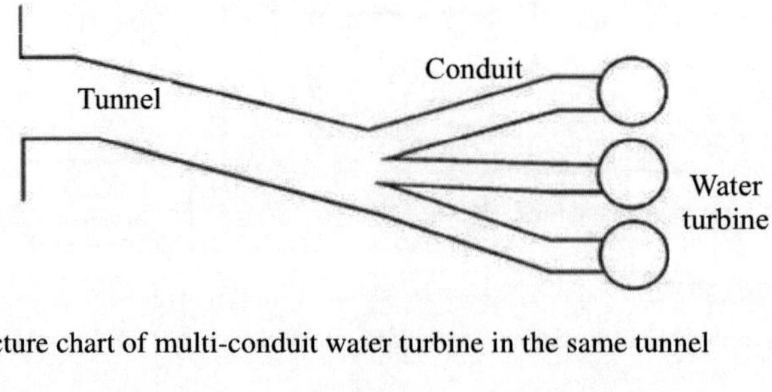

Fig. 3.67 Structure chart of multi-conduit water turbine in the same tunnel

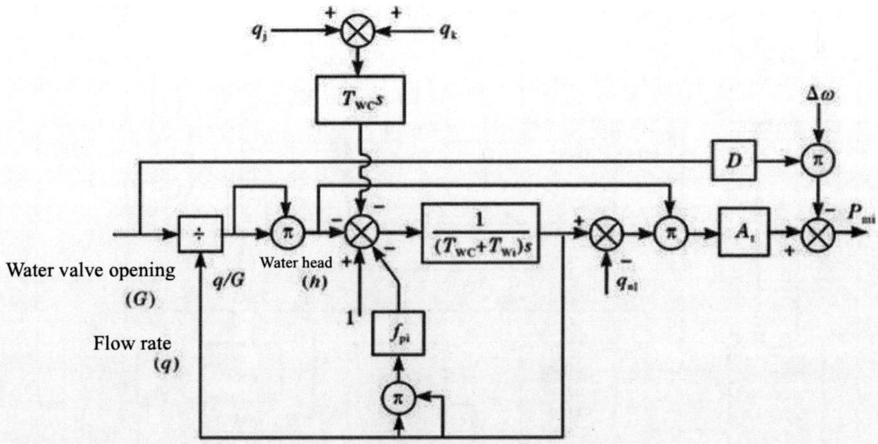

Fig. 3.68 Model of multi-conduit water turbine in the same tunnel

The calculation formula is as shown below

$$(T_{wi} + T_{wc}) = \frac{dq_i}{dt} = (1 - h_i) - T_{wc}\frac{d(q_j + q_k)}{dt} - f_{pi}q_i^2 - f_{pc}q_c^2 \quad (3.84)$$

wherein, T_{WC} is the water time constant of the public tunnel; T_{Wi} is the water time constant of the conduit; f_{pc} is the water head loss coefficient of the public tunnel; f_{pi} is the water head loss coefficient of the conduit; q_i is the water flow rate of the conduit; $q_c = \sum q_i$ is water flow rate of the public tunnel.

3.6.3 Dynamic Model of Pressurized Water Reactor Nuclear Power Plant

Pressurized water reactor nuclear power plant is mainly composed of nuclear reactor, pressurizer, steam generator, turbine generator and its auxiliary equipment. Nuclear fuel fission occurs in the reactor to release nuclear energy. Nuclear energy is converted into thermal energy, which is carried out by the coolant. Nuclear energy enters the steam generator through the hot wire and is transferred to the working medium. After preheating and evaporation, it is converted into saturated steam to drive the turbine generator.

Figure 3.69 shows a basic structural block diagram of model of pressurized water reactor nuclear power plant [9].

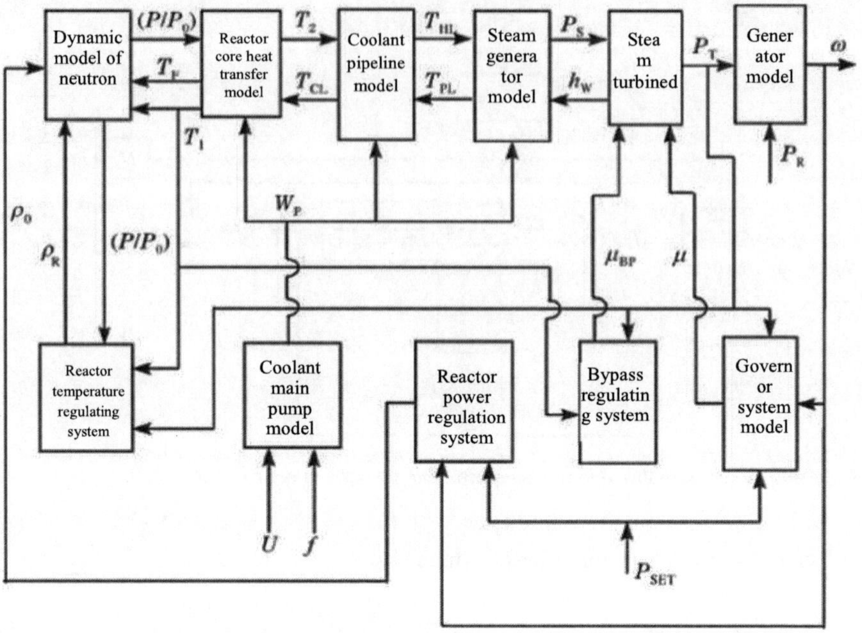

Fig. 3.69 Basic block diagram of pressurized water reactor nuclear power plant

3.7 Auto Generation Control

Various control measures in the power system include local control and centralized control, and their influence on voltage stability is very important. Local control measures, especially in power plants, are usually automatic and have fast response speed. The centralized control measures are generally to set up a control or dispatching center in each power company, which issues slower automatic and manual control commands to power plants and substations. The most basic centralized automatic control is auto generation control, commonly known as AGC.

3.7.1 AGC Control Mode

For a large-scale interconnected system, the process of realizing automatic regulation of active power and frequency, keeping the balance of generated power and load power of the whole system, and keeping the system frequency constant or the deviation from the rated frequency within the allowable range, is auto generation control.

AGC is used to: ① maintain the system frequency at the rated value or within the allowable frequency bias range; ② control the net interchange power in the area at the planned value; and ③ realize economic dispatch control (EDC) among units within the jurisdiction under the premise of safe operation.

AGC is mainly divided into the following three basic control modes.

1. Flat frequency control (FFC)

Under FFC mode

$$\mathrm{ACE} = B\Delta f \tag{3.85}$$

This mode is regulated according to the frequency bias, and the regulation ends in case of $\Delta f = 0$. Its ultimate goal is to maintain the system frequency without controlling the interchange power on the tie line. It is usually only suitable for small systems with close connection among power plants, and is very difficult in regulating large systems.

2. Flat tie line control (FTC)

Under FTC mode

$$\mathrm{ACE} = \Delta P \tag{3.86}$$

This regulation mode only maintains the interchange power on the tie line constant and does not control the system frequency. It is usually suitable for small areas and connections to large systems. It can ensure the transmission power of the tie line, while the frequency is maintained by regulating the generator output simultaneously by two adjacent systems.

3. Tie line load frequency bias control (TBC)

Under TBC mode

$$\mathrm{ACE} = \Delta P + B(\Delta f) \tag{3.87}$$

wherein, ΔP is the difference between the actual value and the planned value of the interchange power on the tie line; B is the frequency bias coefficient, which is usually set as the frequency response characteristic coefficient value β of the area.

TBC control mode forms area control error (ACE) signal by linear combination of power bias and frequency bias on the tie line. The area is controlled by ACE signal. The ultimate goal is not only to eliminate frequency bias ($\Delta f = 0$), but also to eliminate power bias ($\Delta P = 0$). TBC control mode is an ideal control mode, which is widely used in modern large-scale interconnected systems.

Compared to FFC and FTC control modes, TBC control mode has the following characteristics:

(1) During normal operation, each area performs its own control tasks. It is stipulated that the load changes occurring in each area are balanced by regulating

the generated power in that area, that is, the generated power changes in each area are determined according to the load changes in that area. At this time, the net interchange power transmitted by the tie line is maintained at the planned value, all areas share the task of system frequency regulation, and the system frequency is maintained at the normal value.

(2) In case of an accident or emergency, if one area or several areas in the system cannot perform its (or their) control tasks, as long as the whole system is still in synchronous state, the normal area can provide emergency power support to the accident area. At this time, the net interchange power transmitted in the area is allowed to deviate from the planned value, and the frequency can be allowed to deviate from the normal value, so as to ensure that sufficient and effective support is provided to the accident area through the tie line, thus restoring the whole system to normal as soon as possible.

(3) The *B* coefficient is usually set as the natural frequency response characteristic β value of the area. Under the condition of $\Delta f = 0$, all areas are controlled for operation in the mode equivalent to the FTC mode. If the interchange power of tie line is set, all areas are controlled for operation in the mode equivalent to FFC mode.

3.7.2 AGC Model

AGC model can be divided into two parts according to variable sources: network side (Fig. 3.70) and generator side (unit side) (Fig. 3.71). It can realize tie line frequency control and economic dispatch algorithm.

The description of variables and parameter symbols in Fig. 3.70 is as follows:

f is the tie line frequency; PTSC is the planned load curve (it can be manually input, and the initial steady-state value is taken if it is not input); f_0 is the frequency reference value; ACE is the total frequency bias signal; K_P is proportional gain; K_I is the integral gain; ACEi is area control error (MW).

In Fig. 3.71, the typical parameters and their calculation descriptions are as follows:

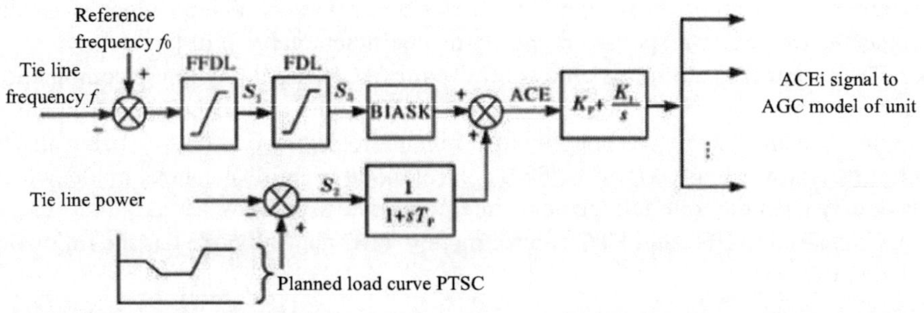

Fig. 3.70 Network side model of AGC

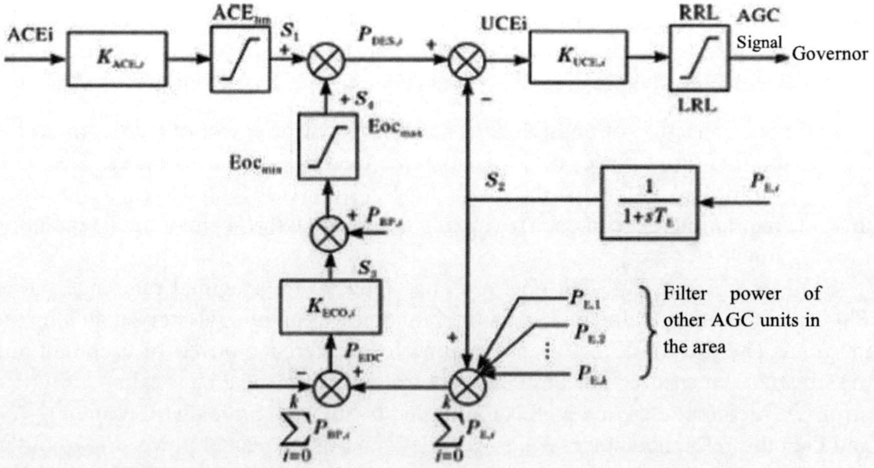

Fig. 3.71 Generator side (Unit Side) model of AGC

$K_{\mathrm{ACE},i}$ is the distribution coefficient of area ACE; $\mathrm{ACE}_{\mathrm{lim}}$ is the ACE regulation limit, calculated according to the actual regulative range, $-\mathrm{ACE}_{\mathrm{lim}} \leq K_{\mathrm{ACE,i}} \cdot \mathrm{ACEi} \leq \mathrm{ACE}_{\mathrm{lim}}$.

For the description of the generator side AGC model variables and parameters, the following per-unit values are all relative to the maximum generated power of this unit:

$P_{\mathrm{BP},i}$ is the unit reference power, and the initial power of the generator is taken in the program, i.e., the generated power P_0 in the power flow calculation; $P_{\mathrm{DES},i}$ is the expected power signal of the controlled unit; UCEi is the error control signal of the controlled unit; $\sum_i^k P\mathrm{BP}, i$ is the total base capacity of the adjustable units in the area; $\sum_i^k P\mathrm{E}, i$ is total actual output for the adjustable units in the area.

RRL and LRL are the upper and lower limits of the unit output change rate, MW (p.u.)/min. The reference value of p.u. value is the maximum generated power of this unit. Thermal power units are rated at 1–15% per/min according to different types; 5% rating/min for nuclear power units; 30–150% rating/min for hydroelectric units and 0.5–2.5% rating/s. LRL is a value less than or equal to 0.

It can be seen from the above that the AGC inputs include bus frequency in this area, interchange power of tie line and output of each regulating unit. The values obtained by using power bias amount and the frequency bias amount of the tie line respectively multiplying by the bias factor BIASK are added to obtain the area power control error signal. ACE signal has the following two functions in the control process:

(1) The sum of the increment of area frequency and the bias amount of tie line can be used to judge the area where power increase or decrease occurs, i.e., the area where power imbalance occurs.

(2) The output of the controlled unit is regulated by modifying ACE control signal, so as to re-balance power generation and load, restore the frequency to the normal value, and restore the power of the tie line to the initial set value.

For interconnection of multiple areas, the interchange power of the tie line is the sum of the measured power of all tie lines on the boundary of the area. The ACE signal is multiplied by the distribution factor K_{ACE} and distributed to each controlled unit; the regulation range of each regulating unit is limited by symmetrical regulation limiting links.

Each AGC controlled unit operates according to the specified reference power $P_{BP,i}$, and the initial value of $P_{BP,i}$ is the initial power of generator calculated by the program. The difference between the sum of the reference power of each unit and the sum of the measured power of each controlled unit in the area is called economic error. The economic error signal is multiplied by the economic distribution $K_{ECO,i}$, and then the reference power $P_{BP,i}$ is added, and the expected power $P_{DES,i}$ of the controlled unit is obtained by adding the economic limiting link and ACE signal. The sum of economic distribution factors of all units participating in secondary regulation in the area should be 1.

The final power requirement signal of each controlled unit actually consists of three parts: $P_{AGC} = P_{E,i} + K_{ACE,i} \cdot \mathrm{ACE} + P_{ECO}$, i.e., actual power $P_{E,i}$ of generator, frequency control signal $K_{ACE,i} \cdot \mathrm{ACE}$ of tie line, and economic error power P_{ECO} distributed by areas. Finally, the signal is sent to each controlled unit.

In the simulation calculation, the planned power of the tie line depends on the initial power value of the tie line. If the planned power needs to be modified, it needs to be preset. The setting method is as shown in Fig. 3.72. The change rate of the tie line power is $(P_2 - P_1)/(T_2 - T_1)$. The change rate remains unchanged during time T_1—T_2. It will take some time for P_1 to increase to P_2.

In actual AGC control, the reference power $P_{BP,i}$ and the economic distribution coefficient $K_{ECO,i}$ are calculated every 4–10 min. Therefore, in the whole process of

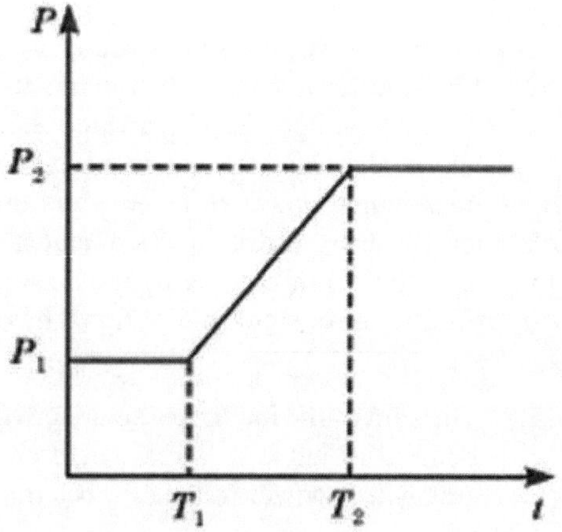

Fig. 3.72 Tie line power setting

simulation calculation, $K_{\mathrm{ECO},i}$ remains unchanged after the original data are input, while $P_{\mathrm{BP},i}$ maintains the initial power of generator unchanged.

In addition, if the unit does not participate in economic allocation, it can be realized by setting the calculation mode at AGC unit side to zero (0: no economic allocation; 1: economic allocation), i.e., setting the economic allocation factor $K_{\mathrm{ECO},i}$ of the corresponding unit as zero, so that ACE signal is directly sent to each regulating unit, which can simulate the way that the dispatching center directly controls ACE without EDC.

References

1. Expert Group on Mathematical Models of Excitation System (1991) Mathematical models of excitation system for calculating power system stability. Proc Chin Soc Electr Eng 5:65–75
2. IEEE Task Force on Excitation Limiters (1995) Recommended models of overexcitation limiting devices. IEEE Trans Energy Convers 10(4):706–713
3. IEEE Task Force on Excitation Limiters (1995) Underexcitation limiter models for power system stability studies. IEEE Trans Energy Convers 10(3):524–531
4. IEEE Task Force on Load Representation for Dynamic Performance (1995) Standard load models for power flow and dynamic performance simulation. IEEE Trans Power Syst 10(3):1302–1313
5. Tang Y, Zhang H, Hou J et al (2007) A synthesis load model with distribution network. Power Syst Technol 31(5):33–38
6. Tang Y, Zhang HB, Zhang DX et al (2009) A synthesis load model with distribution network for power system simulation and its validation. In: Power engineering society general meeting, Calgary
7. Zhejiang University Teaching and Research Group DC Power Transmission Research Section (1982) Water Resources and Electric Power Press, Beijing
8. Zhao W (2004) HVDC transmission engineering. China Electric Power Press, Beijing
9. IEEE Working Group on Prime Mover and Energy Supply Models for System Dynamic Performance Studies (1991) Dynamic models for fossil fueled steam units in power system studies. IEEE: Trans Power Syst 6(2):753–761
10. IEEE Working Group on Prime Mover and Energy Supply Models for System Dynamic Performance Studies (1992) Hydraulic turbine and turbine control models for system dynamic studies. IEEE Trans Power Syst 7(1):167–179

Chapter 4
Static Analysis of Voltage Stability

4.1 Basic Principle of Static Analysis of Voltage Stability

The static analysis method of voltage stability is to capture the system state on different time sections along the time track. By assuming that the differential of the system state variables on the section is zero, the differential-difference-algebraic equations describing the nonlinear dynamic power system are simplified to pure algebraic equations. Since the early voltage instability accidents are mostly the medium and long-term voltage instability processes, which have a long time span and are usually caused by slow load growth or system output power transfer, the static analysis method has become the mainstream of voltage stability research for a long period of time.

The research of static voltage stability has been accumulated for a long time and abundant research results have been obtained. As this analysis method can reflect the voltage stability level of the system to a certain extent, and can give the voltage stability index and the sensitivity information of the state variables and control variables, it is convenient for the monitoring and optimization adjustment of the system, and can better meet the requirements of practical application.

In the static analysis method, the dynamic characteristics of the dynamic elements in the power system are ignored, and the required system element model is relatively simple, which is generally a steady state or quasi steady state model. With the deep understanding of voltage stability, more and more static analysis models consider the actual regulation and limitation characteristics of the system [1, 2], making the static analysis method more consistent with the reality of the power system.

In the static voltage stability analysis of the power system, the active power of the generator is determined by the output power of the generator. In addition, due to the excitation regulation, the bus voltage amplitude of the generator can be maintained within a certain range. Therefore, for generator nodes, the active power P and voltage amplitude U can be given, while the voltage phase angle θ and reactive power Q are to be determined. As a result, generators are usually represented as PV node. However, in the process of power flow calculation, the reactive power output of generator nodes

© Science Press 2021

Y. Tang, *Voltage Stability Analysis of Power System*, Power Systems,
https://doi.org/10.1007/978-981-16-1071-4_4

may exceed the allowable range in order to maintain a given voltage. Especially when the reactive power supply of the power system is not sufficient, the reactive power of *PV* node is often out of bounds. Therefore, it is necessary to monitor the reactive power of *PV* node. When the reactive power is beyond the given range, the common way is to fix the reactive power output of the node on the limit value, and convert the node from *PV* node to *PQ* node. Of course, if the active power output and reactive power output of some generators are given and the voltage amplitude and phase angle are to be calculated, it will be treated as a *PQ* node.

In addition, in the power flow calculation, it is usually necessary to specify a balance node, the voltage amplitude and phase angle of the node are given, and the active power and reactive power are to be calculated, also known as $V\theta$ node.

Load is an integral part of a power system, and the selection of a load model is the basis of voltage stability research, which greatly affects the analysis results of voltage stability. Since the power system load is composed of different kinds of loads, not only the composition changes at any time, but also the load composition of each node is different, it is difficult to accurately obtain the load model of each node. As a result, load models are often simplified according to different research contents and purposes. In the static voltage stability analysis based on power flow, only the static characteristic model of the load is usually considered.

The static analysis method of voltage stability is based on the power flow equation (or extended power flow equation) of the system. The critical point of voltage stability is the point where the system physically reaches the maximum transmission power and the point where the Jacobian matrix of the system power flow is singular mathematically. Therefore, in essence, static voltage stability analysis is a problem to study whether there is a feasible solution to the power flow equation. In this chapter, the static analysis system of voltage stability is constructed by introducing the research results of model, method, criterion and engineering application of voltage stability static analysis method.

4.2 Research on Transmission Capability of Power Systems

With the continuous development of the power system, it is urgent to make full use of the transmission capability of the existing transmission network to reduce the operating cost to the maximum extent and improve the economic benefits of the system operation. However, with the increase of transmission power, the capacity of a power grid to bear the disturbance is gradually reduced, which brings high security risks to the system. Therefore, for a large-scale interconnected power system, how to accurately determine the transmission capability of the power system and satisfy the demand of electric load to the maximum under the constraint of safety and reliability has become an urgent research topic of the modern power system.

4.2.1 PV *Curve*

The voltage instability of a power system often occurs in the period when the system is close to the maximum transmission power. In the critical state, if the load increases by a small amount, the system voltage will drop sharply, resulting in system voltage instability. *PV* curve is a basic tool of static voltage stability analysis based on voltage stability mechanism, in which *P* can be expressed as the total load of a certain area, or the transmission power of a transmission section or regional tie-line, and *U* is the key bus voltage. By establishing the relationship between load and node voltage, *PV* curve can visually and continuously show the process of system voltage reduction and even collapse with the increase of load. At the same time, by calculating the *PV* curve of each node in the system, two important parameters about the voltage stability of the system can be obtained: the critical voltage and the power limit of the load point, which can be used to indicate the voltage stability margin of the system and to represent the ability of each load node to maintain the voltage stability.

From Fig. 4.1:

$$\dot{U} = \dot{E} - (R_s + jX_s)i \tag{4.1}$$

The power absorbed by the load can be calculated by using the following formula (let $\dot{E} = E\angle 0^\circ$, $\dot{U} = U\angle\theta$):

$$\dot{S} = \dot{U}\frac{\overset{*}{E} - \overset{*}{U}}{R_s - jX_s} = (U\cos\theta + jU\sin\theta)\left(\frac{E - U\cos\theta + jU\sin\theta}{R_s - jX_s}\right) \tag{4.2}$$

P and Q can be obtained respectively

$$P = \frac{UE(R_s\cos\theta - X_s\sin\theta - U^2R_s}{R_s^2 + X_s^2} \tag{4.3}$$

$$Q = \frac{UE(R_s\sin\theta - X_s\sin\theta - U^2R_s}{R_s^2 + X_s^2} \tag{4.4}$$

It can be obtained by transposition

$$P + \frac{U^2R_s}{R_s^2 + X_s^2} = \frac{UE(R_s\cos\theta - X_s\sin\theta}{R_s^2 + X_s^2} \tag{4.5}$$

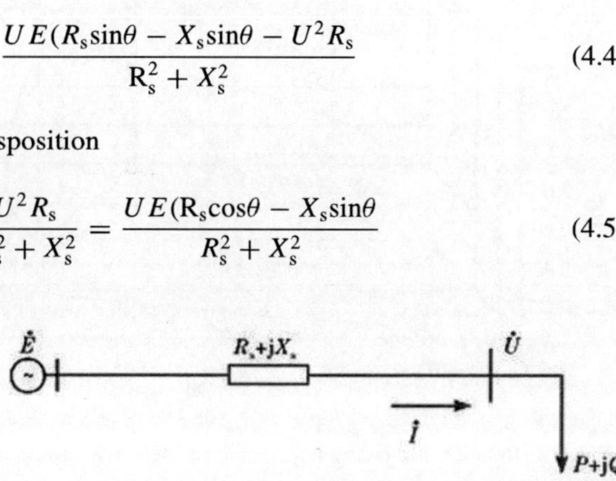

Fig. 4.1 Schematic diagram of single-machine single-load system

$$Q + \frac{U^2 X_s}{R_s^2 + X_s^2} = \frac{UE(R_s \sin\theta - X_s \cos\theta}{R_s^2 + X_s^2} \tag{4.6}$$

It can be obtained by eliminating θ from the two equations above

$$\left(P + \frac{U^2 R_s}{R_s^2 + X_s^2}\right)^2 + \left(Q + \frac{U^2 X_s}{R_s^2 + X_s^2}\right)^2 = \frac{U^2 E^2}{R_s^2 + X_s^2} \tag{4.7}$$

It can be obtained by multiplying both sides by factor $(R_s^2 + X_s^2)^2$, i.e., Z_s^4

$$\left(U^2\right)^2 + (2PR_s + 2QX_s - E^2)U^2 + Z_s^2\left(P^2 + Q^2\right) = 0 \tag{4.8}$$

The voltage solution can be obtained:

$$U = \sqrt{\frac{E^2}{2} - PR_s - QX_s \pm \sqrt{\left(PR_s + QX_s - \frac{E^2}{2}\right)^2 - Z_s^2\left(P^2 + Q^2\right)}} \tag{4.9}$$

For simplicity, if the transmission resistance R_s is ignored, then

$$U = \sqrt{\frac{E^2}{2} - QX_s \pm \sqrt{\frac{E^4}{4} - X_s^2 P^2 - X_s E^2 Q}} \tag{4.10}$$

In (P, Q, U) space, we can make the above relation surface, as shown in Fig. 4.2. The upper part of the surface corresponds to the plus sign solution of the Eq. (4.10), which is the high voltage solution; the lower part corresponds to the minus sign solution, which is the low voltage solution. The "waist line" of the surface is the unique

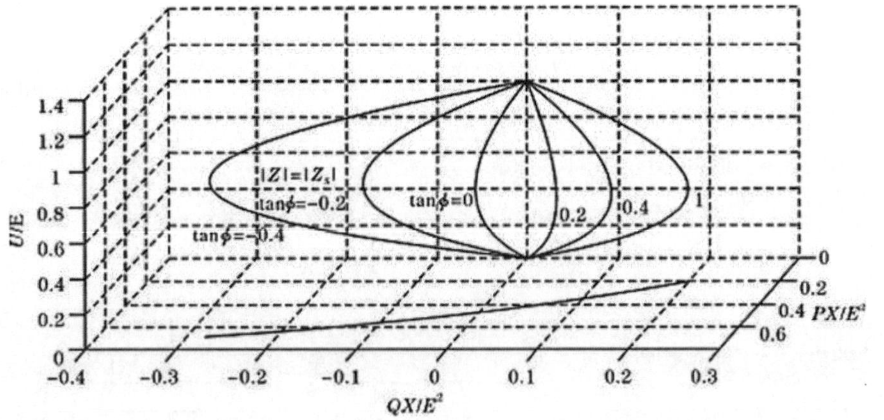

Fig. 4.2 Relationship between active power, reactive power, and voltage of load

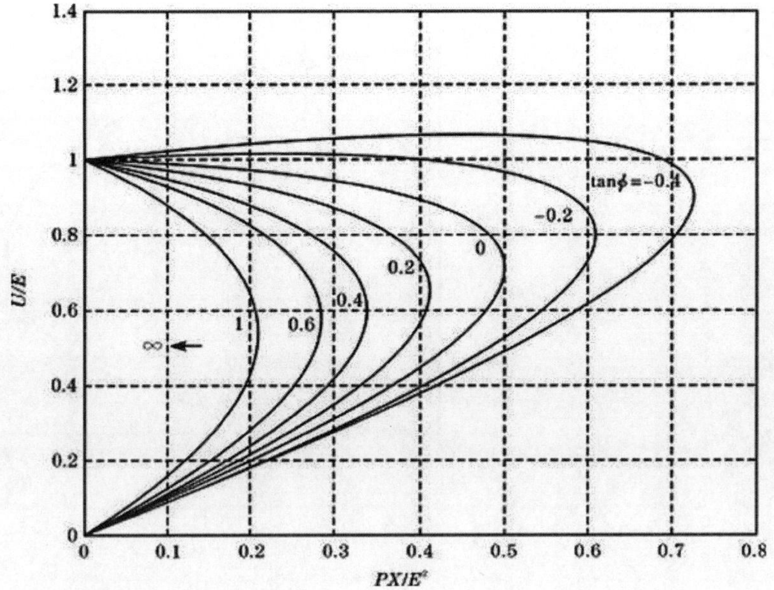

Fig. 4.3 Standardized *PV* curve

solution of the voltage, where the module value of the load equivalent impedance is equal to that of the system equivalent impedance.

The curves in solid lines in the surface intersects with $Q = P\tan\phi$. The projection of these curves on (*P*, *U*) plane is known as *PV* curve, or nose curve, as shown in Fig. 4.3, which plays an important role in the analysis of voltage stability.

The vertex ($|Z| = |Z_s|$) of *PV* curve corresponds to the load capacity limit state of the system, that is, the critical point of voltage stability. The upper half segment of *PV* curve is the high voltage solution or feasible solution, which is the balance point of stable operation of the system. The lower half segment is the low voltage solution or infeasible solution, which is the balance point of the system instability. When the system load increases gradually, the system operating point will transition from the upper half segment of *PV* curve to the lower half segment, and the system will lose stability at the inflection point, which is the critical point of voltage stability.

When the system runs in the lower half segment of *PV* curve, the voltage will be uncontrollable and unstable. As shown in Fig. 4.4, both curves 1 and 2 are *PV* curves of the system, except the power supply potential *E* corresponding to curve 2 is larger than that of curve 1, other conditions are the same. It is assumed that the system is originally running at point A on the lower half segment of curve 1. If the power supply potential suddenly rises to the corresponding value of curve 2, the system operating point will transition to point B on the lower half segment of curve 2 according to the nature of constant load power. It can be concluded that when the system runs in the lower half segment of the curve, the rise of power supply voltage will lead to a drop in load node voltage, and the voltage control will lose causality.

Fig. 4.4 Voltage control

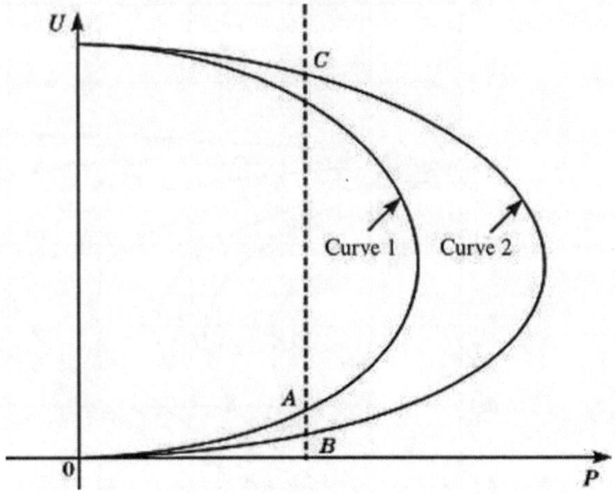

4.2.2 VQ Curve

If the surface in Fig. 4.2 is projected on (U, Q) plane, a group of *VQ* curves can be obtained, as shown in Fig. 4.5. It shows the corresponding relationship between U and Q when P is constant in the case of a single power supply single load system. It can be seen from the figure that the critical voltage value at high load level is very high, so the reactive power margin is small and the voltage stability is poor. The right side $\frac{dQ}{dU} > 0$ of the curve is voltage stability, representing the normal operation area of the system, and its slope can represent the voltage stability of the bus; the left side

Fig. 4.5 *VQ* curve of single power supply single load system

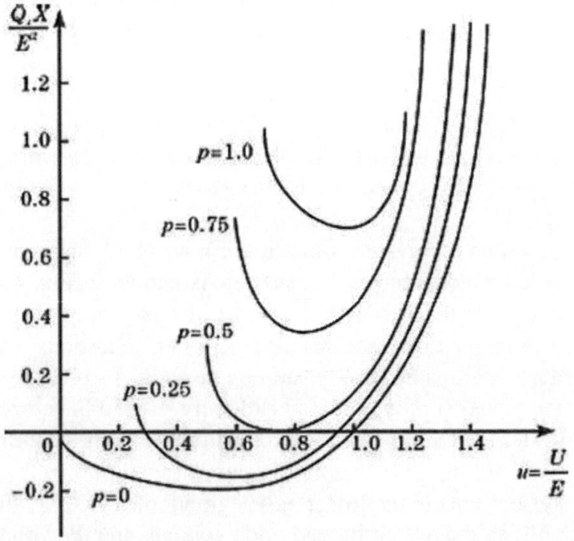

$\frac{dQ}{dU} < 0$ is voltage instability, representing the unstable operation area of the system; the bottom $\frac{dQ}{dU} = 0$ of the curve is the critical point of voltage stability.

For a large-scale power system, *VQ* curve can be obtained by a series of power flow calculations. It represents the relationship between the critical bus voltage and the reactive power of the bus. It is assumed that in the power flow calculation program, the observation bus is equipped with a virtual synchronous condenser, which is not limited by reactive power. It is *PV* node. By setting different terminal voltage values, the corresponding reactive power output value is obtained. The voltage is taken as an independent variable and as x-coordinate, the reactive power as y-coordinate. The capacitive character is positive, and the inductive character is negative. The *VQ* curve can be obtained [3, 4].

VQ curve can reflect the reactive power margin of the bus. If the lowest point of the curve is above the *U*-axis and the curve does not intersect with the *U*-axis, it indicates that the system has insufficient reactive power. If there is no reactive power injection, the system cannot operate. Additional reactive compensation is required to avoid voltage collapse, as shown in curves $p = 0.75$ and $p = 1.0$, Fig. 4.5. The vertical distance between the lowest point and the *U*-axis represents a negative reactive margin, indicating the reactive power to be injected at least during system operation. If the lowest point of the curve is below the *U*-axis, there will be two intersections between the curve and the *U*-axis, as shown in curves $p = 0$ and $p = 0.25$. The vertical distance between the lowest point and the *U*-axis represents the reactive margin of the system.

Through the *VQ* curve, the voltage stability characteristics and reactive power compensation requirements can be understood. Moreover, the calculation speed of the curve is fast and the convergence is good, which can reflect the demand for reactive power compensation, reactive power margin and voltage stability. Therefore, *VQ* curve is widely used in the field of voltage stability analysis. However, *VQ* curve also has its own limitations. When the system scale is large, the calculation of *VQ* curve for each power level and each fault of many nodes requires a large amount of computation. In addition, *VQ* curve can only provide the local compensation needs, not the overall optimal compensation scheme.

4.2.3 Solution of the Curve—Continued Power Flow [4–6]

Since the *PV* (*VQ*) curve can provide safety information such as active power transmission margin, reactive power reserve and voltage stability level, it is very significant to accurately solve the *PV* curve for the research of power system static voltage stability.

As the Jacobian matrix near the "nose" of the *PV* curve is nearly singular, the power flow equation is often ill conditioned, resulting in the difficulty in convergence of conventional power flow calculation and the inability to determine the critical point of voltage instability. Continue power flow (CPF) is a powerful tool to solve the *PV*

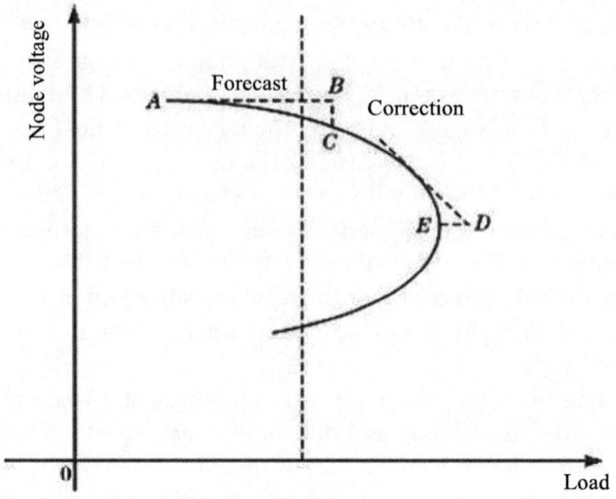

Fig. 4.6 Prediction-correction diagram of CPF

curve. By continuously updating the power flow equation, it makes the power flow equation have solutions at all possible load states, no matter whether in stable balance point or unstable balance point, and overcomes the problem of poor convergence of power flow in the operating state near the stability limit.

1. Basic principle

CPF includes two important steps: prediction and correction. From the initial stable operating point, with the slow change of load, the next operating point is estimated and corrected along the *PV* curve drawn until the complete *PV* curve is drawn, as shown in Fig. 4.6. The specific steps are as follows:

(1) Starting from the known solution A, estimate the solution B of load growth mode in a tangential direction, that is, the prediction step;
(2) Solve the accurate solution C by using the conventional power flow (the system load value is constant), which is the correction step;
(3) Repeat the above steps as the load increases further.

If the load estimation obtained exceeds the maximum load of the accurate solution, the correction calculation based on load value will not converge, as shown in point D in the figure. At this time, the correction calculation based on voltage is adopted to obtain the accurate solution E. Therefore, when approaching the voltage stability limit, in order to determine the accurate maximum power, the load increment should be gradually reduced during the continue prediction.

2. Mathematical foundation

Compared with the conventional power flow, CPF takes the increase of load as a parameter and introduces it into the power flow equation, which is expressed as

$$F(\theta, U) = \lambda K \tag{4.11}$$

wherein, θ is the phase angle of bus voltage; U is the vector of bus voltage; λ is the load parameter; K is the vector of load change of each bus.

λ value meeting the following requirements should be given to solve the Eq. (4.11):

$$0 < \lambda < \lambda_{cr} \tag{4.12}$$

wherein, $\lambda = 0$ is the basic load condition, λ_{cr} is the critical load.

Then Eq. (4.11) can be rewritten as:

$$F(\theta, U, \lambda) = 0 \tag{4.13}$$

(1) Prediction step

It is a step to estimate the solution of a changed state variable (θ, U, λ) by linear approximation. Take the total differential for Eq. (4.13):

$$F_\theta \mathrm{d}\theta + F_U \mathrm{d}U + F_\lambda \mathrm{d}\lambda = 0 \tag{4.14}$$

i.e.,

$$\begin{bmatrix} F_\theta, F_U, F_\lambda \end{bmatrix} \begin{bmatrix} \mathrm{d}\theta \\ dU \\ d\lambda \end{bmatrix} = 0 \tag{4.15}$$

wherein, $[F_\theta,\ F_U]$ is the Jacobian matrix of the conventional power flow equation; $\begin{bmatrix} \mathrm{d}\theta \\ dU \\ d\lambda \end{bmatrix}$ is the tangent vector required for the prediction step.

Since the load parameter λ is introduced and a state variable is added, an additional equation is needed to solve the equations. This can be satisfied by assuming that a component of the tangent vector is $+1$ or -1, which is called a continuous parameter. Equation (4.15) changes to

$$\begin{bmatrix} F_\theta, F_U, F_\lambda \\ e_k \end{bmatrix} \begin{bmatrix} \mathrm{d}\theta \\ dU \\ d\lambda \end{bmatrix} = \begin{bmatrix} 0 \\ \pm 1 \end{bmatrix} \tag{4.16}$$

wherein, e_k is the row vector, except that the Kth element is 1, the other elements are 0.

The load parameter λ is selected as the continuous parameter, and the corresponding tangent vector component is set to $+1.0$. In the subsequent prediction step, the state variable with the largest change rate near the given solution is selected as the continuous parameter, and the sign of the corresponding tangent vector component is determined by the sign of the slope.

Once the tangent vector is determined, it can be predicted by the following equation:

$$\begin{bmatrix} \theta \\ U \\ \lambda \end{bmatrix} = \begin{bmatrix} \theta_0 \\ U_0 \\ U_0 \end{bmatrix} + \sigma \begin{bmatrix} \mathrm{d}\theta \\ dU \\ d\lambda \end{bmatrix} \tag{4.17}$$

wherein, subscript "0" is the solution of the state variable before the start of the prediction step; σ is the step length, so that the predicted value at the next point has a power flow solution. If a power flow solution cannot be obtained in the correction step under a given step length, the step length should be reduced.

(2) Correction procedure

x_k is the kth component of the state variable, and η is the predicted value of $\mathrm{x_k}$. $\mathrm{x_k}$ is selected as the predicted value, thus an equation is added to obtain the extended power flow equation:

$$\begin{bmatrix} F(\theta, U, \lambda \\ x_k - \eta \end{bmatrix} = 0 \tag{4.18}$$

The predicted values $\left[\theta', U', \lambda'\right]$ are substituted into the Eq. (4.18) as initial values, which can be solved by the improved Newton–Raphson iteration method. Due to the introduction of additional equations, the Jacobian matrix at the critical point is non-singular, and a complete *PV* curve can be obtained.

The tangent component of λ means that $\mathrm{d}\lambda$ is positive in the upper half segment of *PV* curve, negative in the lower half segment, and 0 at the critical point. Therefore, whether the system reaches the critical point of voltage stability can be judged from $\mathrm{d}\lambda$.

In addition, the choice of continuous parameters is very important to the convergence of numerical calculation, otherwise the solution will be divergent. For example, if the load parameter is selected as the continuous parameter, the solution will not converge near the critical point when the predicted value exceeds the maximum load. By this time, if the voltage amplitude is selected as the continuous parameter, the solution may converge. If the increase of load is selected as the continuous parameter, the correction step is the vertical line on the *PV* plane; if the voltage amplitude is selected as the continuous parameter, the correction step is the horizontal line on the *PV* plane.

CPF has strong robustness and flexibility. It is an ideal method to solve the convergence problem near the critical point. This method has been widely used in power system static stability evaluation and maximum available transmission capacity calculation since it was proposed. However, due to the large amount of calculation and the long time of calculation, the conventional power flow calculation method should be combined with CPF to achieve fast and accurate results. The specific method is to gradually increase the load from the basic working condition, and use conventional power flow methods, such as the Newton–Raphson method or quick decoupling method (*PQ* decomposition method) to calculate the power flow solution until the point where the calculation does not converge. From this point, CPF is used to solve the power flow. CPF is only necessary to solve the exact solution close to the critical point. As a result, the solving speed is greatly improved by this method. Reference [6] proposes an improved CPF, in which *PQ* decomposition method is used to solve the power flow, Lagrange quadratic interpolation technology is used in the prediction process, and local parametric method is used in the correction process to solve the correction equation. Compared to the traditional CPF based on the Newton–Raphson method in the polar coordinate form, the calculation speed is faster, the memory is less and the accuracy is higher with the improved CPF.

(3) Case study

Taking IEEE 30-node system as an example, the wiring diagram of the system is shown in Fig. 4.7. Please refer to Tables 4.1 and 4.2 for the initial parameters of generator reactive power output and node load.

The *PV* curve is solved by the CPF with prediction and correction functions and step length based on voltage drop. The Bus 30 *PV* curve is shown in Fig. 4.8, and "+" indicates that the corresponding generator reaches the reactive power limit.

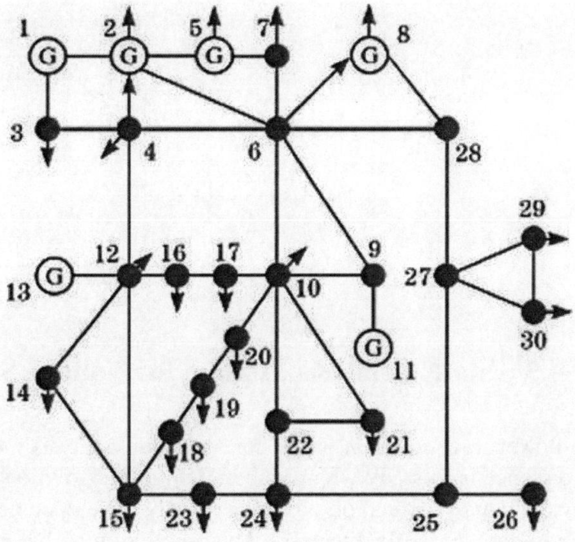

Fig. 4.7 Wiring diagram of IEEE 30-node system

Table 4.1 Parameters of generator reactive power output of IEEE 30-node system

Generator node No	Q_{min}	Q_{max}
2	−20	60
5	−15	62.5
8	−15	50
11	−10	40
13	−15	45

Table 4.2 Initial parameters of node load of IEEE 30-node system

Node No	P_d/MW	Q_d/Mvar	Parallel admittance
2	21.7	20.7	
3	2.4	1.2	
4	7.6	1.6	–
5	94.2	19	
7	22.8	10.9	
8	30	30	
10	25.8	2	19
12	11.2	7.5	
14	6.2	1.6	
15	8.2	2.5	
16	3.5	1.8	
17	9	5.8	–
18	3.2	0.9	
19	9.5	3.4	
20	2.2	0.7	
21	17.5	11.2	
23	3.2	1.6	
14	8.7	6.7	4
26	3.5	2.3	
29	2.4	0.9	–
30	10.6	1.9	

4.3 Static Analysis Method for Voltage Stability

Power flow equation is the core of static analysis method for voltage stability. Its essence is to study whether there is feasible solution for power flow equation. The static analysis method has been greatly developed and is mature for its simplicity, rapidity and effectiveness. This section mainly introduces the common static analysis methods of voltage stability, including nonlinear programming method,

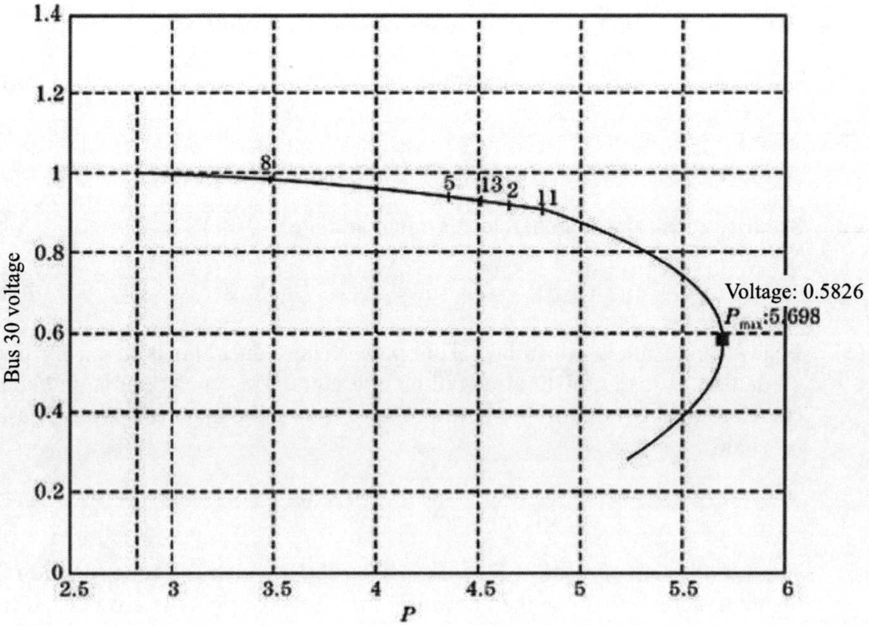

Fig. 4.8 Bus 30 *PV* curve

singular value decomposition method, eigenvalue analysis method, sensitivity analysis method, bifurcation analysis method, and describes and analyzes the principle, application and research status of these analysis methods.

4.3.1 Nonlinear Programming Method

Mathematical programming with nonlinear constraints or objective functions is an important branch of operation research. Nonlinear programming is used to study the extremum problem of an *n*-ary real function under constraints of a set of equations or inequations, and at least one of the objective functions and constraints is a nonlinear function of unknown quantity. Nonlinear programming method is widely used in power systems, and it is also a method to find the critical point of static voltage stability. It transforms the calculation of critical point into the optimization problem of nonlinear objective functions, taking the maximum apparent power of the total load or the maximum active power of any load node as the objective function:

$$\text{Max}\left|S_{(t)}\right| \tag{4.19}$$

Constraints can be considered in the following aspects.

(1) Active power and reactive power constraints of generator output

$$P_i^{\min} \le P_i \le P_i^{\max} \tag{4.20}$$

$$Q_i^{\min} \le Q_i \le Q_i^{\max} \tag{4.21}$$

Security constraint of generator terminal voltage

$$U_i^{\min} \le U_i \le U_i^{\max} \tag{4.22}$$

(2) Load change mode constraint: This part of the constraint is given by load prediction or load growth arranged by operators. This constraint is unique to calculate voltage stability limit. For M load nodes, the apparent power phasor is set as

$$S = [S_1, S_2, \ldots, S_M] \tag{4.23}$$

The variation coefficient of S_i is defined as $\beta_i(t)$, which is a time function. If the total increment of the load power of the system at time t is given as $\Delta S(t)$, then the relationship between the total increment of apparent power of system load and the variation of each load is

$$S_i(t) = S_i^0 + \beta_i(t)\Delta S(t) \tag{4.24}$$

wherein, S_i^0 is the apparent power of the ith load node in the current state; $S_i(t)$ is the apparent power of the ith load node at time t; βi(t) is the distribution rate of the ith load at time t.

From Eq. (4.24), the load variation constraint can also be rewritten as follows:

$$\beta_i(t)\sum_{j\in J_{\mathrm{L}}} S_j(t) - S_i(t) = \beta_i(t)\sum_{j\in J_{\mathrm{L}}} S_j^0 - S_i^0 \tag{4.25}$$

Take

$$C_i(t) = \beta_i(t)\sum_{j\in J_{\mathrm{L}}} S_j^0 - S_i^0 \tag{4.26}$$

Then:

$$\beta_i(t)\sum_{j\in J_{\mathrm{L}}} S_j(t) - S_i(t) = C_i(t) \tag{4.27}$$

(3) Generator participation constraint: When the load increases, the active power and reactive power output of the generator must be increased to maintain the

power balance of the system. The active output participation factor phasor of the generator is defined as γ, and the ith component is γ_i. It can consider the optimal distribution of generator output in the system in the determination of γ, but at the same time, it should consider the limitation of generator adjustment speed in the period of load increase.

(4) Other constraints: In the model, the limitation of regulation and regulation capacity of OLTC tap, line security constraint, tie-line power limitation, and line disconnection security constraint will affect the calculation of voltage stability limit point, so these should be considered in the model.

Knowing how to deal with a large number of inequation constraints is the key to the success of the algorithm. In the middle and late 1980s, there was a breakthrough in interior point method in the field of mathematical programming. Its reliability, super-linear convergence and insensitivity to the scale of the problem make it the best linear and nonlinear programming algorithm at present [7–12]. Reference [13] extends the zero characteristic root method for directly solving the voltage collapse critical point, converts the critical point calculation into a nonlinear programming problem, and uses the predictor–corrector primal–dual interior-point method to solve. This method can consider various inequation constraints, and therefore has strong robustness. Reference [14] deals with inequality constraints by interior point method, which can provide reference for accident prevention and control, as well as for dispatch optimization. The algorithm has strong robustness and practical value. However, due to the higher order of the modified equation, the required storage space is large and the calculation speed is slow. Based on the Refs. [14, 15] introduces parameterized power flow equation so as to avoid morbidity caused by singularity of Jacobian matrix near the critical point, and can precisely determine the voltage stability critical point. A new algorithm for voltage stability critical point based on interior point theory is proposed. This algorithm adopts a new data structure, which makes the coefficient matrix of the modified equation and node-admittance matrix have the same structure and greatly improves the calculation speed. In addition, the interior point method can also be used for voltage stability evaluation for AC/DC systems [16, 17].

The non-linear programming method can effectively avoid singularity of power flow Jacobian matrix and unsolvability of power flow when voltage stability limit is approaching. However, with the expansion of the system scale, the number of constraint equations would increase sharply resulting in a large amount of calculation, greatly increases difficulty in solving non-linear programming and limited calculation scale of the method.

4.3.2 Singular Value Decomposition Method

Singular value decomposition is a powerful computational tool in matrix problems and has been applied in many fields [18–25]. In this method, the stability of the studied system is determined by the determinant symbol of the power flow Jacobian matrix. The minimum singular value of the power flow Jacobian matrix can be used as an index of the static voltage stability, and also can represent the distance between the operating point and the static voltage stability limit. Singular value decomposition method is applied to power system analysis based on the following theorems and definitions [18].

Theorem 4.1 *Assume that* A $\in \mathbf{R}^{m\times n}$. *Then there would be orthogonal unit matrices U and V, resulting in*

$$V^{\mathrm{T}}AU = \begin{bmatrix} \Sigma & 0 \\ 0 & 0 \end{bmatrix} \tag{4.28}$$

wherein, $\Sigma = \mathrm{diag}\,(\delta_1, \delta_2, \ldots, \delta_r)$, *and* $\delta_{\max} = \delta_1 \geq \delta_2 \geq \cdots \geq \delta_r = \delta_{\min} \geq 0$.

Definition 4.1 Assume A $\in \mathbf{R}^{m\times n}$, then there would be a singular value decomposition formula (4.28), so $\delta_1, \delta_2, \ldots, \delta_n$ ($\delta_{r+1} = \delta_{r+2} = \cdots = \delta_n = 0$) are treated as singular values of *A*, the column vectors of *U* are treated as right singular vectors of *A*, and the column vectors of *V* are treated as left singular vectors of *A*.

The right singular value vector of *A* is the orthogonal unit right eigenvector of $A^{\mathrm{T}}A$, and the left singular value vector of *A* is the orthogonal unit left eigenvector of $A^{\mathrm{T}}A$. If there are *n* singular values for *A*, A^{T} also has *n* singular values. The non-zero singular values for A and A^{T} are the same, and the number of non-zero singular values is the rank of *A*.

Singular value decomposition is applied to matrix *A*;

$$A = V\Sigma U^{\mathrm{T}} \tag{4.29}$$

wherein, Σ has a zero singular value, and the relationship between left and right singular vectors is as follows:

$$Au_i = \delta_i v_i, i = 1, 2, \ldots, \min\{m, n\} \tag{4.30}$$

$$Au_i = 0, i = min\{m, n\} + 1, \ldots, n \tag{4.31}$$

$$A^{\mathrm{T}}v_i = \delta_i u_i, i = 1, 2, \ldots, \min\{m, n\} \tag{4.32}$$

$$A^{\mathrm{T}}v_i = 0, i = min\{m, n\} + 1, \ldots, m \tag{4.33}$$

Therefore, for linear equation set $AX = b$, A $\in \mathbf{R}^{n\times n}$, *A* is non-singular, while $b \in \mathbf{R}^n$. After singular value decomposition of matrix *A*, the solution *X* of the system can be written as

$$X = A^{-1}b\left(V\Sigma U^{\mathrm{T}}\right)b = \sum_{i=1}^{n} \frac{U_i V_i^{\mathrm{T}}}{\delta_i} b \tag{4.34}$$

It can be seen that if the singular value δ_i is small enough, small changes in matrix A or vector b will cause big changes in X.

Theorem 4.2 *Assume* $A \in \mathbf{R}^{m \times n}$. *If the singular value* $\delta_{\max} = \delta_1 \geq \delta_2 \geq \cdots \geq \delta_n = \delta_{\min} \geq 0$, *then*

$$\|A\|_F^2 = \delta_1^2 + \delta_2^2 + \cdots + \delta_n^2 = \sum_{i=1}^{n} \delta_i^2 \tag{4.35}$$

Let $E_j = v_j u$T j ($j = 1, 2, \ldots, r$), $A = \delta_1 E_1 + \delta_2 E_2 + \cdots + \delta_r E_r$.
If

$$A' = \delta_1 E_1 + \delta_2 E_2 + \cdots + \delta_{r-1} E_{r-1} \tag{4.36}$$

It can be proven that the A' is the matrix with rank of $r - 1$ closest to A in terms of the norm F of matrix A, similarly, $A'' = \delta_1 E_1 + \delta_2 E_2 + \cdots + \delta_{r-2} E_{r-2}$ is the matrix with rank of $r - 2$ closest to A, and so on.

Jacobian matrix, i.e., $J \in \mathbf{R}^{m \times n}$, is an important matrix in static analysis method for voltage stability of power system. Applying the above singular value decomposition theory to this matrix, it can be obtained that Jacobian matrix is non-singular under normal circumstances, $\delta_{min} > 0$; Jacobian matrix is singular when the system reaches the static stability limit, $\delta_{\min} = 0$.

Among the numerous matrices that can reduce the order of Jacobian matrix, the reduced-order matrix J' constructed under Eq. (4.36) is the closest to the original matrix J in terms of the norm, and

$$\|J - J'\|_F = \delta_{min} \tag{4.37}$$

As you can see, $\delta_{\min}$ can reflect that how close the Jacobian matrix is to singularity.

The singularity degree of the matrix can also be expressed by the conditional numbers, and the numbers of the non-singular matrix are

$$\mathrm{con}(A) = \frac{\delta_{max}}{\delta_{min}} \tag{4.38}$$

When matrix A is singular, $\delta_{\min} = 0$ and con(A) is infinite; if con(A) is close to 1, A is far from singularity, and the larger con(A) is, the closer the singularity would be. Therefore, A is considered to be more singular than B if con(A) > con(B).

1. Power system model

Assuming that the total nodes, excluding the balance node, in the system of the power system is n, and the number of nodes with adjustable voltage is m, the power flow equation of the system can be described as

$$\begin{cases} P(\theta, U) = 0 \\ Q(\theta, U) = 0 \end{cases} \tag{4.39}$$

wherein, P = $[P_1, P_2, \ldots, P_n]$ represents active balance; $Q = [Q_1, Q_2, \ldots, Q_n]$ represents reactive balance; and θ and U represent the angle and amplitude of the node voltage. Linearize the equations into

$$\begin{bmatrix} \Delta P \\ \Delta Q \end{bmatrix} = J \begin{bmatrix} \Delta\theta \\ \Delta U \end{bmatrix} = \begin{bmatrix} J_{P\theta} & J_{PU} \\ J_{Q\theta} & J_{QU} \end{bmatrix} \begin{bmatrix} \Delta\theta \\ \Delta U \end{bmatrix} \tag{4.40}$$

wherein, matrix J is a complete Jacobian matrix; sub-matrices $J_{P\theta}$, J_{PU}, $J_{Q\theta}$ and J_{QU} are the sub-matrices of Jacobian matrix formed by partial differential of power flow equation.

(1) *PQ* can be decoupled

That is, the active power P is related to the angle θ and independent of the voltage U; the reactive power Q is related to voltage U and independent of angle θ, then there are

$$\begin{bmatrix} \Delta P \\ \Delta Q \end{bmatrix} = \begin{bmatrix} J_{P\theta} & 0 \\ 0 & J_{QU} \end{bmatrix} \begin{bmatrix} \Delta\theta \\ \Delta U \end{bmatrix} \tag{4.41}$$

$$\Delta P = J_{P\theta} \Delta\theta \tag{4.42}$$

$$\Delta Q = J_{QU} \Delta U \tag{4.43}$$

Equation (4.42) can be used to describe the steady state rotor angle problem when the system satisfies the decoupling condition. If $J_{P\theta}$ is non-singular, the rotor angle of the system is stable. Equation (4.43) can be used to describe the sensitivity of node voltage to the injected reactive power, and can be treated as an index to measure the stability of system voltage.

(2) *PQ* cannot be decoupled

When the system is under heavy load, the interaction between reactive power and angle cannot be ignored, and *PQ* cannot be decoupled. Under this circumstance, a simplified Jacobian matrix can be introduced. Assuming that $\Delta P = 0$ in Eq. (4.41), the following equation can be obtained

$$\begin{bmatrix} 0 \\ \Delta Q \end{bmatrix} = \begin{bmatrix} J_{P\theta} & 0 \\ 0 & J_{QU} \end{bmatrix} \begin{bmatrix} \Delta\theta \\ \Delta U \end{bmatrix} \tag{4.44}$$

The simplified Jacobian matrix can be defined as:

$$J_R = \left[J_{QU} - J_{Q\theta} J_{P\theta}^{-1} J_{PU} \right] \tag{4.45}$$

Equation (4.44) can represent the linear relationship between the node voltage amplitude and the slight increase in reactive power. According to the block matrix theory (Schur Formula), the determinant values for the complete Jacobian matrix J can be calculated by using the following equation:

$$\det(J) = \det(J_{P\theta})\det(J_R) \tag{4.46}$$

It shows that the complete Jacobian matrix J is also singular if the matrix $J_{P\theta}$ or J_R is singular. Since $J_{\text{P}\theta}$ can reflect the steady state rotor angle stability problem of the system, $J_{P\theta}$ is non-singular (i.e., $J_{P\theta} \neq 0$) if the system has no rotor angle stability problem. Therefore, only when the reduced-order Jacobian matrix J_R is singular, is the complete Jacobian matrix J singular, and the singularity of J_R can be used to indicate the voltage stability of the system.

It can be concluded from the singular value analysis of matrix J

$$J = V\Sigma U^{\text{T}} = \sum_{i=1}^{2n-m} V_i \delta_i U_i^{\text{T}} \tag{4.47}$$

wherein, singular value vectors V_i and U_{i} are the ith column of normalized matrices V and U, $\sum$ is the diagonal matrix of positive real singular value δ_i, $\delta_1 \geq \delta_2 \geq \cdots \geq \delta_{2n-m}$. If one of the singular values approaches 0 and the system approaches the collapse point, the system response is determined by the minimum singular value δ_{2n-m} and its corresponding singular vectors V_{2n-m} and U_{2n-m}. And the following equation can be obtained

$$\begin{bmatrix} \Delta\theta \\ \Delta \text{U} \end{bmatrix} = \delta_{2n-\text{m}}^{-1} U_{2n-m} V_{2n-\text{m}}^{\text{T}} \begin{bmatrix} \Delta P \\ \Delta Q \end{bmatrix} \tag{4.48}$$

wherein, $U_{2n-m} = [\theta_1, \ldots, \theta_{\text{n}}, U_1, \ldots, U_{n-m}]^{\text{T}}$, $V_{2n-m} = [P_1, \ldots, P_n, Q_1, \ldots, Q_{n-m}]^{\text{T}}$, and meet the standardization:

$$\sum_{i=1}^{n} \theta_i^2 + \sum_{i=1}^{n-m} U_i^2 = 1 \tag{4.49}$$

$$\sum_{i=1}^{n} P_i^2 + \sum_{i=1}^{n-m} Q_i^2 = 1 \tag{4.50}$$

Take

$$\begin{bmatrix} \Delta P \\ \Delta Q \end{bmatrix} = V_{2n-m} \tag{4.51}$$

Substituting Eqs. (4.49) and (4.50) into Eq. (4.48), we can get the equation below

$$\begin{bmatrix} \Delta\theta \\ \Delta U \end{bmatrix} = \frac{U_{2n-m}}{\delta_{2n-m}} \tag{4.52}$$

From Eq. (4.52), it can be seen that the minimum singular value is very small when the system is close to the voltage collapse point. As a result, small power fluctuations may result in large changes in voltage.

2. Application of singular value decomposition in voltage stability analysis

As previously mentioned, the singularity of the matrix can be measured by the conditional numbers. With the change of operation conditions, the minimum singular value of Jacobian matrix is consistent with the change of conditional numbers. Therefore, the minimum singular values of the complete Jacobian matrix and the reduced-order Jacobian matrix can be used as indexes of voltage stability. In voltage stability analysis, the application of left singular value vector V_{2n-m} and right singular value vector U_{2n-m} is described as follows:

(1) The maximum element value in V_{2n-m} corresponds to the most sensitive direction of active power and reactive power injection changes, so the left singular value vector can obtain the most dangerous load and the changing pattern of power generation in the system; in Eq. (4.51), V_{2n-m} provides a typical mode of power injection change at the node; the left singular value vector can also provide the influence of regional section power flow on voltage stability and can select the weak transmission line of the system.
(2) The maximum element value in U_{2n-m} corresponds to the most sensitive node voltage, so the right singular value vector can be used to identify weak nodes and critical areas in the system. For example, the index for the definable node strength degree is $LC_j = \frac{U_j}{\delta_{2n-m}}(j \in \{1, 2, \cdots n-m\})$; in Eq. (4.52), U_{2n-m} provides typical modes for node voltage and angle change.

3. Examples [19]

Take IEEE 39-Bus System as an example. The system wiring diagram is shown in Fig. 4.9.

Please refer to Table 4.3 for calculation results of LC_{j} while keeping the load of the generator and other nodes unchanged, increasing the load for Node 7 and adopting the load growth mode of constant power factor. It can be seen that with the increase of load, the *LC* index for most *PQ* nodes increases. Among them, the *LC* index in the area formed by Nodes 5–8 exceeds 7.5, while the LC index of Node 7, the weakest node which is very close to the voltage critical point, even exceeds 9.5. It can also be found in Table 4.4 that the actual voltage level in the weak area has dropped a lot. It shows that with the increase of load, some weak nodes will form a weak area, in which the voltage will drop to different degrees resulting in a threat to the stability of the area. However, the weak nodes and weak areas can be well identified by using the minimum singular values and indexes, which is helpful

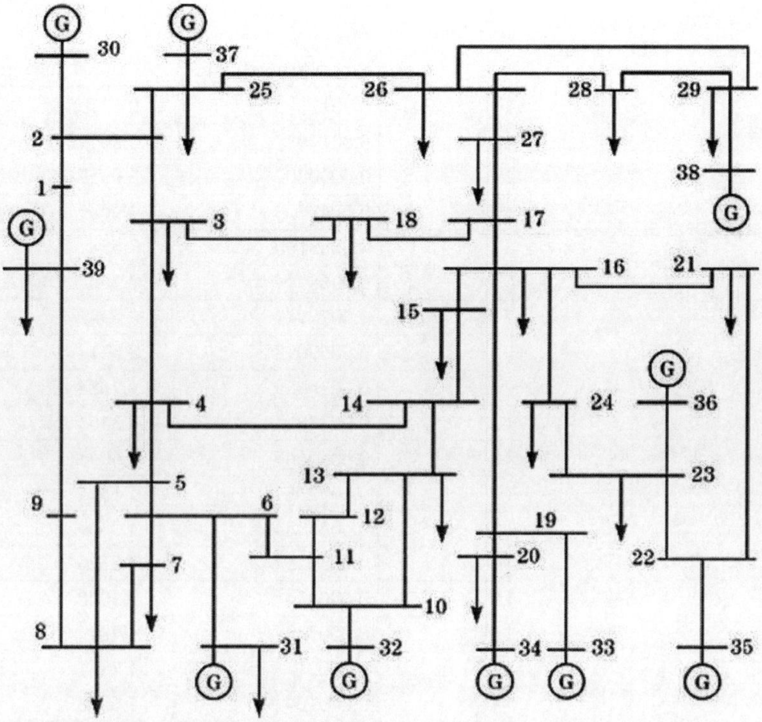

Fig. 4.9 Wiring diagram of IEEE 39-node system

for operators to make corresponding operations according to these indexes so as to maintain stable operation of the system.

4.3.3 Eigenvalue Analysis

The eigenvalue decomposition theory is introduced into the study of voltage stability, and eigenvalue decomposition is adopted in the Jacobian matrix J of power flow equation:

$$J = \sum_{i=1}^{m} \sigma iuiv_i^{\mathrm{T}} \tag{4.53}$$

Substituting the linearized form of power flow, we can obtain:

Table 4.3 Calculation results of *LC* index for IEEE 39-bus system

Node No	*LC* index	
	Working condition	Critical behavior
1	0.00126	0.33198
2	0.00149	0.80413
3	0.0000202	2.0357
4	0.00433	4.9085
5	0.0068166	7.5581
6	0.006891	7.7603
7	0.0070741	9.5921
8	0.0069098	8.6714
9	0.0030969	2.7684
10	0.0043845	4.3464
11	0.0050609	5.4041
12	0.0052357	5.1574
13	0.0045384	4.3309
14	0.0039257	4.0585
15	0.000694	2.1068
16	0.000384	1.284
17	0.00021	1.3831
18	0.000164	1.6358
19	0.00036	0.44744
20	0.000194	0.25665
21	0.000842	0.88988
22	0.000677	0.45963
23	0.00076	0.47835
24	0.000594	1.1578
25	0.0017933	0.59442
26	0.0001273	0.74632
27	0.0000349	1.0484
28	0.000336	0.38082
29	0.000356	0.2598

Table 4.4 Actual voltage levels of nodes in weak area

U_5	U_6	U_7	U_8
0.7323	0.72954	0.66297	0.68969

$$\begin{bmatrix} \Delta P \\ \Delta Q \end{bmatrix} = [U] \begin{bmatrix} \sigma_1 & & & & \\ & \sigma_2 & & & \\ & & \cdot & & \\ & & & \cdot & \\ & & & & \cdot \\ & & & & & \sigma_m \end{bmatrix} [V]^{\mathrm{T}} \begin{bmatrix} \Delta\theta \\ \Delta U \end{bmatrix} \tag{4.54}$$

wherein, m is the order of Jacobian matrix for power flow equation; σ_i is the eigenvalue of the ith order; v_i and u_i respectively represent corresponding left and right eigenvectors. Equation (4.54) can also be expressed as

$$\begin{bmatrix} \Delta\theta \\ \Delta U \end{bmatrix} = \sum_{i=1}^{m} \sigma^{-1} u_i v_i^{\mathrm{T}} \begin{bmatrix} \Delta P \\ \Delta Q \end{bmatrix} \tag{4.55}$$

From Eq. (4.55), it can be seen that if there is an eigenvalue close to zero, any small power change will cause a large change in the state variable.

In eigenvalue analysis, the right eigenvector u_m corresponding to the minimum eigenvalue σ_m reflects the most sensitive direction of active and reactive perturbations under the minimum mode. When the direction of power perturbation is consistent with u_m, the change caused in the state quantity is the largest. Therefore, the right eigenvector corresponding to the minimum mode eigenvalue reflects the direction in which the system is most prone to instability to some extent.

Eigenvalue analysis can help identify weak links and areas of voltage stability [26–33]. In Eq. (4.55), $\Delta Q = e_k$, if $\Delta P = 0$, wherein e_k is a unit column vector with the kth element being 1 and other elements being 0. It is assumed that the active power injection quantity of the system remains unchanged and unit reactive power is only injected at the kth node, the change of system state quantity resulting from which is

$$\begin{bmatrix} \Delta\theta \\ \Delta U \end{bmatrix} = \sum_{i=1}^{m} \frac{u_i}{\sigma_i} v_{n+k-1,i} \tag{4.56}$$

wherein, $v_{n+k-1,i}$ is the $n + k - 1$th element of the v_i, the left eigenvector. Therefore, the voltage sensitivity of the kth node is

$$\frac{\mathrm{d}U_k}{\mathrm{d}Q_k} = \sum_{i=1}^{M} \frac{1}{\sigma_i} u_{n+k-1,i} v_{n+k-1,i} = \sum_{i=1}^{m} \frac{P_{n+k-1,i}}{\sigma_i} \tag{4.57}$$

wherein, $p_{ki} = u_{ki}v_{ki}$ is defined, which is called the participation factor of the kth state variable to the ith characteristic mode σ_i. From Formula (4.57), it can be calculated that:

(1) The participation factor p_{ki} reflects the relative contribution degree of the ith characteristic mode to the voltage sensitivity of the kth node. The larger p_{ki} is,

the higher the voltage sensitivity of the kth node would be determined by the mode σ_i.

(2) Comparing the contribution of the same mode σ_i to the voltage sensitivity of different nodes, we can find out the main nodes with strong correlation to the characteristic mode σ_i. If the system is unstable under this mode, the nodes with strong correlation constitute the instability area for the system under this unstable mode.

(3) If the minimum mode $\sigma_{\min} > 0$, the nodes with strong correlation constitute the area with the worst stability degree or the most prone to instability in the whole system.

It can be seen that the eigenvalue analysis method is essentially an extended analysis of the traditional sensitivity criterion in Jacobian matrix. Since the above model contains both active component and voltage phase angle, it can better reflect the reality of the system than the voltage sensitivity index, as the decoupling characteristics of the system do not exist any longer when the system is operating near the critical point.

According to the contribution degree to the node voltage sensitivity, the nodes with strong correlation to the minimum characteristic mode can be found according to the order of size of the participation factors. These nodes are defined as key nodes forming an area with poor system stability, namely, the weak area, which are often treated as the best location for reactive compensation.

Modal analysis can determine the key load buses, key lines and key units of the system. The key load buses can identify the relatively weak areas in the system for the stability calculation personnel, provide the basis for taking measures and further stability analysis, and also help the planning personnel to determine the location for reactive compensation devices in the system during the planning phase. Key units and key lines can provide important information to operators and mode calculation personnel for stability analysis.

4.3.4 Sensitivity Analysis

The sensitivity analysis method is based on the power flow equation of the power system, and studies the voltage stability of the system through the differential relation between some physical quantities, which will help people have a deeper understanding of the influence of system elements, control modes, and faults, etc. on the voltage stability [34–44].

The power flow equation of the power system can be expressed as

$$f(x, u, \lambda) = 0 \tag{4.58}$$

wherein, x is the state variable, such as the voltage amplitude and angle of the nodes; u is the control variable, such as power supply voltage, active power generation, and reactive compensation, etc.; λ is the parameter, such as active and reactive power of the load.

From the linearized equations, we can obtain

$$\Delta f = \frac{\partial f}{\partial x}\Delta x + \frac{\partial f}{\partial u}\Delta u + \frac{\partial f}{\partial x}\Delta\lambda \quad (4.59)$$

Therefore, Δx can be

$$\Delta x = -\left[\frac{\partial f}{\partial x}\right]^{-1}\frac{\partial f}{\partial u}\Delta u - \left[\frac{\partial f}{\partial x}\right]^{-1}\frac{\partial f}{\partial \lambda}\Delta\lambda = S_{xu}\Delta u + S_{x\lambda}\Delta\lambda \quad (4.60)$$

wherein, S_{xu} is defined as the sensitivity of the state variable x to the control variable u; $S_{x\lambda}$ is defined as the sensitivity of state variable x to the change of parameter λ.

In the sensitivity analysis of voltage stability, the common sensitivities include: the sensitivity of reactive and active power generation to voltage, and the sensitivity of reactive power generation to active or reactive power load.

Since the reactive power generation of the generator is a function of state variable x, control variable u, and parameter λ, it can be expressed as

$$q = q(x, u, \lambda) \quad (4.61)$$

From the linearized Eq. (4.61), we can obtain

$$\Delta q = \frac{\partial q}{\partial x}\Delta x + \frac{\partial q}{\partial u}\Delta u + \frac{\partial q}{\partial \lambda}\Delta\lambda \quad (4.62)$$

Substituting Eq. (4.60) into Eq. (4.62), we can obtain

$$\Delta q = \left(-\frac{\partial q}{\partial x}\left[\frac{\partial f}{\partial x}\right]^{-1}\frac{\partial f}{\partial u} + \frac{\partial q}{\partial u}\right)\Delta u - \frac{\partial q}{\partial x}\left[\frac{\partial f}{\partial x}\right]^{-1}\frac{\partial f}{\partial \lambda}\Delta\lambda = S_{qu}\Delta u + S_{q\lambda}\Delta\lambda \quad (4.63)$$

Similarly, S_{qu} is defined as the sensitivity of reactive power generation q to control variable u; $S_{q\lambda}$ is defined as the sensitivity of reactive power generation q to the change of parameter λ, including the sensitivity of q to the change of active and reactive power load. The change in reactive power generation of generator j caused by the load change of Node i is

$$\Delta Q_{ji} = \frac{\partial Q_{ji}}{\partial Q_i}\Delta Q_i + \frac{\partial Q_{ji}}{\partial P_i}\Delta P_i = S_{qp}(j, i_q)\Delta Q_i + S_{qp}(j, i_p)\Delta P_i \quad (4.64)$$

Therefore, the corresponding change in reactive power generation of the whole system caused by the load change of Node i is

$$\Delta Q_T = \sum_{j=1}^{n_G}\left[S_{qp}(j, i_q)\Delta Q_i + S_{qp}(j, i_q)\Delta P_i\right]$$
$$= S_{Qqi}\Delta Q_i + S_{Qpi}\Delta P_i \quad (4.65)$$

wherein, n_G is the total node numbers in the system that generate reactive power generation; S_{Qqi} is the sensitivity of the total reactive power generation to the reactive power load change of Node *i*; S_{Qpi} is the sensitivity of the total reactive power generation to the active power load change of Node *i*.

The sensitivity $S_{vq}(i)$ of the voltage at Node *i* to reactive power injection at the same node is the slope of the *VQ* curve, which can be expressed as

$$S_{vq}(i) = \frac{\partial U_i}{\partial U_i} = -\left[\frac{\partial f}{\partial x}\right]^{-1}_{U_i, Q_i} \quad (4.66)$$

When the system voltage is instable, $S_{vq}(i)$ would tend to infinity.

Similarly, the sensitivity $S_{vp}(i)$ of the node voltage to active power injection has the same characteristics, which can be defined as

$$S_{vp}(i) = \frac{\partial U_i}{\partial P_i} = -\left[\frac{\partial f}{\partial x}\right]^{-1}_{U_i, P_i} \quad (4.67)$$

Otherwise, the simplified Jacobian matrix J_R can be obtained by assuming the active power of the node is constant, that is, ΔP is constant, and then the linear relationship between the node voltage amplitude and the slight increase in reactive power can be obtained. The *i*th element in J_R is the *VQ* sensitivity of Node *i*. Positive value indicates stable operation. The smaller the value is, the more stable the system would be. Critical stability is achieved at infinite. Negative value indicates unstable operation.

4.3.5 Bifurcation Analysis

The equation describing the differential dynamic behavior of the power system can be expressed as $\dot{x} = f(x, u)$, wherein x is the state variable and u is the variable parameter. Under steady state conditions, the power system operates under equilibrium conditions $(\dot{x}) = f(x, u) = 0)$. In small disturbance range, there are two basic problems:

(1) Is the balance point (x, u_0) stable?
(2) What will happen to the stable balance point of the system when there are changes in the active and reactive power load?

The former problem is a problem related to motion stability and can usually be judged by Lyapunov stability theory; and the latter problem is a structural stability problem. The structural stability problem is the property that the topological structure of the dynamical system remains unchanged after being subjected to small disturbances. Bifurcation theory is a powerful tool to analyze structural stability [45–58].

If a dynamical system is structurally unstable, any small disturbance will cause sudden changes in topological structure of the system, which is called bifurcation. Bifurcation is a qualitative change of system state, such as disappearance of equilibrium (SNB) or changing from equilibrium state to oscillation (Hopf bifurcation). The power system can virtually be regarded as a non-linear dynamical system with parameters. If a certain parameter changes continuously in the system, a critical point may be reached. The system presents a sudden jump from one state to another, which may be different from the original state in performance and value.

The problem of static stability is whether the dynamical system can maintain stable operation at a balance point. There are two types of static instability: oscillatory type and aperiodic type. Static bifurcation can be divided into saddle node bifurcation, pitchfork bifurcation and transcritical bifurcation. Most of the recorded faults in voltage instability accidents happened show that the voltage instability mode of power system is monotonous. It is generally believed that this instability mode is related to saddle node bifurcation in the system [50, 51], so saddle node bifurcation is mainly discussed in this section.

Saddle node bifurcation refers to the bifurcation that occurs when the real part of eigenvalue in balance equation changes from negative to positive with the change of parameters. At the saddle node bifurcation, the system has zero eigenvalue, which corresponds to the singularity of the Jacobian matrix. The eigenvector corresponding to zero eigenvalue contains valuable information about bifurcation properties, system response and control effectiveness. The left eigenvector indicates the state variable has a significant impact on the zero eigenvalue, that is, in order to correct the bifurcation characteristics of the system and obtain the expected dynamical behavior, the states should be controlled for effectiveness so as to stabilize the voltage. The right eigenvector indicates the new state direction of the system evolution due to saddle node bifurcation. The most dangerous disturbance mode that causes saddle node bifurcation and system voltage instability can be determined by using the information related to the vector. Methods for solving saddle node bifurcation can be divided into direct method and continuation method [52].

1. Direct method

The study of direct method is more mature and the method is widely used in practice. It searches for static bifurcation point of the system directly from the current operating state. The method is characterized by small amounts calculation and overcomes the ill condition of the Jacobian matrix. However, it is difficult for operators to fully understand the capability requirements of the system to maintain the voltage level from the current state to the bifurcation state due to the little information obtained, which is a disadvantage. Direct method can be divided into one-parameter direct method and multi-parameter direct method.

(1) One-parameter direct method

In 1979, Seydel proposed to transform the solving of static bifurcation point problem on the equilibrium solution manifold to solving the following equations [52] by introducing two nontrivial vectors u and v:

$$\begin{cases} f(x, \mu) = 0 \\ f_x(x, \mu)v = 0 \text{ or } u^{\mathrm{T}} f_x(x, u) = 0 \\ f_x(x, \mu)v = 1 \text{ or } u^{\mathrm{T}} f_u(x, u) = 0 \end{cases} \tag{4.68}$$

wherein, x is the state variable, $x \in \mathbf{R}^n$; μ is the bifurcation parameter, $\mu \in \mathbf{R}$; u and v are the left and right eigenvectors corresponding to the zero eigenvalue of Jacobian matrix respectively, $\mu, v \in \mathbf{R}^n$. Iterative method can be used to solve Eq. (4.68) to obtain static bifurcation points.

References [53–55] have made lots of research and analysis on reducing the equation dimensions, simplified calculation and calculation difficulties caused by the singularity of power flow Jacobian matrix at bifurcation points. But these methods have damaged the sparsity of the original equations to a certain extent, so overall consideration should be taken in practical application.

One-parameter direct method cannot take all kinds of inequality constraints into consideration. Type judgment for bifurcation points and determination of new bifurcation direction cannot be made owning to the small amount of information obtained. In conclusion, the calculation results still have certain limitations.

(2) Multi-parameter direct method

In 1981, Jarjis and others proposed a multi-parameter direct method, the basic idea of transforming the problem of bifurcation points into a non-linear optimization problem [56] by defining the following vector functions.

Vector function

$$\varphi(x, v, \mu) = \begin{bmatrix} f(x, \mu) \\ f_x(x, \mu)v \\ \|v\| - 1 \end{bmatrix} = 0 \tag{4.69}$$

Solve the following optimization problems:

$$\begin{cases} \min\|\mu\| \\ \text{s.t.} : \varphi(x, v, \mu) = 0 \end{cases} \tag{4.70}$$

Lagrangian function can be constructed

$$L = \|\mu\|^2 + \lambda^{\mathrm{T}}\varphi(x, v, \mu) \tag{4.71}$$

wherein, λ is the Lagrange multiplier; μ is the bifurcation parameter vector with random direction, $\mu \in \mathbf{R}^n (n \geq 1)$.

The bifurcation point for saddle node closest to the current operating point can be found by applying non-linear optimization technology, and the bifurcation point found at this time should be on the bifurcation hypersurface. This situation is more in line with the actual situation of the power system. The application range for this method is wider. However, except for the similar limitations with the one-parameter method, the calculation amount is much larger.

2. Continuation method

Continuation method was applied to non-linear problems in the 1930s, and was used in voltage stability analysis for power systems until the 1990s as a method to track the equilibrium solution manifold. At present, most researches are mainly focused on tracking one-parameter equilibrium solution manifold. Due to the complexity of the power system itself, the tracking of multi-parameter manifold is too complicated. Although there were reports on tracking two-dimensional parameter equilibrium solution manifold, the tracking method of higher dimensional equilibrium solution manifold has not been seen. Undoubtedly, the tracking of multi-dimensional parameter equilibrium solution manifold will be one worthy of further research [57, 58].

(1) Single parameter continuation method

The basic idea of the single parameter continuation method is to obtain the extended power flow equation by parameterizing the conventional power flow equation, and then assume that the initial value of power flow (x_0, μ_0) is known. Starting from the known point, obtain the approximation $(\tilde{x}_1, \tilde{\mu}_1)$ of the upper and lower points of the solution curve with tangent method or interpolation method under the given step control strategy through the prediction link, and finally get the accurate value (x_1, μ_1) of the lower point through the correction link, so as to cycle until obtaining the bifurcation point.

The extended power flow equation is as follows:

$$\begin{cases} f(x,\mu) = g(x) - \mu b = 0 \\ P(x,\mu) = 0 \end{cases} \tag{4.72}$$

wherein, $g(x)$ is the conventional power flow equation; b is the direction vector; μ is the bifurcation parameter. The second equation is a parametric equation, which has two main methods: arc length parameterization and local parameterization. With the introduction of parametric equation, the Jacobian matrix of equations (4.72) is not singular at the bifurcation point, which overcomes the defection of singularity of the Jacobian matrix of conventional power flow equation at the bifurcation point, and the non-convergence of power flow calculation caused by the ill condition of Jacobian matrix near the bifurcation point of conventional power flow equation.

If arc length parameterization is adopted, Eq. (4.72) can be rewritten as

$$\begin{cases} f(x,\mu) = g(x) - \mu b = 0 \\ \sum_{i=1}^{n}\left\{[x_i - x_i(s)]^2\right\} + [\mu - \mu(s)]^2 - (\Delta s)^2 = 0 \end{cases} \tag{4.73}$$

wherein, s is the arc length parameter; Δs is the arc length variation.

In the application of continuation method, the prediction methods mainly include the tangent method and secant method, which are often used in combination. The tangent method is used for the first point prediction; the secant method is used in later points; while the arc length method is used in correction. The following measures should be taken for step control: during correction, if the iteration does not converge after the pre-specified number of times, the step should be reduced to half of the previous one and re-corrected; if it converges after a few iterations, the step of the next iteration should be twice of the current one; if it converges at an appropriate number of times, the step of the next iteration should remain unchanged.

(2) Multi (two) parameter continuation method

The basic idea of this method is to obtain the static bifurcation point (SNB) of the single parameter with continuation method. Then, starting from the bifurcation point, continue to solve the nonlinear equations representing the saddle node bifurcation with continuation method, so as to solve and trace the two-dimensional bifurcation boundary of the system. The advantage of continuation method is that it can trace out the equilibrium solution manifold of given parameter interval and conveniently consider all kinds of inequality constraints. The results obtained contain very useful engineering information with strong robustness.

After obtaining the static bifurcation point of a single parameter, the following nonlinear equations can be solved to trace and calculate the two-dimensional bifurcation boundary of the system:

$$\begin{cases} f(x,\mu) = 0 \\ Aq = 0 \\ \langle q, q\rangle - 1 = 0 \end{cases} \tag{4.74}$$

wherein, A is the augmented matrix of the system; q is the unit vector containing the right eigenvector corresponding to the zero eigenvalue; μ is the bifurcation parameter vector, and $\mu \in \mathbf{R}^2$.

On the basis of the two-dimensional bifurcation boundary, the higher-dimensional parameter bifurcation boundary can be obtained gradually, but the representation and calculation method of the boundary will be more complex.

The disadvantage of multi (two) parameter continuation method is that the calculation is complex and large, it has strict requirements on several key links such as prediction, correction and step control, it is not easy to derive a new branch direction and the location of the bifurcation point cannot be accurately determined. So, there is no online application yet.

4.4 Static Analysis Index for Voltage Stability

So far, researchers have put forward a variety of steady state analysis indexes of voltage stability from different perspectives. Generally, these indexes can be divided into two categories: state indexes and margin indexes. Both of these indexes can give a measure of the distance between the current operating point and the voltage collapse point of the system. However, the state index only uses the information of current operating state, including impedance modulus index, various sensitivity index, singular value index and eigenvalue index. While the margin index involves the simulation of the transition process and the calculation of the critical point, it starts from the given operating state of the system and approaches the voltage collapse point gradually through the load growth or the transmission power growth according to a certain mode. The distance from the current operating point to the voltage collapse point of the system is taken as the index of the voltage stability degree [59, 60], including L index based on power flow solution and VIPI index based on solution pair of power flow. Compared with the state index, the margin index contains a large amount of information, which can easily consider the influence of various factors in the transition process, such as constraints, active power distribution of generators and load growth mode.

4.4.1 Sensitivity Index

The sensitivity method uses the relationship of some variables in the system to analyze the stability problem. This kind of method can not only give the index of voltage stability, but also easily identify the strength of each node in the system from the useful information it provides, as well as the corresponding countermeasures to be taken. Therefore, the sensitivity method has been widely used and studied. However, several sensitivity methods do not involve the static and dynamic characteristics of the load, the reactive power constraints of the generator, the economic dispatch control among the generators and other constraints. Therefore, the calculation results usually have many corrections and deviations. The most common sensitivity criteria are $\mathrm{d}Q/\mathrm{d}U$ and $\mathrm{d}p/\mathrm{d}U$. In a simple system, all kinds of sensitivity criteria are equivalent to each other and can accurately reflect the ultimate capacity of system transmission power. However, when they are extended to a complex system, due to the common constraints of various factors, they are no longer consistent with each other, which may not accurately reflect the ultimate transmission capacity of the system, or even may be contrary to the simple system.

The differential algebraic equation representing the mathematical model of the power system is as follows:

$$\dot{X} = F(X, Y, P) \tag{4.75}$$

$$0 = G(X, Y, P) \tag{4.76}$$

wherein, F represents the dynamic part of the system, including the dynamic characteristics of generator and exciter; G represents the steady part of the system, including the generator stator and internal and external network model; X is the state variable; Y is the algebraic variable; P is the parameter. If it is near a stable operating point $(X(P_0), Y(P_0))$, then the partial derivative of the parameter P and the value of the matrix at the balance point is obtained, and

$$\frac{\partial F}{\partial X}\frac{\partial X}{\partial P} + \frac{\partial F}{\partial Y}\frac{\partial Y}{\partial P} + \frac{\partial F}{\partial P} = 0 \tag{4.77}$$

$$\frac{\partial G}{\partial X}\frac{\partial X}{\partial P} + \frac{\partial G}{\partial Y}\frac{\partial Y}{\partial P} + \frac{\partial G}{\partial P} = 0 \tag{4.78}$$

The sensitivity expressions of algebraic variables and state variables can be obtained:

$$\frac{\partial Y}{\partial P} = -\left[\frac{\partial G}{\partial Y}\right]^{-1}\left[\frac{\partial G}{\partial X}\frac{\partial X}{\partial P} + \frac{\partial G}{\partial P}\right] \tag{4.79}$$

$$\frac{\partial X}{\partial P} = A^{-1}\left[\frac{\partial F}{\partial Y}\left[\frac{\partial G}{\partial Y}\right]\frac{\partial G}{\partial P} - \frac{\partial F}{\partial P}\right] \tag{4.80}$$

where

$$A = \left[\frac{\partial F}{\partial X} - \frac{\partial F}{\partial Y}\left[\frac{\partial G}{\partial Y}\right]^{-1}\frac{\partial G}{\partial X}\right] \tag{4.81}$$

Substitute Eq. (4.80) into Eq. (4.79), the parameter sensitivity of algebraic variables can be obtained. From the above Eqs. (4.79)–(4.81), the sensitivity of any algebraic and state variables of the system to parameters can be obtained.

The sensitivity index can be mathematically divided into state variable sensitivity and algebraic variable sensitivity, but also into node sensitivity, branch sensitivity and generator sensitivity physically [4].

The voltage stability of the system can be judged directly through the sensitivity index. Its essence is to reflect the critical state of the system at the transmission power limit. For $\mathrm{d}U_\mathrm{L}/\mathrm{d}Q_\mathrm{L}$, when the reactive demand Q_L decreases (or increases), the voltage U_L at this point rises (falls), the system voltage is stable; for $\mathrm{d}Q_\mathrm{L}/\mathrm{d}Q_\mathrm{G}$, when the reactive power load demand Q_L increases (or decreases), the generator reactive output Q_G increases (or decreases), the system voltage is stable.

In addition, by using the sensitivity index and sorting the absolute value of the sensitivity value, we can find the weak nodes or areas of the system voltage stability, so as to determine the location of the reactive compensation installed in the system. In addition, it is also a common method to identify the weak bus of the system through the comprehensive discriminant of sensitivity index and the rate of change of sensitivity [34–44].

4.4.2 Singular Value/Eigenvalue Index

From the previous analysis, the static stability of the system can be judged by the determinant symbol of the power flow Jacobian matrix. The minimum singular value of the power flow Jacobian matrix can be used as an index of the static voltage stability, and it can also represent the distance between the operating point and the static voltage stability limit. The maximum element in the right singular vector indicates the most sensitive voltage amplitude, that is, indicating voltage key point of the system; the maximum element in the left singular vector indicates the most sensitive direction of power injection, i.e., the key generator in the system [18–29].

Similarly, the minimum eigenvalue of the Jacobian matrix is used to measure the static voltage stability margin of system operating point. The maximum element of the right eigenvector related to the minimum eigenvalue corresponds to the voltage key point in the system, and the maximum element of the left eigenvector indicates the most sensitive direction of power change.

In practice, it is found that with the increase of the system, the eigenvalues and singular values become more similar. When the dimension of the equation increases, the minimum eigenvalue and singular value of the matrix will decrease. If the generator of the system reaches the reactive power limit and changes from *PV* node to *PQ* node, the eigenvalue jumps and the linearity is not good. Usually, at the critical point, the change rate of singular value and eigenvalue is very large, especially when several generators reach the limit.

Through some fast algorithms, for example, using power flow to calculate *LU* decomposition results, and then using approximate algorithm to calculate the minimum singular value quickly; for the eigenvalue, only m minimum eigenvalues can be calculated. If the corresponding mode of the maximum eigenvalue is considered as voltage stability, the eigenvalues larger than m need not be calculated, which greatly reduces the calculation amount.

4.4.3 Index Based on the Power Flow Solution

1. Voltage stability index L based on general power flow solution

Kessel first proposed the L index [59], and Refs. [60, 61] ignored the real part of the node admittance matrix and simplified the index. L index is a quantitative index that represents the distance between the actual state and the stability limit. Its value is between 0 and 1, which is used to describe the existence of voltage solution. When applying to multi-node system, the method divides all nodes into generator node set G and load node set L, in which the generator node is PV node with constant voltage amplitude. For all load nodes $j \in \mathrm{L}$ in the system, L index can be calculated according to the following formula, wherein the maximum L value and the proximity of 1 represent the divergence proximity of the system.

$$L = \max_{j \in L} L_i = \max\left\{1 - \frac{\sum_{j \in L} F_{ji} U_i}{U_j}\right\} \tag{4.82}$$

wherein, F_{ji} is the load participation factor. Matrix F is a submatrix of H matrix obtained by partial inversion for node admittance matrix Y used in power flow calculation:

$$\begin{bmatrix} U^{\mathrm{L}} \\ I^{\mathrm{G}} \end{bmatrix} = H \begin{bmatrix} I^{\mathrm{L}} \\ U^{\mathrm{G}} \end{bmatrix} = \begin{bmatrix} Z^{\mathrm{LL}} & F^{\mathrm{LG}} \\ K^{\mathrm{GL}} & Y^{\mathrm{GG}} \end{bmatrix} \begin{bmatrix} I^{\mathrm{L}} \\ U^{\mathrm{G}} \end{bmatrix} \tag{4.83}$$

wherein, U^{L} and I^{L} are voltage vector and injection current vector of load node respectively; U^{G} and I^{G} are voltage vector and injection current vector of generator node respectively.

The application of L index is illustrated by taking the New England 30-node system as an example. The system wiring diagram is shown in Fig. 4.10, and the calculation results are shown in Fig. 4.11. Among them, L_i is calculated for multiple nodes as a function of reactive command. The results show that calculation of L_i contains a lot of valuable information. For example, the stability of all nodes belonging to the same critical area indicates that the characteristic curve of L_i is similar. If the difference between the two adjacent L_i indexes is too large, it means that the line is overloaded. From a large number of calculation experience, it can be taken that L_i is equal to 0.2 as the threshold value. If the L_i of a node exceeds 0.2, the area surrounding the node is the critical area [62].

With a simple structure and fast calculation speed, L index can be used for on-line voltage stability analysis. However, the index is based on the traditional power flow model, and it cannot provide any information about the branches and the participation of generators and cannot consider the load model and the reactive limit of generators, etc., so there is still severe limitation. Reference [63] considered the load model and proposed an improved L index.

2. Indication of adjacent voltage collapse based on power flow solution pair VIP I [4, 64]

The power flow equation usually presents multiple solution pairs. The number of solution pairs decreases with the operating point approaching the collapse point. There is only 1 pair of solutions near the collapse point, and 2 solutions at the collapse point become one solution. Figure 4.12 shows the change of multiple power flow solutions of the New England 30-node system when only Q_{11} is increased.

The voltage instability proximity index (VIPI) uses this solution pair to predict the proximity of voltage instability. The index is defined according to the power flow

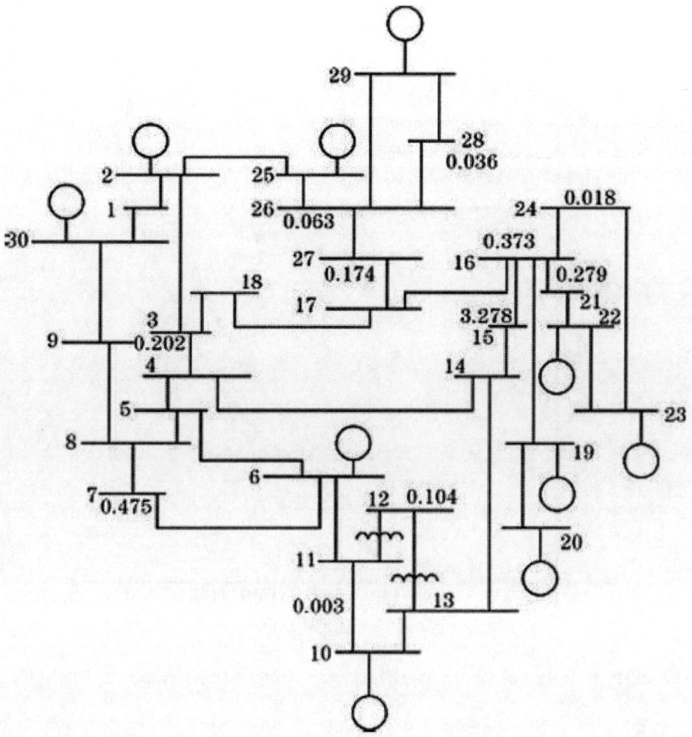

Fig. 4.10 Schematic diagram of New England 30-node system

equation in the following rectangular coordinates:

$$y_s = y(x) = (Ax)x + Bx + C \tag{4.84}$$

wherein, x is the voltage vector in rectangular coordinates; y_s is the node injection vector; $y(x)$ is the quadratic function of x; A is the three-dimensional Hessian matrix; B is the constant equation; C is the constant vector.

Let x_1 and x_2 indicate high voltage (operable) solution and low voltage solution respectively, and define two vectors, i.e., a and b:

$$a = (x_1 + x_2)/2 \tag{4.85}$$

$$b = (x_1 - x_2)/2 \tag{4.86}$$

At the voltage collapse point, $x_1 = x_2$, that is, $b = 0$.

VIPI is defined as the angle between the vectors Y_s and $Y(a)$, measured in degrees, Y_s is the node injection vector, $Y(a)$ is the singular vector of the node injection space:

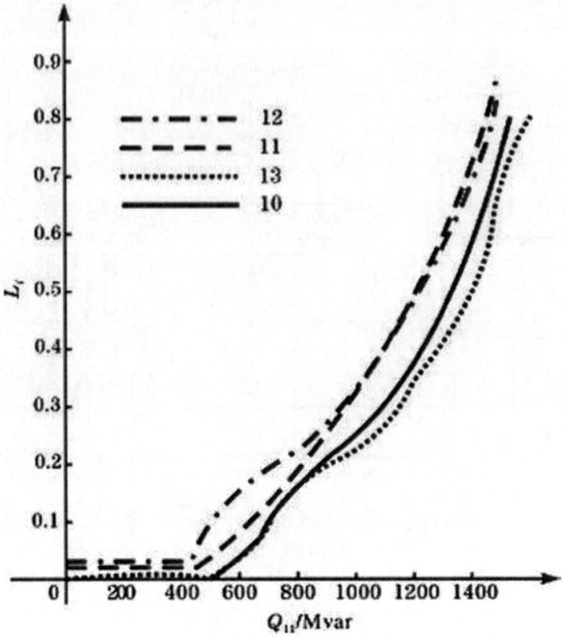

Fig. 4.11 Stability Index L_i of nodes 10–13 as a function of reactive power load Q_{11}

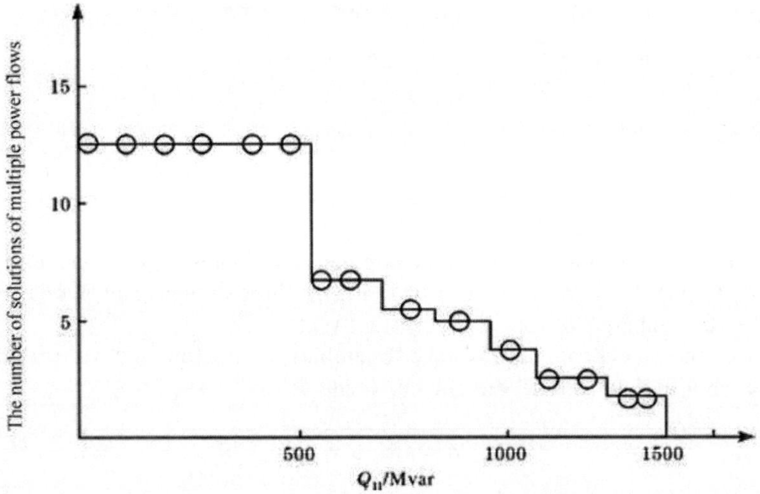

Fig. 4.12 Change of solution number of system power flow with Q_{11} added

$$\text{VIPI} = \theta = \cos^{-1}\frac{\left|Y(a)^{\text{T}}Y_{\text{s}}\right|}{|Y(a)||Y_{\text{s}}|}(^{\circ}) \tag{4.87}$$

4.4.4 Local Index

With the continuous expansion of the power system, if the voltage stability analysis of all nodes in the system is conduced, it will pay a high calculation cost. However, the focus of research is only on the influence of load growth of a node or area on the voltage stability of the system. Therefore, when accuracy allows, the local voltage stability analysis can be carried out by equivalent methods. For example, for a single bus or line, at present, voltage stability evaluation and weak node identification are mainly based on the local information of monitored bus and the information of adjacent bus, so as to quickly apply to online analysis.

References [65, 66] put forward the index of local load margin. The theory assumes that the load of all nodes (except the studied nodes i) remains unchanged and the load power factor of node i remains unchanged, by increasing the load of node i, the relative value of the distance from the starting load (expressed by P_{0i}) to the peak point of PV curve (expressed by $P_{\text{max}i}$) is the local load margin index $P_{\text{Lmg}i}$:

$$P_{\text{Lmg}i} = \frac{P_{\max i} - P_{0i}}{P_{\max i}} \tag{4.88}$$

The $P_{\text{Lmg}i}$ value is between 0 and 1, and the load margin is 0 at the collapse point. $P_{\text{Lmg}i}$ is for specific nodes. When it is necessary to evaluate the whole power system, if every node of the system is calculated, it will take a lot of time. Therefore, in order to effectively estimate $P_{\text{Lmg}i}$, it is possible to maintain the principle of equal power flow before and after equivalence, forming the equivalent system of node i. According to the sensitivity obtained from the relevant part of the Jacobian matrix and hypothesis that node i increases by multiple of basic load, the local load margin of node i is calculated.

References [67, 68] put forward the local safety index of load node voltage stability. Only local quantities are used for the index. Voltage safety is judged by the condition of solution of simplified two-node power flow equation, and line stability factor is constructed by the condition of solution of real part and imaginary part of power flow equation; safety index by real part of voltage and safety index by voltage amplitude are constructed by rectangular coordinate form of power flow equation and solution condition of power flow.

4.4.5 Impedance Modulus Index

For the system shown in Fig. 4.13, the voltage solution can be obtained from Eq. (4.9):

$$U_L = \sqrt{\frac{E^2}{2} - P_L R_s - Q_L X_S \pm \frac{1}{4}\sqrt{(2P_L R_s + 2Q_L X_s - E^2)^2 - 4Z_s^2(P_L^2 + Q_L^2)}} \tag{4.89}$$

So, when

$$(2P_L R_s + 2Q_L X_s - E^2)^2 - 4Z_s^2(P_L^2 + Q_L^2) = 0, \tag{4.90}$$

the voltage has a unique solution:

$$U_L = \sqrt{\frac{E^2}{2} - P_L R_s - Q_L X_s} \tag{4.91}$$

At this time, the load impedance Z_L meets

$$Z_L^2 = R_L^2 + X_L^2 = \frac{U_L^4}{P_L^2 + Q_L^2} \tag{4.92}$$

Substituting Eq. (4.91) into Eq. (4.92), we can obtain

$$Z_L^2 = \frac{\left(\frac{E^2}{2} - P_L R_s - Q_L X_s\right)^2}{P_L^2 + Q_L^2} = \frac{Z_s^2(P_L^2 + Q_L^2)}{P_L^2 + Q_L^2} = Z_s^2 \tag{4.93}$$

Then

$$|Z_L| = |Z_s| \tag{4.94}$$

Through the condition of the power flow solution, it can be obtained that when the equivalent impedance of the load node is equal to the equivalent impedance of the node network (the equivalent impedance of the system Thevenin), the transmission

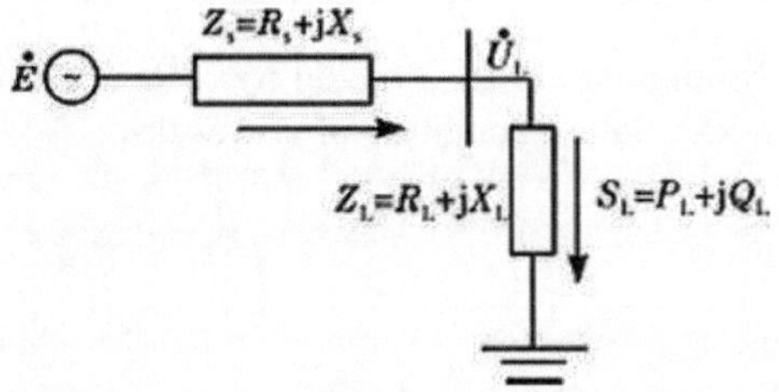

Fig. 4.13 Thevenin equivalent system

power of the network reaches the limit. Therefore, the load impedance and Thevenin equivalent impedance of the system can be monitored at the load node. When the load impedance is greater than the Thevenin equivalent impedance, the system voltage is stable; when the load impedance is less than the Thevenin equivalent impedance, the system voltage is unstable; when the two are equal, it is the critical point of voltage stability. This method has good linearity, but how to get the Thevenin equivalent parameters of the system accurately becomes the key restriction to the development of this method [69–71].

4.4.6 Energy Function Index

The energy function index (TEF) is based on Lyapunov stability theory. As an index of voltage stability, the scalar function is proved to be directly related to the area surrounded by the nose curve under some assumptions. For a balance point, the energy function is expressed as

$$\begin{aligned}\mathrm{TEF} = {} & \frac{1}{2}\sum_{i=1}^{n}\sum_{j=1}^{n} B_{ij}U_i^0U_j^0\cos\left(\theta_i^0-\theta_j^0\right)\frac{1}{2}\sum_{i=1}^{n}\sum_{j=1}^{n} B_{ij}U_i^1U_j^1\cos\left(\theta_i^1-\theta_j^1\right)\\ & -\sum_{i=1}^{n}P_i(P_0)\cos\left(\theta_i^1-\theta_i^0\right)-\sum_{i=1}^{n}\int_{U_i^0}^{U_i^1}\frac{Q_i(U,P_0)}{U}\mathrm{d}U\\ & +\sum_{i=1}^{n}\sum_{j=1}^{n} G_{ij}U_i^0U_j^0\cos\left(\theta_i^0-\theta_j^0\right)\left(\theta_i^1-\theta_i^0\right)\\ & +\sum_{i=1}^{n}\sum_{j=1}^{n} G_{ij}U_i^0U_j^0\cos\left(\theta_i^0-\theta_j^0\right)\left(U_i^1-U_i^0\right)\end{aligned}\tag{4.95}$$

wherein, G_{ij} and $\mathrm{B_{ij}}$ are the real part and imaginary part of the ij_{th} element in the node admittance matrix; $P_i(p)$ and $Q_i(p)$ are the active and reactive power injected into node k respectively; $U_i^0\angle\theta_i^0$ and $U_j^0\angle\theta_j^0$ are the node voltage vectors of node i and j at the balance point ($\mathrm{Z_0}$, P_0); and $U_i^1\angle\theta_i^1$ and $U_j^1\angle\theta_j^1$ represent the voltage vectors of the same parameter value p_0 at the other balance point Z_1, which is related to the "closest" unstable balance point.

The expression of TEF provides a measure of the "energy distance" between two balance points. When the system reaches the collapse point, two solutions Z_0 and Z_1 are combined into one solution, namely, $\mathrm{Z_0} = \mathrm{Z_1}$. At the collapse point, the TEF value becomes zero. This characteristic can be used to predict that a system draws near voltage collapse. One of the disadvantages of the energy function index is that it is not easy to include more complex system model, and the calculation of the second balance point Z_1 is difficult, especially for heavy load systems [4].

4.4.7 Second-Order Index

All the above-mentioned indexes are first-order indexes. The linearity of first order indexes is usually poor, and only when the system is close to the collapse point, its changes are obvious and cannot express the degree of approaching the collapse point accurately. Berizzi and others [72] put forward the concept of second-order index, which overcomes the weakness of bad linearity by using the additional information in the first order index.

The second-order index is based on "quadratic" function, which is characterized by linear ratio of parameter changes, such as function

$$f(\lambda) = (a - b\lambda)^{-\frac{1}{c}} \tag{4.96}$$

Then:

$$\frac{f(\lambda)}{\mathrm{d}f/\mathrm{d}\lambda} = ac/b - c\lambda \tag{4.97}$$

Then the second-order index can be defined as:

$$l = \frac{1}{l_0}\frac{f(\lambda)}{\mathrm{d}f/\mathrm{d}\lambda} \tag{4.98}$$

If the system encounters generator capacity limit or other limits, the change of $f(\lambda)$ will be partially offset by the high value of $\mathrm{d}f/\mathrm{d}\lambda$, so the second-order index greatly improves the linearity of the first order index. If the limitation of the system is not considered, the second-order index is completely linear. Only two points need to be calculated to get the critical point of the system, which can be used as an approximate estimation method.

Because of the nonlinearity of power system, it is almost impossible to find the better linear index directly in the first order index. Therefore, the key of the problem is to find the function with quadratic condition and construct the second-order index. So far, the following second-order indexes are proposed: test function, minimum singular value, reduced-order Jacobian matrix, and power loss sensitivity. These "quadratic" functions, strictly speaking, are only close to "quadratic", but they have met the basic analysis requirements.

1. Test function

The test function is based on a cluster of scalar t_{lk}, and is defined as:

$$t_{lk} = \left| e_l^{\mathrm{T}} J J_{lK}^{-1} e_l \right| \tag{4.99}$$

wherein, J corresponds to the system Jacobian matrix; e_l is the l-th unit vector; J_{lk} represents an operation on the Jacobian matrix J, that is, remove l lines and replace them by e_k^{T} line, that is

$$J_{lk} = \left(I - e_l e_l^{\mathrm{T}}\right)J + e_l e_k^{\mathrm{T}} \tag{4.100}$$

wherein, I is the unit matrix. If $l = k = c$, c is the largest element or critical element in the right eigenvector of Jacobian matrix, then the test function is "critical":

$$t_{cc} = \left|e_c^T J J_{cc}^{-1} e_c\right| \tag{4.101}$$

When the state variable of the system changes, the critical test function t_{cc} is expressed as the quadratic form of load margin $\Delta\lambda$, that is

$$t_{cc} = m\Delta\lambda^{\frac{1}{c}} \tag{4.102}$$

wherein, m is a scalar constant; $c = 2$ or 4. Generally, this quadratic form is only shown in the test function of the bus in the key area, while in other nodes, it is not sensitive to the change of parameters and has no quadratic characteristics. Therefore, when using the test function, the key problem is how to determine the key nodes. In engineering, the following methods are usually adopted:

(1) The key nodes are determined by the network structure, operators' experience and simulation calculation results;
(2) Combined with eigenvalue analysis or singular value analysis, the key nodes are determined by the maximum elements of the right eigenvector or right singular vector corresponding to the minimum eigenvalue of Jacobian matrix;
(3) The maximum nodes of participation coefficient are defined by load growth;
(4) Combined with sensitivity method, the key nodes of the system are determined by weak nodes.

2. Minimum singular value

References [73, 74] show that the minimum singular value of the Jacobian matrix of the system changes linearly when the load power increases, as shown by the dotted line in Fig. 4.14. Therefore, second-order index with good linearity can be constructed based on the minimum singular value, as shown in the solid line in Fig. 4.14.

By decomposing singular value of Jacobian matrix,

$$J = U\Sigma V^{\mathrm{T}} = \sum_{i=1}^{n} \sigma_i u_i v_i^T \tag{4.103}$$

wherein, U and V are orthogonal matrices, u and v are left and right singular vectors, and Σ is diagonal matrix, whose diagonal element is the singular value σ_i of the Jacobian matrix. When the state variable changes by Δx, the new Jacobian matrix is

$$[J]_{X+\Delta X} = [J]_X + [f_{XX}]_X \Delta X \tag{4.104}$$

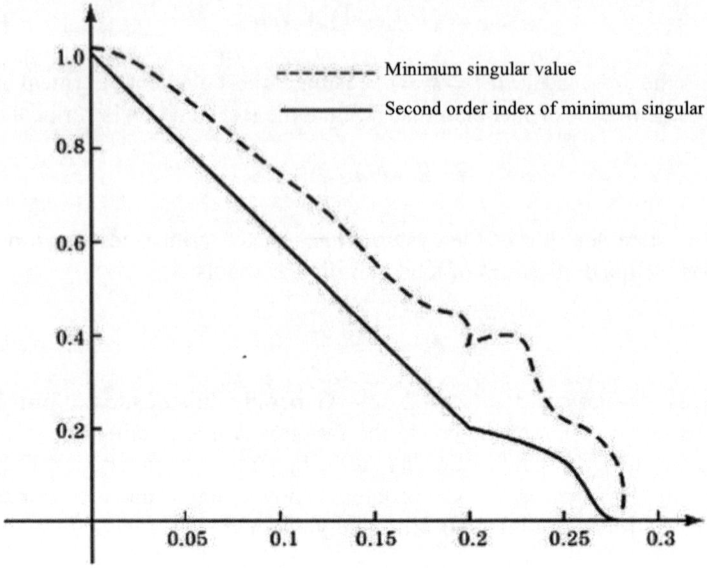

Fig. 4.14 The minimum singular value and its second-order index

wherein, [*fxx*] is the Hessian matrix of the power flow equation. If Δp is the change of power injection vector, the state variable is

$$\Delta x = -J^{-1}\Delta p \tag{4.105}$$

Substituting Eq. (4.105) into Eq. (4.104), we can obtain

$$[J]_{X+\Delta X} = (U + \Delta U)(\Sigma + \Delta \Sigma)(V + \Delta V)^{\mathrm{T}} \tag{4.106}$$

By substituting Eqs. (4.103)–(4.105) into Eq. (4.106) and neglecting disturbed quadratic and cubic terms, we can get

$$U\Delta\Sigma V^{\mathrm{T}} + \Delta U\Sigma V^{\mathrm{T}} + U\Sigma\Delta V^{\mathrm{T}} = -f_{XX}J^{-1}\Delta p \tag{4.107}$$

Both the left and right singular matrices are orthogonal before and after the disturbance, so

$$(U + \Delta U)(U + \Delta U)^{\mathrm{T}} = I \tag{4.108}$$

$$(V + \Delta V)(V + \Delta V)^{\mathrm{T}} = I \tag{4.109}$$

At the same time, if $M = U^{\mathrm{T}}\Delta U$, $N = V^{\mathrm{T}}\Delta V$, then $M = M^{\mathrm{T}}$, $N = N^{\mathrm{T}}$, and the diagonal elements of M and N are all 0. By expanding Eq. (4.107), we can get that the rth singular value of J changes to

$$\Delta\sigma_r = -\left|U^{\mathrm{T}} f_{XX} J^{-1} \Delta p V\right|_{rr} \tag{4.110}$$

For the minimum singular value, we can get

$$\Delta\sigma_{\min} = c^{\mathrm{T}}\Delta p \tag{4.111}$$

wherein, $c = -V_1^{\mathrm{T}}\left(U_1^{\mathrm{T}} f_{XX} J^{-1}\right)$ and U_1 and V_1 are the left and right singular vector of the minimum singular value. Through Eq. (4.110), the change of the minimum singular value can be calculated when the power changes. The parameter c can reflect the voltage sensitivity of the node. The installation of reactive compensation device in the node indicated by the maximum parameter c will obtain better effect than that in other nodes.

3. Reduced-order Jacobian matrix

At a given load node, the following simplified equation can be obtained:

$$\begin{bmatrix} \Delta P_1 \\ \Delta Q_1 \end{bmatrix} = D' \begin{bmatrix} \Delta\theta \\ \Delta U_1 \end{bmatrix} \tag{4.112}$$

wherein, $D' = D - CA^{-1}B$, A, B, C, and D are the corresponding submatrices of Jacobian matrix. It can be seen that the matrix D' can be defined at all operating points. As long as there are non-zero elements in the right eigenvector corresponding to the zero eigenvalue of the Jacobian matrix at the collapse point, the matrix A at the collapse point is also non singular. So, $\det D' = \frac{\det J}{\det A}$ just goes to zero at the collapse point.

4. Power loss sensitivity

The voltage operation level of a power system is closely related to the power loss of the transmission network. A typical feature of system voltage collapse is the sudden increase of system power loss. When the system reaches the critical point with the increase of load, the power loss sensitivity approaches infinity. Therefore, the power loss sensitivity can be used as a sign of voltage collapse.

The power loss in the network should be equal to the sum of the injected power of all nodes in the system:

$$\dot{S}_{\text{loss}} = \sum_{i=1}^{n}\sum_{j=1}^{n} \dot{U}_i\left(Y_{ij}\dot{U}_j\right)^* \tag{4.113}$$

If active power and reactive power are described separately, then

$$\begin{cases} P_{\text{loss}} = f(\delta, U) \\ Q_{\text{loss}} = h(\delta, U) \end{cases} \tag{4.114}$$

We can get

$$\begin{cases} \frac{\mathrm{d}P_{\text{loss}}}{\mathrm{d}P} = \frac{\partial P_{\text{loss}}}{\partial \delta}\frac{\partial \delta}{\partial P} + \frac{\partial P_{\text{loss}}}{\partial \theta}\frac{\partial \theta}{\partial P} \\ \frac{\mathrm{d}P_{\text{loss}}}{\mathrm{d}Q} = \frac{\partial P_{\text{loss}}}{\partial \delta}\frac{\partial \delta}{\partial Q} + \frac{\partial P_{\text{loss}}}{\partial \theta}\frac{\partial \theta}{\partial Q} \end{cases} \tag{4.115}$$

Expressed in matrix form, then

$$\begin{bmatrix} \frac{\mathrm{d}P_{\text{loss}}}{\mathrm{d}P} \\ \frac{\mathrm{d}P_{\text{loss}}}{\mathrm{d}Q} \end{bmatrix} = -\begin{bmatrix} \frac{\partial \delta}{\partial P} & \frac{\partial U}{\partial P}\frac{1}{U} \\ \frac{\partial \delta}{\partial Q} & \frac{\partial U}{\partial Q}\frac{1}{U} \end{bmatrix}\begin{bmatrix} \frac{\partial P_{\text{loss}}}{\partial P} \\ \frac{\partial P_{\text{loss}}}{\partial U} \end{bmatrix} = -\left[J^{\mathrm{T}}\right]^{-1}\begin{bmatrix} \frac{\partial P_{\text{loss}}}{\partial P} \\ \frac{\partial P_{\text{loss}}}{\partial U} \end{bmatrix} \tag{4.116}$$

where

$$\begin{cases} \frac{\partial P_{\text{loss}}}{\partial \delta_i} = 2\sum_{i=1}^{n} U_i G_{ij} U_j \sin \delta_{ij} \\ \frac{\partial P_{\text{loss}}}{\partial U_j} U_j = 2\sum_{i=1}^{n} U_i G_{ij} U_j \cos \delta_{ij} \end{cases} \tag{4.117}$$

Through Eq. (4.116), $\frac{\mathrm{d}P_{\text{loss}}}{\mathrm{d}P}$ and $\frac{\mathrm{d}P_{\text{loss}}}{\mathrm{d}Q}$ of power loss sensitivity can be obtained; similarly, $\frac{\mathrm{d}Q_{\text{loss}}}{\mathrm{d}P}$ and $\frac{\mathrm{d}Q_{\text{loss}}}{\mathrm{d}Q}$ of reactive power loss sensitivity can be obtained. It can be seen from the formula that the power loss sensitivity index is closely related to the Jacobian matrix. When the system is close to voltage collapse, the determinant of Jacobian matrix approaches 0, then the power loss sensitivity approaches infinity. Therefore, the power loss sensitivity can be used to judge the voltage stability of the system, but the linearity of the index is very poor. In order to overcome the shortcomings of the poor linear characteristics of power loss sensitivity index, it is found that the second-order index composed of the second derivative of the power loss sensitivity has good linearity. The second derivative of power loss sensitivity can be obtained by finite differentiability:

$$\frac{\mathrm{d}^2 P_{\text{loss}}}{\mathrm{d}P^2} = \frac{\frac{\mathrm{d}P_{\text{loss}}}{\mathrm{d}\lambda}(\lambda + \Delta\lambda) - \frac{\mathrm{d}P_{\text{loss}}}{\mathrm{d}\lambda}(\lambda)}{\Delta\lambda} \tag{4.118}$$

wherein, $\Delta\lambda$ is usually 5 times of the power flow error.

4.5 Engineering Application

4.5.1 Engineering Indexes in China

At present, the guidelines and standards for voltage and reactive power of China's power grid mainly include:

Guide on Security and Stability for Power System (DL 755—2001);

Technical Guideline on Voltage and Reactive Power of Power System (SD 325—1989);

Regulations on the Administration of Voltage and Reactive Power in Power System (NYD [1988] No. 18);

Technical Principle for Reactive Compensation Configuration in Power System of State Grid Corporation (GJDWS [2004] No. 435).

These guidelines and standards involve little of the static voltage stability evaluation. In the *Guide on Security and Stability for Power System* revised in 2001, *PV* and *VQ* curve methods can be used to simulate the long-term voltage instability with slow change process to obtain the voltage stability margin, determine the key buses, key lines and key units of the system, and determine the relative weak area of voltage stability in the system, and provide basis for further analysis of operation and planning personnel.

For the calculation and analysis of static voltage stability, the method of gradually increasing load (according to the circumstances, the load can be increased proportionally according to the method of maintaining constant power factor, constant power or constant current) can be used to solve the critical point of voltage instability (expressed by $\frac{\mathrm{d}P}{\mathrm{d}U} = 0$ or $\frac{\mathrm{d}Q}{\mathrm{d}U} = 0$), so as to estimate the voltage stability margin of the current operating point.

The static stability criterion is

$$\frac{\mathrm{d}P}{\mathrm{d}U} > 0 \tag{4.119}$$

or

$$\frac{\mathrm{d}Q}{\mathrm{d}U} > 0 \tag{4.120}$$

wherein, $P = f(\mathrm{U})$ is the static voltage characteristic of the active power of the system, and the static voltage characteristic of the active load is considered as constant power; $Q = f(\mathrm{U})$ is the static voltage characteristic of the reactive power of the system, and the static voltage characteristic of the reactive power load is considered as constant power.

The static voltage stability reserve factor is

$$K_{\mathrm{V}} = \frac{U_{\mathrm{Z}} - U_{\mathrm{C}}}{U_{\mathrm{C}}} \times 100\% \tag{4.121}$$

wherein, U_Z and U_C are the normal voltage and critical voltage of the bus respectively.

The static voltage stability reserve standards:

(1) Under normal operating modes, for the different power system, the static voltage stability reserve coefficient (K_V) is 10–15%;
(2) Under post-accident operation mode and special operation mode, K_v should not be less than 8%.

1. Selection of indicators

The static voltage stability analysis method generally evaluates the voltage stability of the system comprehensively by calculating various voltage stability safety indexes, and can determine the relative weak links of the system. The relatively practical static voltage stability safety index in engineering application should meet the following requirements:

(1) The physical meaning is clear, and the comparability between different load buses and different operation modes in the same operation mode is strong, which is convenient for comparative analysis and comprehensive evaluation;
(2) The variation of static voltage stability index should have good linearity, so as to reflect the margin between operating point and critical point intuitively;
(3) The critical point of static voltage stability should have the accurate meaning theoretically;
(4) According to the static voltage stability index, the relatively weak bus and weak area of the system can be determined, which is convenient for taking control and strengthening measures;
(5) From the perspective of engineering application, the calculation speed should be fast to ensure that there are certain calculation errors on the basis of certain accuracy.

The state index of voltage stability only uses the information of current operating state, and the calculation is relatively simple, but generally there is nonlinearity. The calculation of the margin index involves the simulation of the transition process and the calculation of the critical point. It has a large amount of information and can take into account the occurrence of various restrictions, but it needs to realize the set transition process. Both of the indexes have their own advantages and disadvantages, which can be used according to the actual situation in the actual analysis of power system.

Voltage amplitude, critical voltage, active power margin and reactive power margin are the static voltage stability safety index with relatively clear physical meaning, which are also concerned by system planning and operation personnel in engineering practice. Therefore, they are widely used in voltage stability evaluation of actual system. However, the power system is a large-scale nonlinear system, and the state variables present strong nonlinear variations under the action of various disturbance variables and control variables. Various static voltage stability evaluation indexes only reflect the voltage stability level of the system from different sides, which have their own limitations in voltage stability evaluation of actual system.

(1) Indexes of voltage amplitude and critical voltage

Voltage amplitude index is usually used to measure voltage stability level and voltage quality. The understanding of the voltage instability starts from the analysis of the characteristics of the large drop of voltage level. In the operation of the actual system, the operators often focus on the bus voltage at the central point. When the dynamic reactive power reserve of the system is sufficient, the operator can control the central point voltage in a reasonable range by conventional voltage control means. Conventional voltage adjustment means mainly include voltage regulation at generator terminal, voltage regulation of parallel reactive power compensation equipment, and adjustment of OLTC tap, etc. As long as the system has enough reactive power reserve and flexible adjustment means, the voltage can always be controlled within the reasonable operating range; otherwise, in the heavy power flow mode, if the reactive power reserve is insufficient, with the increase of load level, the voltage at the center point can no longer be maintained, the voltage amplitude index can reflect the decrease of the system voltage stability level. Voltage from controllable state to uncontrollable state is an important feature of voltage stability deterioration. However, due to the nonlinear characteristics of voltage variation, the voltage amplitude does not reflect the static voltage stability level and reactive power support capacity of the system.

The critical voltage refers to the voltage when the system is in critical state. For a normal operating point, the critical voltage of the bus refers to the bus voltage when the system transits to the critical state according to a certain load growth mode. In the actual system, the load growth mode is uncertain. For the single load bus power growth, the critical voltage changes with the changes of increasing of load power factor. In general, the higher the power factor of the increasing load, the higher the critical voltage, the lower the power factor of the increasing load and the lower the critical voltage. The critical voltage is also related to the sensitivity of the bus voltage. The closer the bus to the power supply, the smaller the voltage sensitivity and the higher the critical voltage; while, the load bus at the terminal or far from the power supply center, the critical voltage is usually low. Due to the uncertainty of the critical voltage and the nonlinearity of the voltage, the magnitude of the critical voltage cannot reflect the magnitude of voltage stability margin of the system. Therefore, the critical voltage obtained under the same calculation conditions reflects the voltage support capacity of the studied bus to a certain extent, but the numerical value itself does not explain the actual problems.

Therefore, in the evaluation of the actual system, there is severe limitation to use voltage amplitude index and critical voltage index to evaluate the voltage stability of the system. The reasonable voltage distribution is a necessary condition for high voltage stability level, not a sufficient and necessary condition.

(2) Indexes of active power margin and reactive power margin

The active power margin changes with different load growth modes and different adjustment principles of active power output. Therefore the calculation of this index is uncertain. The different load growth modes and active output adjustment principles have different physical meanings. Therefore, it is necessary to evaluate the operation of the actual system.

When the load growth is limited to a single load bus, the calculated active power margin is the active power margin of single load bus. In the actual system, even in a very small period of time, the load growth can not be limited to a single bus growth mode. When the load growth is limited to an area, the calculated active power margin index is the area power margin. The area power margin index is calculated based on the assumption of synchronous load growth. However, in the operational process of the actual system, the load growth mode and active output adjustment do not necessarily change in accordance with the mode assumed by the calculation. However, as the static voltage stability index, the active power margin can give quantitative evaluation of the stability reserve of the system under the current operating state, which provides important reference basis for the operation and planning personnel.

When solving the reactive power margin, the active power output of generator changes slightly with the increase of reactive power load, mainly due to the change of active power loss caused by the increase of reactive power. Therefore, the solution of reactive power margin is certain certainty. However, the reactive power margin is not the physical quantity concerned by the operators. The transmission capability, load level and active power reserve coefficient of the system are the most concerned by the operators. Actually, the transmission capacity of active power is closely related to the reactive power support capacity of the system. The active power margin and reactive power margin reflect the voltage stability level of an actual system from different sides. When the system transits from the normal operating state to the critical state of voltage collapse, the variation trend of both is basically the same. The magnitude of reactive power margin will also affect the active power transmission capability of the system, and the both are consistent. In the voltage stability evaluation of the actual system, it can combine the certainty of reactive power margin with the intuition of active power margin.

In the study process of voltage stability, it should be noted that the reactive reserve of the studied system has a decisive influence on the voltage stability level of the system. For the network with close electrical connection, the magnitude of reactive reserve has a strong correlation with the voltage stability of the study area, but for the power grid structure with weak connection, the reactive reserve cannot directly reflect the voltage stability level of the study area. For buses in the same area, the stability margins of buses near the power supply and those at the far-end load are quite different.

There are many indexes used to evaluate the static voltage stability level of the system, but in actual systems, active power margin and reactive power margin indexes are usually selected, because the calculation process of power margin indexes simulates the transition process from the operating point to the critical point, and takes into account the influence of various nonlinear factors in the load aggravating process of the actual system. The physical meaning is clear and theses indexes can reflect the stability reserve margin of the current operating state of the system. In practical applications, as operators are more concerned about the influence of the percentage of active power load growth on voltage stability in load buses or areas, power margin coefficient (active power margin coefficient or active power reserve coefficient) is often used to express the strength of voltage stability level in addition to calculating active power margin and reactive power margin index in calculation and analysis, and power margin coefficient K_p is defined as

$$K_\mathrm{p} = \frac{P_\mathrm{max} - P}{P} \times 100\% \tag{4.122}$$

The power margin coefficient is based on P_max determined by PV curve. Different load power growth modes correspond to different P_max, so there are different power margin coefficients K_p.

The static voltage stability index most concerned by dispatching operators is usually the area power reserve coefficient. The calculation formula of active power reserve coefficient in areas is as follows

$$\text{Active power reserve coefficient in areas} = \frac{\text{Power limit} - \text{initial power}}{\text{Initial power}} \times 100\% \tag{4.123}$$

This index describes the distance between the current operating point of the system and the voltage collapse point under the condition of synchronous increase of all loads in the designated area, which enables the operators to have a more intuitive understanding of the current operating condition of the system in the form of a percentage.

The calculation formula of active power reserve coefficient of single-load bus is the same as the formula (4.123), except that it describes the distance between the current operating point and the voltage collapse point under the condition that the load (in the simulation data, the power grid of 110 kV and below is usually regarded as a load, and the load of the single bus here is the equivalent load of the low-voltage power grid) of single bus increases according to a certain mode. By comparing magnitude of this index for different bus loads, it can be judged that the bus with the first voltage collapse in the power grid is in a weak area of the grid voltage stability.

(3) Sensitivity index

The purpose of sensitivity analysis is to determine various state indexes of the operating point. The state indexes are mainly related to the state quantity of the current operating point or the change relationship between the state quantity and the disturbance variable. Sensitivity index can actually be used as an index reflecting the voltage stability margin, and the index of the critical point has been determined, for example, $\mathrm{d}U/\mathrm{d}Q$ sensitivity index is zero at the critical point, and the minimum singular value or eigenvalue index is zero at the critical point, etc.

As the solution process of the state index does not reflect the transition process of an actual system from the operating point to the critical point, and does not consider the influence of various restrictive factors, the physical meaning is not very clear, and it has severe limitation as an independent voltage stability evaluation index. However, the solution of the state safety index is relatively simple and convenient, and the sensitivity indexes on different buses in the same operating state are comparable, which can be used as auxiliary analysis indexes to help operators to determine the relatively weak load buses and weak areas of the system, so as to take timely measures and conduct effective monitoring.

Combining the advantages and disadvantages of various evaluation indexes, the method of combining active power margin index and reactive power margin index is commonly used in engineering to evaluate the static voltage stability level of the system.

2. Comparison of K_V and K_P indexes

According to the *Guide on Security and Stability of Power System* issued in 2001, the steady state rotor angle stability reserve coefficient K_V calculated according to the reactive voltage criterion is 10–15%; under the post-accident operation mode and special mode, K_V should not be less than 8%. In this section, the relationship between K_V and K_p indexes is compared. For convenience of comparison, the power margin index method is abbreviated as K_p method and the voltage margin index method is abbreviated as K_V method.

(1) The load power factor is 0.90

Taking the initial active power of the load as the power reference value, the rated voltage as the voltage reference value, and increasing the load with a constant load power factor of 0.9, the PV curve obtained is shown in Fig. 4.15, and the points marked in the figure are the initial load points. The comparison results of K_v and K_p methods are shown in Table 4.5, wherein the area power reserve coefficient is 18%. Table 4.5 converts indexes of $K_v = 10$ and 15% into corresponding K_p index, taking $K_v = 10\%$ as an example to illustrate the conversion process. According to the PV curve of the load point, the critical power is 1.18 p.u., the critical voltage is 0.80 p.u., then the voltage value corresponding to the operating point with $K_v =$ 10% is 0.89 p.u., the power value of the operating point is 1.11 p.u., so the operating point $K_p = (1.18–1.11)/1.11X100\% = 6.3\%$. Accordingly, if $K_v = 15\%$, $K_p = 18\%$ corresponding to operating point can be obtained. Therefore, the steady state rotor angle stability reserve coefficient $K_v = 10–15\%$ specified in the existing *Guide on Security and Stability of Power System* can be converted to $K_p = 6.3–18\%$, i.e., the minimum operating point for static voltage stability of the power grid is $K_p = 6.3\%$.

(2) The load power factor is 0.95

Taking the initial active power of the load as the power reference value, the rated voltage as the voltage reference value, and increasing the load with a constant load power factor of 0.95, the PV curve obtained is shown in Fig. 4.16, and the points marked in the figure are the initial load points.

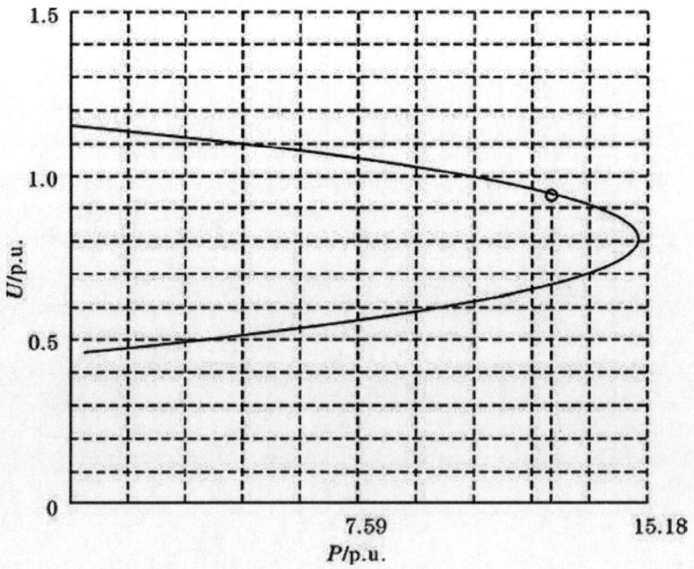

Fig. 4.15 PV curve when load power factor is 0.90

Table 4.5 Comparison of K_v and K_p with load power factor as 0.90

Scheme comparison	Initial power/p.u	Critical power/p.u	Corresponding K_p/%	Initial voltage/p.u	Critical voltage/p.u
$K_p = 18\%$	1.00	1.18	18	0.94	0.80
$K_p = 8\%$	1.09	1.18	8	0.90	0.80
$K_v = 10\%$	1.11	1.18	6.3	0.89	0.80
$K_v = 15\%$	1.00	1.18	18	0.94	0.80

The results of comparison between K_v and K_p are shown in Table 4.6, wherein the area power reserve coefficient is 35%. The steady state rotor angle stability reserve coefficient $K_v = 10$–15% for the method can be converted to $K_p = 19.5$–57%, i.e., the minimum operating point for static voltage stability of the power grid is $K_p = 19.5\%$.

(3) The load power factor is 1.00

Take the initial active power of the load as the power reference value and the rated voltage as the voltage reference value, and increase the load with a constant load power factor of 1.00. The PV curve obtained is shown in Fig. 4.17, and the points marked in the figure are the initial load points.

The comparison results of K_v and K_p methods are shown in Table 4.7, wherein the area power reserve coefficient is 87%.

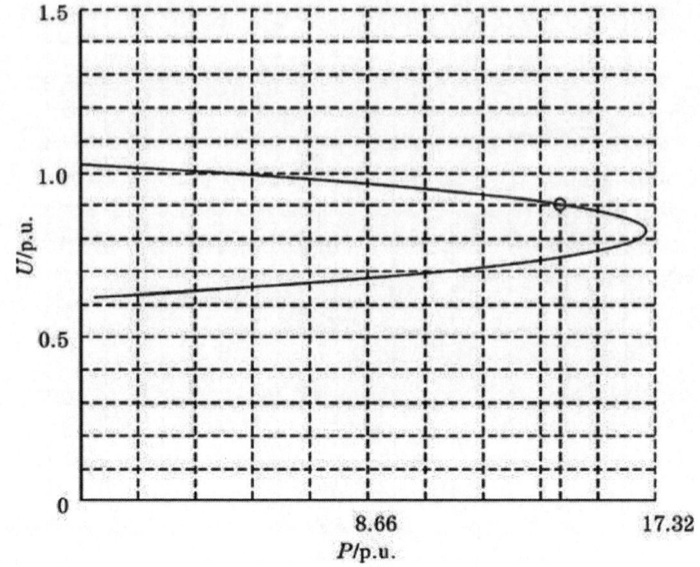

Fig. 4.16 PV curve when load power factor is 0.95

Table 4.6 Comparison of K_v and K_p with load power factor as 0.95

Scheme comparison	Initial power/p.u	Critical power/p.u	Corresponding K_p/%	Initial voltage/p.u	Critical voltage/p.u
$K_p = 35\%$	1.00	1.35	35	0.99	0.87
$K_p = 8\%$	1.25	1.35	8	0.94	0.87
$K_v = 10\%$	1.13	1.35	19.5	0.97	0.87
$K_v = 15\%$	0.86	1.35	57	1.02	0.87

Through the position comparison of operating points $K_p = 8\%$, $K_v = 10\%$, and $K_v = 15\%$ at the above-mentioned different load power factors (0.90, 0.95 and 1.00), we can see that:

(1) When the load power factor is 0.90, K_p values corresponding to operating points $K_v = 10\%$ and 15% are 6.3% and 18% respectively; when the load power factor is 0.95, K_p values corresponding to operating points $K_v = 10\%$ and 15% are 19.5% and 57% respectively; when the load power factor is 1.00, the K_p value corresponding to operating point $K_v = 10\%$ is 146%, and the initial voltage value of operating point $K_v = 15\%$ has exceeded the voltage value at which the load is 0 on PV curve, so its K_p value has no solution.

(2) When the load power factor is smaller, the upper half of the PV curve has more curvature, that is to say, with the load increasing, the voltage drops faster, and the K_p value range corresponding to $K_v = 10$–15% includes the operating point of $K_p = 8\%$; however, when the load power factor is larger, the upper

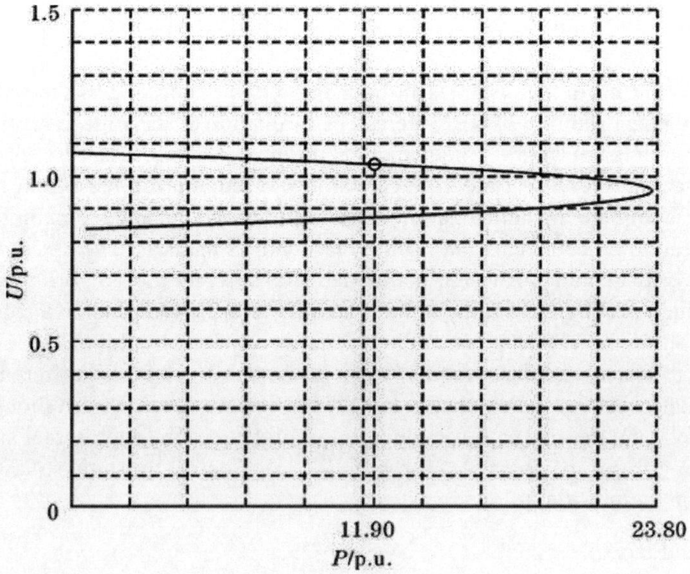

Fig. 4.17 PV curve when load power factor is 1.00

Table 4.7 Comparison of K_v and K_p with load power factor as 1.00

Scheme comparison	Initial power/p.u	Critical power/p.u	Corresponding K_p/%	Initial voltage/p.u	Critical voltage/p.u
$K_p = 87\%$	1.00	1.87	87	1.04	0.95
$K_p = 8\%$	1.73	1.87	8	0.99	0.95
$K_v = 10\%$	0.76	1.87	146	1.05	0.95
$K_v = 15\%$	–	1.87	–	1.12	0.95

half of the PV curve is relatively flat, and the voltage drops slowly with the load increasing. The K_p value range corresponding to $K_v = 10–15\%$ is larger than the operating point of $K_p = 8\%$, especially when the load power factor approaches 1.00, the initial voltage value of operating point $K_v = 15\%$ has exceeded the voltage value at which the load is 0 on the PV curve, i.e., the operating point of $K_v = 15\%$ does not exist.

Therefore, the static voltage stability index obtained by K_v method is more conservative than K_p index. The stable operation range obtained by K_v method and K_p method is not much different when the load power factor is smaller. However, when the load power factor is larger, the operation range of steady state rotor angle stability obtained by K_v method is much less than Kp index, and the transmission capability of the system cannot be fully utilized.

3. Main factors affecting the static voltage stability evaluation

The power margin of the system is the main index of the static analysis method to evaluate the voltage stability of the system. The evaluation of the voltage stability of the actual system is not a purely mathematical problem. It is necessary to fully consider various factors that affect the voltage stability calculation in the actual system, to avoid the complete mathematization of the actual problem, and the main features of things need to be considered and so do reasonable standards. The determination of the static critical point position is mathematically strict for most algorithms, which is not the main problem solved by voltage stability calculation. In the voltage stability calculation of the actual system, the standards required by the calculation conditions are different, and the similarity with the actual system is also different. Various factors that affect the static voltage stability level in the actual system should be fully considered, and the simulation calculation should conform to the actual situation of the power grid as far as possible. The main factors affecting the static voltage stability evaluation include the following aspects.

(1) Load model

In the static voltage stability analysis and calculation, the load is usually simulated by a constant power model, so in the process of load growth, the active and reactive power demands of the load are not affected by bus voltage changes, and the obtained power margin index is relatively conservative. The load of the actual system often includes constant power, constant impedance, constant current model and induction motor model. During the load growth process, with the gradual reduction of bus voltage, the active and reactive power demands of the load are reduced, and the actual power margin is higher than the calculation index obtained when the constant power load model is adopted. However, in the actual calculation of the power grid at present, in order to ensure the safety, the constant power load model is mostly adopted, leaving a certain static voltage stability margin for the power grid.

(2) Load growth mode

In the actual calculation and analysis, the average load is equivalent to 35, 110 or 220 kV, and the capacity of reactive power compensation devices in the distribution network is included in the load. The power factor of this equivalent load cannot accurately reflect the power factor of the actual load. Therefore, in the process of simulating slow load growth, the common constant power factor growth mode cannot accurately reflect the real static voltage stability margin of the system.

(3) Distribution of active output of generator

With the increase of load, the generator's active power output should increase synchronously to meet the demand of load growth. Different adjustment principles of active power output can obtain different critical points, which is also one of the reasons for uncertainty in the solution of active power margin. In the simulation calculation, the generator output that meets the demand of load growth should be reasonably distributed according to the flow direction and capacity of external power received in the studied area, so as to make the simulation calculation conform to the actual situation of the power grid as far as possible.

When the operating state of the system is far away from any critical voltage collapse state, the power margin obtained by different active power output adjustment principles may be quite different. However, when the reactive power standby in the system is insufficient, resulting in a lower reactive power margin of the load center bus, when the operating state of the system is close to the critical operating state, the calculation results will tend to be the same regardless of the load growth mode and the principle of adjusting the active power output of the generator.

(4) Generator reactive capacity limit

The main limiting factor affecting the reactive power output of the generator is usually the temperature rise of the rotor and stator windings, and the maximum reactive power output limit of the generator set is also an important factor affecting the voltage stability margin. In the power flow data, the maximum reactive power output limit of the generator set is generally given according to the rated reactive power of the generator, but the actual reactive power output limit of the generator varies with the active power output and terminal voltage of the generator set. Therefore, a more accurate method should be simulated according to the reactive power limit curve of the generator set, which needs to be drawn according to the nominal data on the nameplate of the generator, the temperature rise test data and all possible limiting factors.

(5) Reasonable division of study area

When calculating the area power margin, it is necessary to simulate the power margin index of the system under the condition that all loads in the area increase synchronously, and the standard of area division will affect its margin index. From the experience of simulation calculation, the larger the study area is, the more prone to voltage instability accidents it is, i.e., the lower the area power margin, under the condition of simulating the synchronous growth of all loads in the area. According to the load coherence, the study area is reasonably subdivided, and the area power margin index obtained by simulation is obviously improved.

In the actual calculation, at present the provincial power grid division or area power grid division is the most calculated, and the voltage stability index calculated by taking the provincial power grid as the study area is generally small, even the area power reserve coefficient index of some provincial power grids does not meet the requirements under the normal operating mode, but in fact there is no voltage stability problem in the provincial power grid; relatively speaking, the voltage stability index calculated by taking the area power grid as the study area is relatively reasonable. Taking a provincial power grid as an example, the provincial power grid is divided into 12 subareas according to the power supply areas. The area power reserve coefficients are 11.1%, 16.9%, 18.0%, 19.3%, 31.2%, 36.0%, 36.2%, 39.2%, 40.6%, 40.6%,

47.3% and 58.2% respectively from small to large ratio. However, if the province is regarded as a power supply area, the calculated area power reserve coefficient is only 3.8%. This is because the load of other power supply areas does not increase synchronously when the load of the power supply area is simulated and studied by the first division method, while the load of all 12 subareas in the whole province is increasing synchronously when the second division method is adopted, and the system operating state deteriorates rapidly and approaches the critical point soon.

(6) Checked fault mode

Referring to foreign static voltage stability evaluation standards, it is not only required to meet a certain power margin index under normal operating mode, but also a certain power margin index under *N*-1 and *N*-2 fault modes. However, it is very tedious and unnecessary to calculate the static voltage stability margin for all *N*-1 and *N*-2 fault modes. Therefore, it is necessary to calculate the static voltage stability margin by reasonably selecting the fault mode that has relatively serious influence on the voltage stability, and it should be avoided to select too minor a fault mode, resulting in misjudgment of the static voltage stability level of the system. Voltage stability problems mostly occur in the receiving power grid, and the fault modes that have significant influence on its static voltage stability margin generally include:

(1) Line faults that may cause a large amount of power flow transfer;
(2) Shutdown of important units in receiving power grid;
(3) Major step-down transformer faults in the receiving power grid, etc.

4.5.2 Engineering Indexes of Other Countries

1. Western Electricity Coordinating Council (WECC)

The WECC is composed of 81 subsystems and 23 members. The power system under its jurisdiction includes 14 states in the western United States, 2 provinces in Canada and one state in Mexico. After two times of massive power outage in 1996, WECC revised its reliability criteria and formulated new standards.

The Commission's voltage stability standards are based on active and reactive power margins. All member systems must give the minimum specified margin on the basis of considering the uncertainty factors. This margin must be larger than the margin standard of performance grade A (performance grade will be described below) under the basic condition of *N*-0. Under normal conditions, when corrective measures are not started, unpredictable load increase or connection power flow is allowed. Each member system must check the corresponding terms according to the standards.

Table 4.8 is WECC voltage stability standard [75]. The system elements in the table include all equipment, such as generators, transmission lines, transformers, and reactive power sources, etc. In order to achieve the purpose of voltage stability

Table 4.8 WECC voltage stability standard

Performance level	Disturbance (1) (2) (3) (4) What causes: Fault or non fault DC disturbance	Active power margin/MW (PV curve method) (5) (6) (7)	Reactive power margin/Mvar (*VQ* curve method) (6) (7)
A	Any element such as: 1 generator One line One transformer A reactive power source DC monopole	≥5%	Worst case (8)
B	Bus	≥2.5%	Worst case (9)
C	Any combination of the following elements: One line and one generator One line and one reactive power source 2 generators 2 lines Two transformers Two reactive power sources DC bipolar	≥2.5%	Worst case (9)
D	A combination of any three or more of the following: Three or more lines in a row All substations Entire power plant including switchyard	>0	>0

analysis, the system can reach another acceptable steady-state operating point after breaking the elements that have the most significant influence on the margin in the preset accident study. At the same time, the margin index of all critical buses under the worst operating conditions of the system should also be calculated. Among them, the worst case refers to the maximum load under the maximum power generation condition, the low load under the minimum power generation condition, and the maximum connection power flow under the worst load condition.

Reactive Power Reserve Work Group (RRWG) recommends that the voltage stability probability load forecasting for the service area should not be less than once every two years. The power generation mode, load level, load characteristics, reactive power capacity of generators, OLTC tap and connection power flow in the study area and adjacent areas are important factors for voltage stability study.

For internal system application, this standard includes: For performance level A accidents, low-pressure load reduction is adopted to achieve the margin specified in Table 4.8; if the load area is radial or local network, and the pre-estimated accidents will not cause system collapse beyond the local area, the margin index in Table 4.8 does not need to be implemented.

(1) Table 4.8 applies to systems in which all elements are put into operation, systems in which one element is out of service, and systems which are readjusted.
(2) For member systems to which this standard applies, low-pressure load reduction is permitted at performance level A.
(3) The element failure list of each performance level is not necessarily different from the disturbance performance of WECC reliability standard.
(4) For the basic case of N-0, the margin index must be larger than that of performance level A.
(5) For each performance level, the active power margin of the maximum operating point on the P-axis must be larger than or equal to the measured peak point of the PV curve.
(6) This standard should be applied to post transient analysis techniques.
(7) Each member system should give due consideration to the uncertainties of the system.
(8) The bus with the least reactive power should have a sufficient reactive power margin. When the worst single fault occurs, the following conditions can be met: ① exceeding 5% of the maximum predicted load; ② exceeding 5% of the maximum allowable tie-line power flow. The worst single fault is to reduce the reactive power margin the most.
(9) The bus with the least reactive power should have a sufficient reactive power margin. When the worst single fault occurs, the following conditions can be met: ① exceeding 2.5% of the maximum predicted load; ② exceeding 2.5% of the maximum allowable tie-line power flow. The worst single fault is to reduce the reactive power margin the most.

Performance level: the performance level defines the minimum active power and reactive power margin required.

Performance level A: In case of any single disturbance in the system, the external system will not be adversely affected, such as loss of load or equipment load beyond the scope of emergency.

Performance level B: When any bus is lost in the system, the system and external system will not be so affected that voltage collapse will occur.

Performance level C: In the absence of system regulation, the system loses any combination of two system elements.

Performance level D: In the absence of system regulation, the system loses any combination of three system elements.

Table 4.9 Table of stability margin values

Section power flow	Minimum voltage margin (K_V)
Normal power flow	0.15
Aggravated power flow	0.15
Forced power flow	0.10

2. Russian power system

The voltage margin parameter (K_V) belongs to the load point parameter, which is calculated by the following formula:

$$K_V = \frac{U - U_{KP}}{U} \tag{4.124}$$

wherein, U is the load point voltage under the studied operation mode; U_{KP} is the limit voltage at this point. When the voltage is lower than this limit value, it will cause the static instability of the motor.

For 110 kV and higher voltage grade lines, the limit voltage at load point should not be less than 0.7 U_{HOM} and 0.7 U_{HOPM}. Wherein, U_{HOM} is the rated voltage of the inspected load point; U_{HOPM} is the voltage of the inspected load point under normal operation mode. In order to check the implementation of the voltage standard margin of check load point in the actual project, the voltage of any node in the power system can be referred to. The reference node voltage value can be determined by power flow calculation.

The stability margin value should not be lower than the value shown in Table 4.9.

Under the above mentioned normalized disturbance, the post fault operation mode should meet the voltage margin of not less than 0.1.

References

1. Zhang Y, Zhou S, Wang L et al (2000) Dynamic element model and its implementation in static voltage stability analysis. Proc Chin Soc Electr Eng 20(3):66–70
2. Kundur P (2002) Power system stability and control. China Electric Power Press, Beijing
3. Taylor CW (2002) Power system voltage stability (trans Wang Weisheng). China Electric Power Press, Beijing
4. Zhou S, Zhu L, Guo X et al (2004) Voltage stability and control of power system. China Electric Power Press, Beijing
5. Ajjrapu V, Christy C (1992) The continuation power flow: a tool for steady state voltage stability analysis. IEEE Trans Power Syst 7(1):416–423
6. Chen H, Chen Z (1996) The application of power flow equation in the study of static voltage stability. J Shanghai Jiaotong Univ 30(9):69–74
7. Li Y, Zhang B, Sun H (2007) The safety constrained optimal power flow (I) theoretical analysis based on the nonlinear interior point method. Proc CSU-EPSA 31(19):7–13
8. Karmarkar NK (1984) A new polynomial time algorithm for linear programming. Combinatorica 4(4):373–395

9. Granville S, Mello JCO, Mello ACG (1996) Application of interior point methods to power flow unsolvability. IEEE Trans Power Syst 11(2):1096–1103
10. Wei H, Sasaki H, Kubokawa J et al (1998) An interior point nonlinear programming for optimal power flow problems with a noval data structure. IEEE Trans Power Syst 13(3):870–877
11. El-Bakry AS, Tapia RA, Tsuchiya T et al (1996) On the formulation and theory of the Newton interior point method for nonlinear programming. J Optim Theory Appl 89(3):507–541
12. Forsgren A, Gill PE, Wright MH (2002) Interior methods for nonlinear optimization. SIAM Rev 44(4):525–597
13. Guo R, Han Z, Wang Q (1999) Nonlinear programming model and algorithm for critical point of voltage collapse. Proc Chin Soc Electr Eng 19(4):14–17
14. Irisarri GD, Wang X, Tong J et al (1997) Maximum loadability of power systems using interior point non-linear optimization method. IEEE Trans Power Syst 12(1):162–172
15. Wei H, Ding X (2002) The critical point algorithm of voltage stability based on modern interior point theory. Proc Chin Soc Electr Eng 22(3):27–31
16. Wang X, Li H, Li B (2007) Voltage stability evaluation of AC and DC system based on interior point method. Proc CSU-EPSA 19(6):72–77
17. Zhang M (2006) Research on voltage stability analysis of AC/DC hybrid system. Master's thesis, Sichuan University, Chengdu
18. Feng Z, Liu Q, Ni Y et al (1992) Voltage static stability analysis of multimachine power system—singular value decomposition method. Proc Chin Soc Electr Eng 12(3):10–19
19. Wu H, Li X, He Y et al (2009) Static voltage stability analysis based on singular value decomposition method considering static load characteristics. Sichuan Electr Power Technol 32(3):5–8
20. Li X, Wang X (2003) A fast voltage stability analysis method based on static equivalence and singular value decomposition. Proc Chin Soc Electr Eng 23(4):1–5
21. Liu F, Yao X (1999) The singular value decomposition method used to analyze the effect of secondary voltage control. Proc CSU-EPSA 23(18):1–4, 8
22. Chen M, Zhang B, Duan X (2006) Research on voltage stability weak nodes based on minimum singular value sensitivity. Power Syst Technol 30(24):36–39, 55
23. Xu Z, Luan Z (1997) Singular values and stability indexes for measuring node voltage stability. Proc CSU-EPSA 21(8):42–44
24. Gao B, Morison GK, Kundur P (1992) Voltage stability evaluation using modal analysis. IEEE Trans Power Syst 7(4):1529–1542
25. Lof PA, Smed T, Andersson G et al (1992) Fast calculation of a voltage stability index. IEEE Trans Power Syst 7(1):54–64
26. Verghese GC, Perez-Arriaga IJ, Schweppe FC (1982) Selective modal analysis with applications to electric power systems. IEEE Trans Power Appar Syst 101(9):3117–3125
27. Liu T, Song X, Tang Y et al (2010) Eigenvalue sensitivity method and its application in small disturbance stability analysis of power system. Power Syst Technol 34(4):82–87
28. Xi Y, Guo Y (1998) A voltage stability algorithm based on eigenstructure analysis. J Tsinghua Univ (Nat Sci) 38(3): 1–5
29. Zhang G, Yang J, Zhang J (2007) An improved eigenstructure analysis method for static voltage stability. Power Syst Technol 31(16):77–82
30. Tan D (1996) A practical calculation and analysis method for voltage stability of power system—brief introduction of eigenvalue analysis method for linear system. Power Syst Technol 20(6):61–62
31. Wu Z, Zhang Y (2001) Analysis of correspondence between eigenvalues and state variables of power system. Proc CSU-EPSA 25(10):23–26
32. Wu J, Zhang B, Chen G (2006) The voltage instability area of AC-DC system is determined by eigenvalue method. Relay 34(7):27–31
33. Nam HK, Kim YK, Shim KS et al (2000) A new eigen sensitivity theory of augmented matrix and its applications to power system stability analysis. IEEE Trans Power Syst 15(1):363–369
34. Yuan J, Duan X, He Y et al (1997) Summary of voltage stability sensitivity analysis methods in power system. Power Syst Technol 21(9):7–10

35. Duan X, Zhang D (1997) Voltage stability sensitivity analysis methods in power system. Proc CSU-EPSA 21(4):9–12
36. Yu Y, Zeng Y, Jia H (2000) The sensitivity criterion of static voltage stability and the evaluation of criterion of dQL/dVL and dPL/dVL. Proc CSU-EPSA 12(3):1–4, 13
37. Luo J, Wu Z, Lian X et al (2010) A new method for calculating the sensitivity of voltage stability margin to line power. Proc CSU-EPSA 4:94–99
38. Long J, Zhou L, Fu K (2009) Voltage stability sensitivity analysis of power system with UPFC. J Guangxi Univ: Nat Sc 34(2):251–255
39. Zhang J, Sun Y (2007) Sensitivity analysis method for selecting emergency control location based on fragile cut set. Power Syst Technol 31(11):21–26
40. Cai G, Zhang Y, Yu T et al (2008) Static voltage stability preventive control based on the unified sensitivity method. High Voltage Technol 34(4):748–752
41. Jiang W, Wang C, Yu Y (2006) A new method for calculating the sensitivity of voltage stability margin to parameters. Proc Chin Soc Electr Eng 26(2):13–18
42. Wang J, Zhang Y, Wang C (2005) Power grid reactive power/voltage assessment method based on sensitivity analysis and optimal power flow. Power Syst Technol 29(10):65–69
43. Zhao H, Zhao Y (2008) Evaluation of power grid vulnerability region based on sensitivity technique. Power Syst Technol 32(14):54–58
44. Ma P, Cai X (2008) Sensitivity method of voltage stability margin evaluation for branch outage contingencies. Proc Chin Soc Electr Eng 28(1):18–22
45. Dobson I, Lu L (1993) New methods for computing a closest saddle node bifurcation and worst cast load power margin for voltage collapse. IEEE Trans Power Syst 8(3):905–913
46. Canizares CA, Alvarado FL (1993) Point of collapse and continuation methods for large AC/DC systems. IEEE Trans Power Syst 8(1):1–8
47. Lu J, Liu CW, Thorp JS (1995) New methods for computing a saddle-node bifurcation point for voltage stability analysis. IEEE Trans Power Syst 10(2):978–989
48. Dobson I (1992) Observations on the geometry of saddle node bifurcation and voltage collapse in electrical power systems. IEEE Trans Circuits Syst I: Fundam Theory Appl 39(3):240–243
49. Alvarado F, Dobson I, Hu Y (1994) Computation of closest bifurcations in power system. IEEE Trans Power Syst 9(2):918–928
50. Zhao X, Zhang X, Su X (2008) Voltage stability studies and bifurcation theory in power systems. Trans China Electrotech Soc 23(2):87–95
51. Chiang HD, Jumeau RJ (1995) A more efficient formulation for computation of the maximum loading in electric power system. IEEE Trans Power Syst 10(2):635–646
52. Seydel R (1979) Numerical computation of branch points in nonlinear equations. Numer Math 33(3):339–352
53. Liu Y, Yan Z, Ni Y et al (2003) An auxiliary-variable-based direct method for computing quadratic turning bifurcation points of power flow equations. Proc Chin Soc Electr Eng 23(5):9–13
54. Zeng J, Han Z (1997) Direct computing method of voltage stability critical point. J Tsinghua Univ (Nat Sci) 37(SI):91–94
55. Ajjarapu V (1995) Application of bifurcation and continuation methods for the analysis of power system dynamics. In: Proceedings of 4th IEEE conference on control applications, Albany, pp 52–56
56. Jarjis J, Galiana FD (1981) Quantitative analysis of steady state stability in power networks. IEEE Trans Power Appar Syst 100(1):318–326
57. CIGRE Task Force 38.02.11 (1995) CIGRE technical brochure: indices predicting voltage collapse including dynamic phenomenon. Electra 159:135–147
58. Zhou S, Jiang Y, Zhu L (2001) Review on steady state voltage stability indices of power systems. Power Syst Technol 25(1):1–7
59. Kessel P, Glavitseh H (1986) Estimating the voltage stability of power system. IEEE Power Deliv PWRD-1(3):346–354
60. Tuan TQ, Fandlno J, Adjsaid N et al (1994) Emergency load shedding to avoid risk of voltage instability using indicators. IEEE Trans Power Syst 9(1):341–351

61. Yu Y, Jia H, Yan X (19999) An improved local voltage stability index Li, which can trace saddle-node bifurcation exactly. Power Syst Technol 23(5):19–23
62. Yu Y, Wang C (1999) Theory and method of power system stability. Science Press, Beijing
63. Canizares C (1998) Voltage stability assessment, procedure and guides. IEEE/PES Power System Stability Subcommittee Special Publication
64. Nagao T, Tanaka K, Takenaka K (1997) Development of static and simulation programs for voltage stability studies of bulk power system. IEEE Trans Power Syst 12(1):273–281
65. Mohamed A, Jasmon GB (1995) A new clustering technique for power system voltage stability analysis. Electr Mach Power Syst 23(4):389–403
66. Sun X, Duan X, He Y (1998) Research on local safety index of load node voltage stability. Proc CSU-EPSA 22(9):61
67. Tiranuchit A, Thomas RJ (1988) A posturing strategy against voltage instability in electric power systems. IEEE Trans Power Syst 3(1):87–93
68. Xu Z, Luan Z, Fan T et al (1997) Singular values and stability indexes for measuring node voltage stability. Proc CSU-EPSA 8:42–44
69. Tang Y, Sun H, Yi J et al (2009) Thevenin equivalent parameter tracking algorithm based on complete differential. Proc Chin Soc Electr Eng 29(13):48–53
70. Tang Y, Lin W, Sun H et al (2010) Analysis of voltage stability criteria considering load characteristics. Proc Chin Soc Electr Eng 30(16):12–18
71. Tang Y, He RM, Ju P et al (2010) Evaluation of dynamic characteristics and safety of electric power receiving-end system. Tsinghua University Press, Beijing
72. Berizzi A, Zeng YG, Marannino P et al (2000) A second order method for contingency severity assessment with respect to voltage collapse. IEEE Trans Power Syst 15(1):81–89
73. Zhang Y, Zhang J, Yuan S (2005) Improved continuation power flow algorithm for obtaining the limit of static voltage stability. Proc CSU-EPSA 17(2):21–25
74. Zhao J, Zhang B (2005) A study on the strategy for improving robustness of continuation power flow computation. Proc Chin Soc Electr Eng 25(22):7–11
75. Western Electricity Coordinating Council (2001) Summary of WECC voltage stability assessment methodology

Chapter 5
Transient (Short-Term) Voltage Stability

5.1 Analysis Method of Transient Voltage Stability

The dynamic behavior of power system can be summarized and reflected by the nonlinear differential-difference-algebraic equations (DDAE). The differential equations partially reflect the dynamic behavior of dynamic elements in the power system, the difference equations partially reflect the discrete actions of system elements, and the algebraic equations partially reflect the interaction between dynamic elements and topological constraints of the network in the power system. In this way, both the disturbance from the dynamic element part and the disturbance from the network part disturb the physical equilibrium of the dynamic elements. The dynamic behaviors of power system are only restricted by the dynamic behavior and mutual relations of system dynamic elements. The research on voltage stability of power system is to abstract the mathematical model reflecting this objective phenomenon from the actual situation of power system, and then return to the actual problems from the mathematical characteristics reflected by the mathematical model and explain them.

The power system may become unstable if there are short-circuit faults, other types of large disturbances or load fluctuations at the edge of steady state rotor angle stability. At this time, the nonlinear characteristics of the power system must be retained in the mathematical description of dynamic behavior for power systems in order to truly reveal the mechanism of voltage stability of the power system and dynamic characteristics of the power system under large disturbance. At present, analysis methods of transient voltage stability mainly include time-domain simulation method, energy function method and nonlinear dynamic method.

5.1.1 Time-Domain Simulation Method

The time-domain simulation method is a method to obtain the changes of voltage and other electrical quantities with time based on the differential-algebraic equations

© Science Press 2021
Y. Tang, *Voltage Stability Analysis of Power System*, Power Systems,
https://doi.org/10.1007/978-981-16-1071-4_5

of power systems, while retaining the nonlinear characteristics of the power system and considering the dynamic action of elements. It is a basic method to analyze the dynamic process of the power system under large disturbance. This method can easily adapt to the power system models, different element models and system faults and operations. It is characterized by detailed mathematical model, simple algorithm principle, provision of time response of system state variables and other advantages. This method is the most effective method of studying stability of power systems. It is now mainly used for dynamic process analysis for the disturbed power system to develop prevention and correction control measures.

Power system is a complex large-scale nonlinear system. It contains a large number of variables based on time constants. Some variables are characterized by rapid change while others are characterized by slow change. Therefore, the power system is a multi-time scale system. From the perspective of electromechanical transients, it contains three groups of variables: rapid change (electromagnetic transient), normal rate (electromechanical transient) and slow change (medium and long-term dynamics). Therefore, the power system is at least a three-time scale dynamic system.

Due to the complexity and multi-time scale characteristics of power system, in order to improve the computational efficiency, the fast dynamics of electromagnetic transients and the slow dynamics of medium and long-term process are usually ignored in electromechanical transient simulation. That is, in the electromechanical transient simulation, it is considered that while the electromagnetic transients have ended and the corresponding variables have decayed, the medium and long-term process has not yet begun to change, which means the medium and long-term dynamic variables remain constant.

According to the purpose of simulation, the mathematical model used by power system simulation software can be linear or non-linear, constant or time-varying, continuous or discrete, of centralized or distributed parameters, and deterministic or random, etc. Some secondary factors are often ignored in development of the mathematical model, so the model developed is often a simplified one.

Common power system simulation software adopts different simulation methods for different dynamic processes. There are mainly electromagnetic transient simulation, electromechanical transient simulation and medium and long-term dynamic simulation. The latest research and development achievements show that multi-time scale full-process dynamic simulation can realize electromagnetic transient-electromechanical transient hybrid simulation and electromechanical transient-medium and long-term dynamic unified simulation [1, 2].

1. Electromagnetic transient simulation

The digital simulation of electromagnetic transient simulates electromagnetic transient between microseconds and several seconds of power system by numerical computation method. The electromagnetic transient simulation should generally consider the distribution characteristics and the frequency characteristics of parameters of transmission lines, and the electromagnetic and electromechanical transients of generators, as well as the nonlinear characteristics of a series of elements (lightning arrester, transformer, and reactor, etc.). Therefore, the mathematical model of electromagnetic transient simulation must establish algebraic, differential or partial differential equations of these elements and systems. The commonly used numerical integration method is implicit integration method.

The electromagnetic transient simulation program mainly focuses on the digital simulation of: ① transients caused by the system external factors, such as lightning overvoltage, etc.; ② transients caused by faults and operations, such as operating overvoltage and power frequency overvoltage, etc.; ③ resonance transients, such as sub-synchronous resonance and ferromagnetic resonance, etc.; ④ control transients, such as the interaction between primary and secondary systems, etc.; ⑤ fast transient and non-sinusoidal quasi-steady processes of power electronic devices, FACTS and HVDC.

As the electromagnetic transient simulation requires not only detailed nonlinear models for dynamic elements of power systems, but also network transients, as well as the description of differential equations, and this restricts the simulation scale of electromagnetic transient simulation program. In general, when conducting electromagnetic transient simulations for large-scale power systems, the power systems should be simplified by equivalence.

At present, the electromagnetic transients program (EMTP) is commonly used. It is characterized by being able to compute the electromagnetic transients of any network with centralized parameter elements and distributed parameter elements. The models and computation methods adopted in the program are verified and compared with the results of field tests. The solution speed is fast and the accuracy can meet the requirements of engineering computation. The EMTP has been revised continuously on the basis of cooperation between multi-countries since 1987. Programs with similar functions to EMTP include EMTPE developed by China Electric Power Research Institute based on EMTP, EMTDC (PSCAD) of Manitoba HVDC Research Center of Canada, Canada, Micro-Tran of University of British Columbia (UBC), British Columbia of Canada, NETOMAC of Siemens, Germany, etc.

2. Electromechanical transient simulation

The electromechanical transient simulation mainly studies the transient stability of the power system after a large disturbance occurs and the steady state rotor angle stability of the power system after a small disturbance occurs. It is a basic method of studying the dynamic behaviors of the power system and its ability to maintain synchronous and stable operation under large disturbances such as short-circuit fault, stoppage or operation of lines, generators and loads, loss of excitation of generators and impact loads, check and analyze the stability of the power system in operation and stability disruption accidents, master the dynamic stability of the power system, and formulate prevention measures of stability disruption.

The mathematical model of power system electromechanical transient simulation can be written as:

$$\begin{cases} \frac{dx}{dt} = f(x, y, t) \\ 0 = g(x, y) \end{cases} \tag{5.1}$$

In the formula, x represents the state variable of the element, determined by various dynamic elements to be considered in electromechanical transient simulation; y represents operation variables of the power system. The differential equation represents the characteristics of dynamic elements of the power system and is the state equations of the systems. The algebraic equation represents the static element characteristics of power system and is the network equation of the system.

The algorithm of electromechanical transient simulation of power system is to solve the differential equations and algebraic equations simultaneously to obtain the time domain solution of physical quantities. The methods of solving differential equations mainly include implicit trapezoidal integration method, improved Euler method and Runge-Kutta method. Among them, implicit trapezoidal integration method has been used increasingly due to the good numerical stability. The methods of solving algebraic equations mainly include Gaussian elimination method for solving linear algebraic equations and the Newton method for solving non-linear algebraic equations. It can be divided into alternate solution and simultaneous solution by the order of solving differential equation and algebraic equation.

The commonly used electromechanical transient simulation programs include PSD-BPA power system analysis program and Power System Analysis Synthesis Program (PSASP) in China. And PSS/E of PTI (USA), ETMSP of EPRI (USA), and programs developed by international electrical companies such as SIMPOW of ABB, PSLF of GE, NETOMAC of SIEMENS with the function of electromechanical transient simulation are commonly used in the world.

3. Medium and long-term dynamic simulation

The medium and long-term dynamic simulation of power systems is a long-term dynamic simulation of the disturbed power system, i.e., the general long-term dynamic stability computation of a power system should take into account the long-term and slow dynamic characteristics of the power system that are not considered in the general transient stability simulation, including the dynamic responses of relay protection system, automatic control system, thermal and hydraulic systems of power plant and nuclear reaction system. The time for the long-term dynamic stability computation ranges from tens of seconds to tens of minutes or even hours.

The long-term dynamic stability computation of a power system is mainly used to analyze the long-term dynamics of power systems (tens of seconds to tens of minutes or even hours). The scope of long-term dynamic simulation and research is mainly as follows:

(1) Perform ex post analysis for complex and severe accidents to understand the essential reason of the accident and study the right anti-accident measures;
(2) Carry out voltage stability analysis to study the mechanism for voltage stability of power system and effective measures for preventing voltage collapse;

(3) Assess the system capacity in enduring extremely serious faults (severe faults beyond the normal design standard) in the planning and design stage to study measures for reducing the frequency of those severe faults and preventing fatalities;
(4) Study accident development processes and give training to operators on their emergency handling capability;
(5) Study and arrange load shedding strategies;
(6) Study the effectiveness for emergency reactive power support;
(7) Study the allocation of spinning reserve and distribution of spinning reserve units;
(8) Study auto generation control strategies;
(9) Study the impact of response from boiler control system (including reactor) and auxiliary equipment of power plant after large disturbance on the operation characteristics of power plant and coordinate the control and protection system of power plant.

The long-term dynamic simulation computation of power systems is also, as the same as the electromechanical transient simulation of power systems, to solve simultaneously the differential equations describing system dynamic element pair and the algebraic equations describing system network characteristics, so as to obtain the time-domain solution for long-term dynamic process of power system. Nevertheless, a power system, with the time constant for long-term dynamic response ranging from tens of milliseconds to more than 100 s, is a typical stiff system and requires adopting the numerical solution of stiff differential equations. Moreover, to avoid too long computation time, the variable step size computation technique is also required.

Presently, the internationally major long-term dynamic stability computation procedures include: the SIMPOW procedure of ABB and the EUROSTAG procedure developed by EDF, etc., LTSP procedure of the US Electric Power Research Institute (EPRI), and EXTAB procedure developed jointly by General Electrical and Tokyo Electric Power Company Holdings, Inc. (TEPCO); moreover, the PSS/E procedure of PTI (USA) and MODES procedure of CEZ have also the long-term dynamic stability computation function. In China, China Electric Power Research Institute has started the study of full dynamic simulation procedure since 1998 and has developed PSD-FDS full dynamic simulation procedure.

4. Multi-time scale full simulation of power systems

With the extensive application of DC power transmission, FACTS and other power and electronic apparatus and other nonlinear elements in a power system, the waveform distortion caused by those elements and their fast transient process have an increasing influence on system electromechanical transient processes, and the mutually independent electromagnetic transient simulation procedures and electromechanical transient simulation procedures of power systems have already been difficult to meet the requirements for simulation by modern power systems. Therefore, it is very necessary to develop the power system simulation technology and software supporting mixed simulation for electromagnetic transient process and electromechanical transient process.

The increasing long-distance power transmission capacity of power systems result in increasingly prominent heavy load problem of power transmission network, and the medium and long-term dynamic stability (including voltage stability problem) of power system after transient stability will be one of the important problems for safe and stable operation of power system step by step, threatening the safe and stable operation of the power system. It will certainly become one of the important items for power system computation and analysis to analyze the long-term dynamic stability problems of power system for avoiding large scope blackout and study effective measures (the third line of defense for safe and stable power system) for preventing accident expansion. Therefore, it is very necessary to develop a full simulation technology and software of power system that can simulate uniformly the electromechanical transient and medium and long-term dynamic processes.

(1) Mixed simulation of electromagnetic transient and electromechanical transient processes

The simulation of electromagnetic transient simulates electromagnetic transient between microseconds and several seconds of power systems by numerical computation method. The characteristics and frequency characteristics of power transmission line distribution parameters, the electromagnetic and electromechanical transient processes of generators, the nonlinear characteristics and DC power transmission of a series of elements (lightning arrester, transformer and reactor) as well as the transient process of FACTS device can be put into consideration for simulation of electromagnetic transient processes. As the electromagnetic transient simulation requires not only detailed nonlinear models for dynamic elements of power system, but also network transients, and this restricts the simulation scale of electromagnetic transient simulation programs. In general, when conducting electromagnetic transient simulation, the power systems should be simplified by equivalence.

Simulation of electromagnetic transient process refers to simulating HVDC systems and FACTS devices with quasi-steady state model by fundamental wave, single phase and phasor simulation technology, and cannot simulate precisely the fast transient characteristics of HVDC, FACTS and other power and electronic apparatus and the waveform distortion characteristics of MOV and other nonlinear elements. It is imprecise to simulate HVDC systems and FACTS devices with quasi-steady state model under asymmetric AC system conditions.

The main idea for mixed simulation of electromagnetic transient and electromechanical transient processes lies in that dividing the large-scale power system into subsystems requiring electromagnetic transient simulation and those requiring only electromechanical transient simulation for accepting electromagnetic transient simulation and electromechanical transient simulation respectively, and handing over electromagnetic transient simulation and electromechanical transient simulation at the boundary of subsystems to improve the simulation precision of electromechanical transient procedures.

(2) Full dynamic (electromechanical transient and medium and long-term dynamic) simulation

Dynamic process of power systems (from electromechanical transient process to medium and long-term dynamic process) is a continuous process, not a discrete separate process. The transient process and the medium and long-term process interact with each and are inseparable. The full dynamic simulation of power system refers to that unifying the electromechanical transient process, the medium-term dynamic process and the long-term dynamic process of power systems organically for simulation.

The time for full dynamic simulation of power systems ranges from several seconds to ten minutes and even several hours, with a large time span. Therefore, the variable step size integral method needs to be used, namely, computing with small-sized steps in the fast stage (electromechanical transient) of the system and computing with large-sized steps in the slow stage (medium and long-term dynamic). The numerical integration method of existing full dynamic simulation procedure applies mostly Gear variable step stiff integration method which has the advantages of uniform models and numerical integration methods that can be used for transient process and medium and long-term dynamic processes, and large step method can be used in the medium and long-term dynamic process for simulation.

Besides being used for the analysis and computation in which the conventional transient stability procedure can be used, the full dynamic simulation of power systems can be applied for the ex post analysis for complex and severe accidents and the study of accident repetition in longer time after disturbance, so as to understand the essential reason of the accident and study the right accident prevention measures, etc. Therefore, studying full dynamic simulation technology and developing and improving full dynamic simulation procedures can not only improve China's study and software development level in the field, but also provide powerful simulation tools for safe operation analysis and decision making of the power grid.

The time-domain simulation method has the following advantages: carefully records the dynamic characteristics of elements, with higher simulation precision; better reflect the full process of voltage instability and provide reliable information for analyzing the voltage collapse mechanism; obtain the preventive and corrective measures for preventing voltage instability. The time-domain simulation method has made undeniable contributions in analyzing the large disturbance voltage stability of power systems and revealing the voltage instability mechanism, but the characteristics of power systems present difficulties for the time-domain simulation method in the following two aspects:

(1) Power system is a high-dimensional strongly nonlinear system covering multiple elements with different time constants and dynamic characteristics. The DAE equations of the power system model is stiff, and in consideration of the numerical stability, convergence of iteration, cumulative errors and other factors of integration by parts, step sizes too large for integration are not

allowed, so heavy computation, long elapsed time and other factors restrict the further application of the time-domain simulation method.

(2) With many operating states and imaginary faults in power systems, another integration is required if the network structure, control parameters or fault settings change. Moreover, we can only draw a conclusion on whether the system is stable or unstable, instead of obtaining stability margin and any other information about system stability through time-domain simulation. So, it is difficult to perform the real-time on-line stability evaluation and control of power systems with the time-domain simulation method.

The above shortcomings limit the application of the time-domain simulation method in voltage stability analysis. At present, the method is mainly used to analyze the influence of dynamic elements on voltage stability, understand the characteristics of voltage collapse and check the correctness of stability criteria.

5.1.2 Energy Function Method

The theoretical basis of the energy function method is the Lyapunov stability theory of power systems. This method approaches the stability problem from the perspective of system energy instead of time domain. Therefore, it can avoid solving nonlinear differential equation set and directly judge system stability through the analysis on the system energy function. Therefore, it is of great significance in can stability analysis of power system and engineering applications.

The energy function method is based on the development of a classical mechanical concept. Reference [3] describes and illustrates the principle of the energy function method with a simple example.

As shown in Fig. 5.1, when the rolling ball system is free from disturbance, the ball is located at the stable equilibrium point (SEP). After being subject to disturbance, the ball is located at the height of h at the end of the disturbance and has the velocity v. The mass of the ball is m, and the total energy V is composed of the sum of

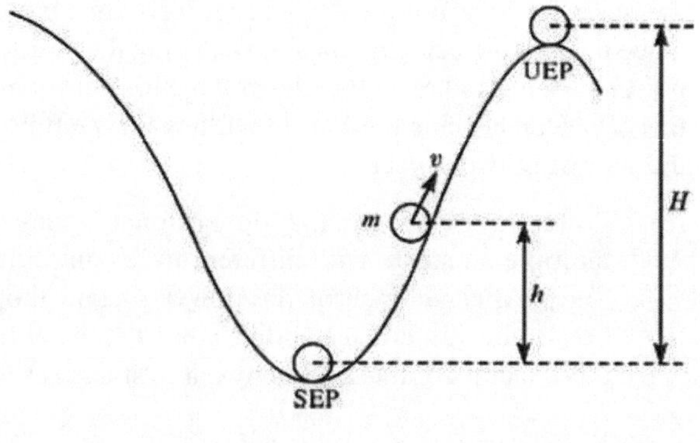

Fig. 5.1 Stability principle of rolling ball system

kinetic energy $\frac{1}{2}\text{mv}^2$ and potential energy *mgh* (g is the acceleration of gravity), i.e., $V = \frac{1}{2}mv^2 + mgh > 0$. In case of friction between the small ball and the wall, the energy will decrease gradually under the action of friction after disturbance. Assuming that the wall height of the container in which the small ball is located is H, when the small ball is located on the wall edge and the velocity is zero, the corresponding potential energy is MGH, which is called unstable equilibrium point (UEP), and the corresponding potential energy is the critical energy V_{cr} of the system, namely, $V_{\text{cr}} = mgH$.

According to the kinematic principle, if the wall friction is ignored, when the total energy V of the small ball is higher than the critical energy V_{cr} at the end of the disturbance, the small ball will eventually roll out of the container and lose its stability; otherwise, if the total energy V of the small ball is less than the critical energy V_{cr}, the small ball will gradually reduce the energy under the action of the friction, and finally remain at SEP; when $V = V_{\text{cr}}$, it is the critical state of the system. Obviously, the stability margin can be determined as per $V_{\text{cr}} - V$.

The core task of the energy function method is to estimate the stability domain of the system's SEP after a fault, which can be divided into two steps in practical application: First, to find the energy function satisfying Lyapunov criterion, and its magnitude shall correctly reflect the severity of the system's instability; second, to make calculations based on the critical energy and make depictions based on the stability domain according to the energy function constructed, so as to obtain the final result. Therefore, the stability or the stability domain can be determined by comparing the transient energy function and critical energy at the end of disturbance.

For the n-unit system, the generator adopts the typical second-order model, neglects the excitation system dynamics, the prime mover and governor dynamics, the network linearity, and the load linearity (constant impedance). Assuming that the load impedance has been included in the node admittance matrix, the generator X'_d has also been included in the node admittance matrix, the system node admittance matrix eliminates the load nodes and network nodes, and there are only the nodes in the generator, the differential equation of the ith generator is described as

$$\begin{cases} M_i \frac{\text{d}\omega_i}{\text{d}t} = P_{\text{m}i} - P_{\text{ei}} \\ \frac{\text{d}\delta_i}{\text{d}t} = \omega_i \end{cases}, i = 1, 2, \ldots, n \tag{5.2}$$

Wherein, ω is the deviation between angular velocity and synchronous speed of rotors; δ is rotor angle; $P_{\text{m}} = \text{const.}$ is mechanical power and P_e is electromagnetic power; and

$$P_{\text{ei}} = \text{Re}\dot{E}_i \overset{*}{I}_i = \text{Re}\left(\dot{E}_i \sum_{j=1}^{n} \overset{*}{Y}_{ij} \overset{*}{E}_j \right)$$

$$= E_i^2 G_{ii} + \sum_{\substack{j=1 \\ j \neq i}}^{n} \left(E_i E_j B_{ij} \sin \delta_{ij} + E_i E_j B_{ij} \cos \delta_{ij} \right) \tag{5.3}$$

Wherein, $\dot{E}_i = E_i \angle \delta_i$, $\dot{E}_j = E_j \angle \delta_j$, $\delta_{ij} = \delta_i - \delta_j$, $Y_{ij} = G_{ij} + \mathrm{j}B_{ij}$ is the element of admittance matrix.

If $E_i E_j B_{ij} = C_{ij}$, $E_i E_j G_{ij} = D_{ij}$ are defined, P_{ei} in Eq. (5.3) is

$$P_{ei} = E_i^2 G_{ii} + \sum_{\substack{j=1 \\ j \neq i}}^{n} \left(C_{ij} \sin \delta_{ij} + D_{ij} \cos \delta_{ij} \right) \tag{5.4}$$

Equations (5.2) and (5.4) are the complete dynamic models of the system.

The kinetic energy of a multi-unit system is defined as

$$V_k = \sum_{i=1}^{n} \frac{1}{2} M_i \omega_i^2 \tag{5.5}$$

During the stable state, $\omega_i = 0$, $V_k = 0$. During fault clearing (disturbance end), the kinetic energy of the system is

$$V_k|_c = \sum_{i=1}^{n} \frac{1}{2} M_i \omega_{ci}^2 = \sum_{i=1}^{n} \int_{\delta_{0i}}^{\delta_{ci}} M_i \frac{\mathrm{d}\omega_i}{\mathrm{d}t} \mathrm{d}\delta_i = \sum_{i=1}^{n} \int_{\delta_{0i}}^{\delta_{ci}} \left(P_{mi} - P_{ei}^{(2)} \right) \mathrm{d}\delta_i \tag{5.6}$$

Wherein, $P_{ei}^{(2)}$ corresponds to the system node admittance matrix in case of a fault, and the calculation formula is shown in Eq. (5.4).

At the same time, the kinetic energy of a multi-unit system is defined as

$$V_p = \sum_{i=1}^{n} \int_{\delta_{si}}^{\delta_{ci}} \left(P_{ei}^{(3)} - P_{mi} \right) \mathrm{d}\delta_i \tag{5.7}$$

Wherein, δ_s is the SEP after the fault, as the reference point of potential energy; $P_{ei}^{(3)}$ corresponds to the system node admittance matrix after fault clearing (disturbance end), and the calculation formula is shown in Eq. (5.4). In this method, the system potential energy during fault clearing is

$$V_p = \sum_{i=1}^{n} \int_{\delta_{si}}^{\delta_{ci}} \left(P_{ei}^{(3)} - P_{mi} \right) \mathrm{d}\delta_i$$

$$= \sum_{i=1}^{n} \int_{\delta_{si}}^{\delta_{ci}} (E_i^2 G_{ii} - P_{mi}) d\delta_i + \sum_{i=1}^{n} \int_{\delta_{si}}^{\delta_{ci}} \sum_{\substack{j=1 \\ j \neq 1}}^{n} C_{ij} \sin\delta_{ij} d\delta_i$$

$$+ \sum_{i=1}^{n} \int_{\delta_{si}}^{\delta_{ci}} \sum_{\substack{j=1 \\ j \neq 1}}^{n} D_{ij} \cos\delta_{ij} d\delta_i$$

$$\stackrel{\text{def}}{=} V_{\text{pos}}|_{\text{c}} + V_{\text{mag}}|_{\text{c}} + V_{\text{diss}}|_{\text{c}} \tag{5.8}$$

Wherein, the first, second and third items on the right are called rotor position potential energy V_{pos}, magnetic potential energy V_{mag} and dissipative potential energy V_{diss} respectively.

In case the total energy of system V is equal to $V_{\text{k}}+V_{\text{p}}$, $V|_{\text{c}}$ is equal to $V_{\text{k}}|_{\text{c}}+V_{\text{p}}|_{\text{c}}$ during fault clearing, and can be calculated as per Eqs. (5.6) and (5.8). In the actual calculation, first use the step-by-step integration method to calculate ω_{c} and δ_{c} in the time domain, and then calculate $V_{\text{k}}|_{\text{c}}$, namely, $\sum_{i=1}^{n} \frac{1}{2} M_i \omega_{ci}^2$, then calculate the potential energy $V_{\text{p}}|_{\text{c}}$ as per Eq. (5.8) (calculate δ_{s} using the Newton-Raphson method—it can be calculated as per the power balance equation $P_{\text{e}i}^{(3)} = P_{\text{m}i}$).

The critical energy of the system can be approximately the potential energy at the UEP (δ_{u}), namely,

$$V_{\text{cr}} \approx V_{\text{p}}|_{\text{u}} = V_{\text{poss}}|_{\text{u}} + V_{\text{mag}}|_{\text{u}} + V_{\text{diss}}|_{\text{u}} \tag{5.9}$$

The calculation formula is similar to $V_{\text{p}}|_{\text{c}}$, as long as δ_{c} is changed to δ_{u}, but the problem is how to calculate δ_{u}.

At present, the calculation methods of critical energy mainly include the relevant unstable equilibrium point (RUEP), potential energy boundary surface (PEBS), and extended equal area criteria (EEAC) [4–6], etc. Once both V_c and V_{cr} are obtained, the stability margin calculation and transient stability analysis can be performed according to the normalized stability.

$$\Delta V \frac{V_{\text{cr}} - V_{\text{c}}}{V_{\text{k}}|_{\text{c}}} \tag{5.10}$$

The accuracy of transient stability analysis with energy function and critical energy in synchronous coordinates is often poor. The research shows that this is mainly caused by the unreasonable observation coordinate system. Therefore, the coordinate of the center of inertia (COI) is usually used in practical analysis.

The equivalent rotor angle δ_{COI} of COI of the system is defined as the weighted average value of each rotor angle, and the weight coefficient is M_i, namely, the inertia time constant of each generator, thus

$$\delta_{\text{COI}} = \frac{1}{M_{\text{T}}} \sum_{i=1}^{n} M_i \delta_i \tag{5.11}$$

Wherein, $M_T = \sum_{i=1}^{n} M_i$.

Similarly, the equivalent velocity ω_{COI} of COI is

$$\omega_{\text{COI}} = \frac{1}{M_{\text{T}}} \sum_{i=1}^{n} M_i \omega_i \tag{5.12}$$

Wherein, ω_i is the deviation from the synchronous speed.

Clearly

$$\frac{\mathrm{d}\delta_{COI}}{\mathrm{d}t} = \omega_{\text{COI}} \tag{5.13}$$

Then, in COI coordinates, the rotor angle and rotor angular velocity of each generator are defined as

$$\begin{cases} \theta_i = \delta_i - \delta_{\text{COI}} \\ \widetilde{\omega}_i = \omega_i - \omega_{\text{COI}} \end{cases} \tag{5.14}$$

Then the motion equation of COI is

$$\begin{cases} M_T \frac{d\omega_{COI}}{dt} = \sum_{i=1}^{n} (P_{mi} - P_{ei}) \overset{def}{=} P_{COI} \\ \frac{d\delta_{COI}}{dt} = \omega_{COI} \end{cases} \tag{5.15}$$

Wherein, P_{COI} is the accelerating power of COI.

The definition of transient energy in COI coordinates is similar to that in synchronous coordinates, namely,

$$\begin{aligned} V =& V_{\text{k}} + V_{\text{P}} \\ =& \sum_{i=1}^{n} \frac{1}{2} M_i \widetilde{\omega}_i^2 + \sum_{i=1}^{n} \int_{\theta_{si}}^{\theta_{ci}} -\left(P_{\text{mi}} - P_{\text{ei}} - \frac{M_i}{M_{\text{T}}} P_{\text{COI}}\right) \mathrm{d}\theta_i \end{aligned} \tag{5.16}$$

During fault clearing, if $\theta_s \rightarrow \theta_c$ is set as a linear path, then

$$\begin{aligned} V_{\text{c}} =& V_{\text{k}}|_{\text{c}} + V_{\text{P}}|_{\text{c}} \\ =& \sum_{i=1}^{n} \frac{1}{2} M_i \widetilde{\omega}_i^2|_c + \sum_{i=1}^{n} (-P_i)(\theta_{ci} - \theta_{si}) \end{aligned}$$

$$-\sum_{i=1}^{n-1}\sum_{j=i+1}^{n} C_{ij}\left(\cos\theta_{ij}^{(c)} - \cos\theta_{ij}^{(s)}\right)$$

$$+\sum_{i=1}^{n-1}\sum_{j=i+1}^{n} D_{ij}\frac{a}{b}\left(\sin\theta_{ij}^{(c)} - \sin\theta_{ij}^{(s)}\right) \tag{5.17}$$

Wherein, $P_i = P_{mi} - E_i^2 G_{ii}$; $\frac{a}{b} = \frac{(\theta_{ci}+\theta_{cj})-(\theta_{si}+\theta_{sj})}{\theta_{ij}^{(c)}-\theta_{ij}^{(s)}}$; the definition of C_{ij} and D_{ij} is the same as that of Eq. (5.4), and the admittance matrix parameters after fault clearing are used. In Eq. (5.17), the first, second, third and fourth items on the right are kinetic energy, position potential energy, magnetic potential energy and dissipative potential energy respectively.

The above is the classical network-based simplified energy function method. When the actual energy function method is applied to the voltage stability study, the structure maintaining energy function method is generally used. In this kind of method, the power network structure is maintained, the load can take into account the static and dynamic characteristics, and the generator can take into account the saliency effect, flux linkage attenuation and excitation control effect. One-axis model or two-axis model as well as Park equation model [1, 7] can be used for the generator.

The energy function method has been used in the study of analysis on transient stability of power systems for decades. With the wide application of computers, great progress has been made in this field. At present, the energy function method is one of the a few methods that can be used to study the problem of domain of attraction of SEP. After the development in recent years, this method has made a series of important progress. Although there is still a certain gap between the analysis accuracy and simulation method, this method is helpful to understand the essence of system stability in theory. The advantages of the energy function method include: ① nonlinearity is considered and the method is applied to large-scale systems; ② it is fast in calculation and it is unnecessary to conduct integration and calculate the rotor angle swing curve step by step, but can judge the stability through the energy criterion; ③ it can give the stability margin. However, the energy function method has two disadvantages: ① the model is simple; ② the analysis results tend to be conservative. Obviously, the energy function method cannot replace the time-domain simulation method, and they are complementary.

5.1.3 Nonlinear Dynamic Method

The power system is a nonlinear dynamics system, and minor changes of the system near the critical point will cause dramatic changes in system state; in addition, it is difficult for the traditional analysis method to describe this problem how the system changes when it exceeds stability limit. To ensure the safety of the power system, people research the new methods of analyzing and controlling nonlinear interaction, such as central manifold theory, bifurcation theory and chaos theory. The most studied one is bifurcation theory.

Bifurcation is a phenomenon of the nonlinear system. If a dynamical system is structurally unstable, any small and proper disturbance will cause sudden changes in topological structure of the system, which is called bifurcation. Bifurcation analysis mainly focus on the fact that the bifurcation method has been used to study the structural stability of the system in the past two decades when dynamic characteristics of solutions described by a set of differential equations are related to the value of parameters contained in the equation, and change as the parameter value changes. Structural stability is that topology of system equilibrium points is constant under small parameters quasi-static changes. Similarly, structural stability is that equilibrium points still keep stability after the system parameters change slowly. The bifurcation theory builds the relationship between system bifurcation and system stability, and develops a whole set of theoretical and numerical methods directed at the system differential-algebraic equation model.

The bifurcation theory mainly includes static and dynamic aspects: static bifurcation studies static equations

$$f(x,\mu) = 0, x \in U \subseteq \mathbf{R}^n, \mu \in J \subseteq \mathbf{R}^m \tag{5.18}$$

The number of solutions abruptly changes with the change of the parameter μ; dynamic bifurcation studies dynamic equations

$$\dot{x} = f(x,\mu), x \in U \subseteq \mathbf{R}^n, \mu \in J \subseteq \mathbf{R}^m \tag{5.19}$$

The topology of solutions abruptly changes with the change of the parameter μ.

The solution of Eq. (5.18) is the equilibrium point of Eq. (5.19), so static bifurcation mainly studies the bifurcation problem of the equilibrium point. The dynamic bifurcation studies not only the bifurcation problem of the equilibrium point but also other bifurcation problems. Actually, the dynamic bifurcation includes static bifurcation.

Nonlinear systems may be in multiple bifurcation forms under the change in parameters. Generally, power system voltage stability is mainly influenced by three types of bifurcation phenomena. The closure of the three bifurcation interfaces constitutes the stability domain of power system [8–10].

(1) Saddle node bifurcation (SNB) is a static bifurcation, which was first proposed by Kwatny. The bifurcation is related to monotonic instability of power systems.
(2) Hopf bifurcation (HB) is a dynamic bifurcation, which was first proposed by Abed. The bifurcation is related to oscillatory instability of power systems.
(3) Singularity induced bifurcation (SIB) is a dynamic bifurcation, which was first proposed by Zaborszky and further discussed by Venkatasubramanian, Beardmore, and others. References [11–13] have supplied this proof by applying the central manifold theory to the ordinary differential equation set of singularity

disturbance corresponding to the differential-algebraic equation set (DAE). Some studies have shown that this bifurcation is also related to monotonic instability of power systems.

1. SNB

Consider a single parameter cluster of ordinary differential equation

$$\dot{x} = f(x, \mu) \tag{5.20}$$

meet the following equilibrium conditions:

$$f\left(x^{*}, \mu\right) = 0 \tag{5.21}$$

An SNB is an intersection of two branches of the equilibrium point. On this bifurcation, the equilibrium point becomes a saddle node. So, the bifurcation is named. In this point, Jacobian matrix f_{x} must be of singularity. So, the requirement of SNB is given by the equilibrium Eq. (5.21) and singularity condition.

$$det\ f_{x}(x^{*}, \mu) = 0 \tag{5.22}$$

Not all points that satisfy the requirement are SNB points. To illustrate the sufficient conditions, we consider a scalar system. For such a system, the sufficient conditions of saddle node bifurcation are as follows:

$$f(x^{*}, \mu) = 0 \tag{5.23a}$$

$$\frac{\partial f}{\partial x} = 0 \tag{5.23b}$$

$$\frac{\partial f}{\partial u} \neq 0 \tag{5.23c}$$

$$\frac{\partial^{2} f}{\partial x^{2}} \neq 0 \tag{5.23d}$$

Equations (5.23c) and (5.23d) are generally called transversal conditions. The previous transversal condition (5.23c) ensures a smooth local function $\mu = h(x)$ at the bifurcation point $\left(\mu_{0}, x_{0}^{*}\right)$. In geometric terms, the equilibrium manifold (5.23a) intersects the line $x = x_{0}^{*}$ transversely. The latter transversal condition (5.23d) implies that this equilibrium manifold is locally maintained on one side of the line $\mu = \mu_{0}$.

In a multi-variable system, two equilibrium points coincide and disappear (or appear simultaneously) at an SNB. One equilibrium point has a positive eigenvalue, and the other equilibrium point has a negative eigenvalue. Both become zero at the bifurcation point.

2. HB

As we know, after the complex eigenvalue crosses the virtual axis on the complex plane due to parameter changes, the SEP becomes unstable. In nonlinear systems, this type of oscillation instability is related to the HB.

In an SNB, an approaching unstable equilibrium point makes domain of attraction of a stable equilibrium point shrink. When the two equilibrium points coincide and disappear, the stability eventually disappears. In an HB, the stability of this equilibrium point disappears because of the interaction between an equilibrium point and a limit cycle. According to the characteristics of this interaction, two types of HB are as follows.

(1) Subcritical HB: an unstable limit cycle exists before this bifurcation, shrinks and disappears. When this unstable limit cycle coincides with a stable equilibrium point at this bifurcation point, after bifurcation, the stable equilibrium point becomes unstable, resulting in a gradually increasing oscillation.
(2) Supercritical HB: a stable limit cycle is generated at the bifurcation point; a stable equilibrium point becomes unstable with the oscillation of increasing amplitude, so that this oscillation is eventually absorbed by a stability limit cycle.

The requirement for an HB is the existence of an equilibrium point with pure imaginary eigenvalues. This condition is not easily built as zero eigenvalue condition. Zero eigenvalue condition is easy as long as the determinant of the system matrix is zero. Most equilibrium points with pure imaginary eigenvalues are HB points. But, similar to SNB, there may be exceptions. After it becomes zero, the real part of this critical eigenvalue pairs does not change the sign, so these points are not HN points.

3. Singular induced bifurcation

The mathematical model of the power system stability problem is usually expressed by DAE:

$$\begin{cases} \dot{x} = f(x, y) \\ 0 = g(x, y) \end{cases} \tag{5.24}$$

Wherein, f refers to the dynamic characteristics of the generator and its excitation system, load, HVDC, and SVC, etc.; g is the power flow model of the network; x is the system state variable (such as generator rotor angle and angular velocity; y means the system algebraic variable (such as node voltage and phase angle).

The equilibrium point E of the system is a point that meets the following:

$$E = \{(x, y) | f(x, y) = 0, g(x, y) = 0\} \tag{5.25}$$

The singularity surface of the system is

$$S = \left\{(x, y) | f(x, y) = 0, \det g_y(x, y) = 0\right\} \tag{5.26}$$

Wherein, $g_y(x, y) = \partial g(x, y)/\partial y$. The point in S is called singularity point. When the dynamic state of the system moves to this point, it is called singularity induced bifurcation. Generally speaking, all kinds of references agree on this criterion of g_y singularity, and the main difference lies in the different models and the different definitions of some singularity points.

In view of the small disturbance linearization equation of Eq. (5.24) at the equilibrium point:

$$\begin{cases} \Delta \dot{x} = f_x \Delta x + f_y \Delta y \\ 0 = g_x \Delta x + g_y \Delta y \end{cases} \tag{5.27}$$

Wherein, $f_x = \partial f/\partial x, f_y = \partial f/\partial y$, $g_x = \partial g/\partial x$, $g_y = \partial g/\partial y$, all values are taken at the current operating point.

When g_y is not singular, the variable Δy is eliminated from Eq. (5.27) to obtain a simplified small disturbance linearization equation:

$$\Delta \dot{x} = \left[f_x - f_y\left(g_y\right)^{-1} g_x\right] \Delta x = A_g \Delta x \tag{5.28}$$

Wherein, $A_g = f_x - f_y\left(g_y\right)^{-1} g_x$ is commonly referred to as the system state matrix, which is the core of small disturbance stability analysis in power systems, and the existence of the solution of Eq. (5.28) also lays the foundation of transient stability simulation in power systems. These methods usually imply the assumption that g_y is not singular.

However, if the assumption that g_y is not singular is excluded, it can be seen from Eq. (5.27) that, when g_y approaches singularity, a small change of Δx will lead to an infinite change of some components in Δy, or a finite velocity change of Δx will lead to an infinite velocity change of some components in Δy. Geometrically, it will be clearer as shown in Fig. 5.2.

The curve in Fig. 5.2 represents the algebraic equation of the system. The system operation must always meet the constraints of the algebraic equation. The differential equation can be visualized as a "tension" on the state variable x. At the normal point, under the combined action of this tensile and algebraic constraint, the motion directions of x and y will be tangent to the algebraic constraint plane; as the current point tends to the SIB point, the tangent plane of the constraint plane tends to be horizontal, and y needs to move more to meet the conditions of the algebraic constraint under the same x amount of movement; when the current point is at the SIB point, the direction in which x is moving is perpendicular to the algebraic constraint plane. At this time, no matter what value Δx takes, Δy is always 0, so that the variables x and y cannot move, and the DAE has no solution. Therefore, the SIB point is sometimes also called the impasse point. The model cannot describe the time behavior of the system after reaching the singularity point because the system described by the DAE model is ill conditioned at the singularity point, which contradicts with "there must be a time solution for any physical system". Therefore, DAE model is no longer suitable for describing the actual power system at this time, because the preconditions for

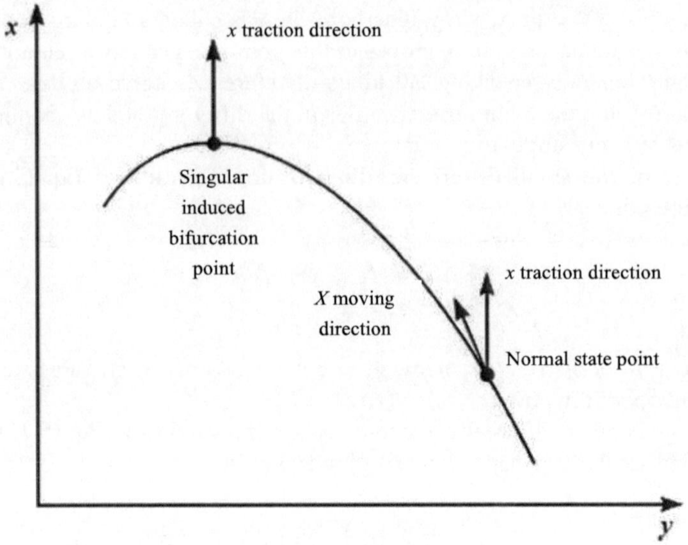

Fig. 5.2 Schematic Diagram of singularity induced bifurcation

describing the power system by DAE model is to ignore the fast dynamic process and fast dynamic stability of the system. A power system is a complex large-scale nonlinear system, which contains a large number of fast dynamic processes with different time constants. It is believed that the fast dynamic will quickly converge to the equilibrium state, that is, the constraint manifold determined by the algebraic equation in DAE model. Therefore, the differential equation describing this dynamic process is simplified to a stable algebraic equation. However, when the system is running near the singularity point, the fast dynamic process plays a leading role, so the DAE model fails and needs to be changed. At this time, the algebraic equation can be processed by introducing a small positive constant ε and using the singularity disturbance method to obtain a differential equation model in the form of singularity disturbance:

$$\begin{cases} \dot{x} = f(x, y) \\ \varepsilon \dot{y} = g(x, y) \end{cases} \tag{5.29}$$

Obviously, the singularity point of the original DAE model will disappear, but the small changes in the state variables of the disturbance system will still lead to large y changes near the original singularity point due to the small constant ε.

Although it is a more reasonable power system model to use singularity disturbance model and other models to describe the dynamic models of elements with differential equation, it is difficult to analyze the model. DAE Eq. (5.24) is a common model for transient stability research of power systems. It can describe the dynamic characteristics of physical system well without singularity, and it is relatively easy to make analysis. Therefore, a more practical approach is to still use Eq. (5.24) in the analysis, and consider the neglected dynamic states after detecting the system state near the singularity point to overcome the singularity.

At present, considerable progress has been made in the study of voltage stability using nonlinear dynamics, especially in the analysis of bifurcation theory. However, the research of nonlinear dynamics in voltage stability is still immature. Compared to the previous two research methods, the basic theoretical framework has not yet been determined and needs to be further researched. Therefore, further research of the theory and phenomenon of nonlinear dynamics is of great significance to further understanding the mechanism of voltage instability and accurately calculating the stability domain of power systems.

5.2 Criterion for Transient Voltage Stability

At present, how to accurately judge the transient voltage instability is still a difficult problem to be solved. The reason is that, on the one hand, the research methods and theories for understanding the transient voltage instability mechanism are not perfect, on the other hand, there is a connection between transient voltage stability and rotor angle stability. In the transient process after a fault, voltage collapse and rotor angle instability often occur at the same time. It is difficult to distinguish which is the primary factor causing system instability.

Because the state variables of all elements in the system change dramatically in the process of transient voltage collapse, the static voltage stability analysis method without consideration of the dynamic process of elements cannot be used to distinguish the transient voltage stability, and the corresponding static voltage stability index cannot be used to distinguish the transient voltage stability of the system. Other methods must be used to accurately distinguish the transient voltage stability.

5.2.1 Engineering Experience Criteria

Among all kinds of transient voltage stability criteria proposed, the engineering experience criteria are widely used at home and abroad, that is, after the system is disturbed, if a (some) bus voltage(s) is (are) lower than a certain voltage level continuously for longer than a certain period of time in transient process, it is regarded as voltage instability, as shown in Fig. 5.3. The voltage instability criteria used in the power systems in China, North America, Brazil and other countries or regions are all engineering experience criteria of this type.

In the past, a common experience criterion of voltage instability in China was based on the magnitude and duration of bus voltage drop at voltage center point, that is, the system or load voltage is regarded to be unstable if the duration (generally 1 s)

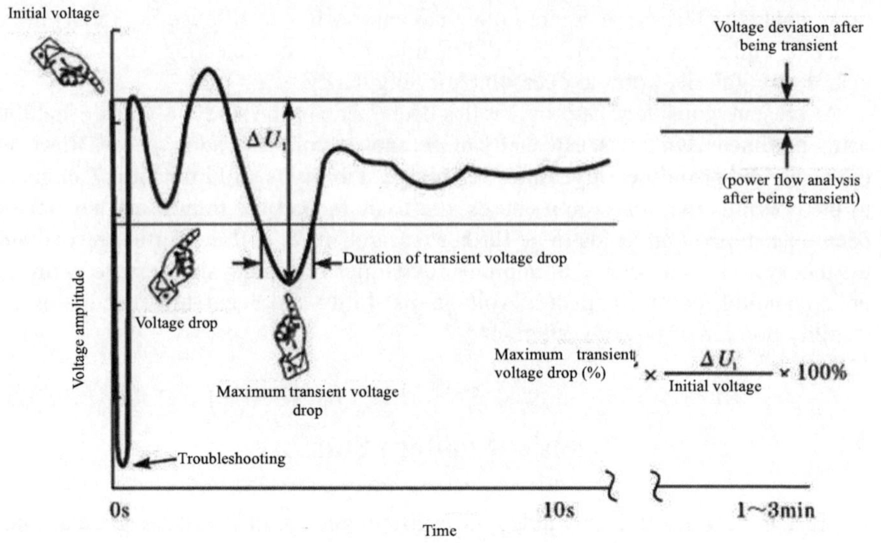

Fig. 5.3 Criterion for transient voltage stability

of bus voltage drop at the voltage center point of the system in a dynamic process is lower than the limit value (generally 0.75 p.u.).

According to the provisions in "executive standard" documents of North American Electric Reliability Council/Western Electricity Coordinating Council (NERC/WECC) issued in April 2003, for the following four types of disturbances (Refer to the Level 4 Classification in *NERC Planning Standard* issued by NERC in 1997 for specific classification.), namely, (A) no emergency; (B) single element loss resulting from single fault; (C) two or more (multiple) elements losses resulting from single or multiple faults; (D) two or more (multiple) elements losses or successive equipment decommissioning operations resulting from extreme faults, the voltage fluctuation is required as follows for different disturbances:

(1) For Type B disturbance: The voltage drop at the load bus or non-load bus shall not exceed 25 or 30%. The time when the voltage drop at the load bus exceeds 20% shall not exceed 20 cycles.
(2) For Type C disturbance: The voltage drop at any bus shall not exceed 30%. The time when the voltage drop at the load bus exceeds 20% shall not exceed 40 cycles.
(3) For Type D disturbance: There are no clear standards for voltage drop/sag.

The Brazilian voltage stability standard is formulated by the National Electrical System Operator (ONS), and the specific contents are as follows: The voltage in the first oscillation cycle after a fault shall not be lower than 60% of the rated value, and the voltage must be maintained at higher than 80% in the subsequent oscillation process. The standard is applicable to any event or switching under the same simulation conditions.

In theory, the engineering experience method cannot be used to accurately distinguish voltage instability. The main reason for this kind of criterion is that there was no method to accurately distinguish transient voltage instability at that time, so, from the perspective of avoiding the loss of load, determining the low voltage level and duration that the load can bear is the criteria for distinguishing transient voltage instability. In addition, based on the experience, it can be regarded that, as long as the voltage is lower than a certain value (which is set as U_c) for over a certain period of time (which is set as T_c), and is no longer capable of returning to the normal or higher voltage level (resulting in load loss) by virtue of the dynamic characteristics of the system itself, the voltage value U_c and the duration T_c can be regarded as the criteria for distinguishing transient voltage instability.

In practical application, the method for determining the specific U_c and T_c is a difficulty of this kind of criterion, because it can be found from a large number of simulation calculations that the voltage drop, recovery, minimum voltage, low voltage duration and others of each bus in the transient process are closely related to the factors of the system such as startup mode, power flow level, voltage level and load type. In some cases, the system can eventually recover to normal or higher voltage level after a process of bearing low voltage for a long time; in other cases, the system's ability to bear low voltage will be weaker. Similarly, some types of loads can bear low voltage levels for extended period of time, while others cannot. In a word, the engineering experience method is a kind of expedient method when there is no accurate voltage stability judgment method. It is based on the previous calculation and operation experience to a great extent. It is very difficult to determine the threshold value of the criterion by theoretical derivation or simulation.

5.2.2 *Theoretical Criteria*

In addition to engineering experience criteria, there are many methods and corresponding theoretical criteria in the process of research and development currently. The author's latest research results are introduced below.

1. Theoretical derivation of voltage stability criteria

According to the definition of voltage stability, the voltage stability is determined according to the power demand of the load and the ability of the system to provide power to the load, in which the dynamic characteristics of the load play a key role. The Thevenin equivalent system shown in Fig. 5.4 is used to make a brief introduction [14].

In Fig. 5.4, $\dot{E}$ and $\dot{Z}_s$ refer to the Thevenin equivalent potential and impedance at the system side, $\dot{U}_L$ refers to the load node voltage (assume that U_L and θ respectively

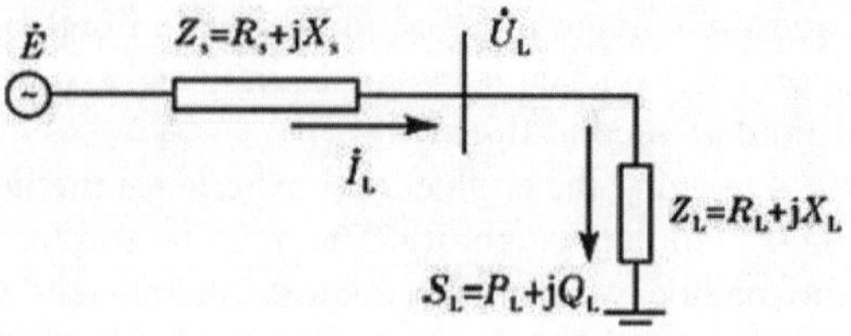

Fig. 5.4 Thevenin equivalent system

are load voltage magnitude and phase angle), and $\dot{I}$, Z_L, P_L, and Q_L respectively refer to the current, load impedance, active and reactive power of load.

(1) Load changes according to constant power factor

Assuming that the Thevenin equivalent potential and impedance at the system side are constant and the power factor of the load impedance is cos ϕ, the load impedance is indicated in the following form:

$$Z_L = R_L + jX_L = R_L + jR_L \tan\phi \tag{5.30}$$

The active power consumed by the load is as follows:

$$P_L = R_L I_L^2 = \frac{R_L E^2}{(R_s + R_L)^2 + (X_s + R_L \tan\phi)^2} \tag{5.31}$$

Calculate P_L extremum under R_L change:

$$\frac{dP_L}{dR_L} = \frac{E^2\left(R_s^2 + X_s^2 - R_L^2 - R_L^2 \tan^2\phi\right)}{\left[(R_s + R_L)^2 + (X_s + R_L \tan\phi)^2\right]^2} = 0 \tag{5.32}$$

We can get

$$R_L^2 + R_L^2 \tan^2\phi = R_s^2 + X_s^2 \tag{5.33}$$

i.e.,

$$|Z_L| = |Z_s| \tag{5.34}$$

By the monotonicity of P_L, it can be proved that, when $|Z_L| = |Z_s|$ is met, the active power of the load is the maximum under the condition that the power factor change of load is changed.

For the system shown in Fig. 5.4, assuming that $E = 1$ p.u., $R_s = 0.01$ p.u., $X_s = 0.5$ p.u. and $\tan\phi = 0.3$. As R_L changes, the change rule of the active power P_L of the load is shown in Fig. 5.4. According to $|Z_L| = |Z_s|$, $R_L = 0.479$ p.u., which also corresponds to the maximum active power of the load shown in Fig. 5.5. Therefore,

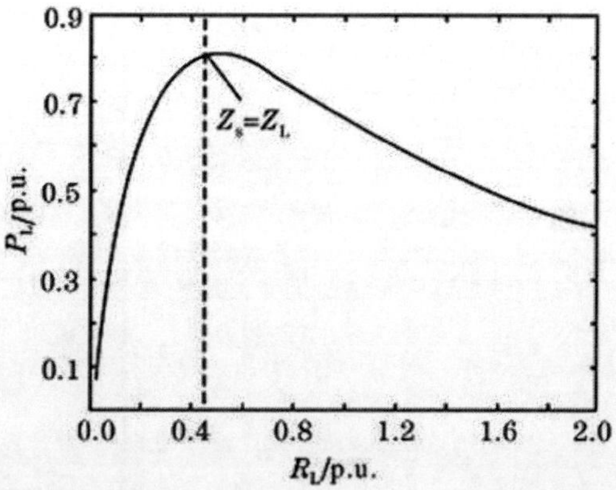

Fig. 5.5 Active power of load varying with constant power factor

in this load model, the $|Z_L| = |Z_s|$ corresponds to the critical point of the voltage stability.

(2) Constant load resistance R_L and changed reactance X_L

Assume that the Thevenin equivalent potential and impedance at the system side are constant, the load resistance R_L is constant, and only the load reactance X_L is changed. Calculate the change extremum of the active power of the load as follows:

$$\frac{\mathrm{d}P_L}{\mathrm{d}X_L} = \frac{-2R_L E^2(X_s + X_L)}{\left[(R_s + R_L)^2 + (X_s + X_L)^2\right]^2} = 0 \tag{5.35}$$

The load reactance under extremum is indicated by X_{LC}, meeting

$$X_{LC} = -X_s \tag{5.36}$$

By the monotonicity of P_L, it can be proved that P_L is the maximum when meeting Eq. (5.36). That is, $X_{LC} = -X_s$, which is the critical point of voltage stability.

For the system shown in Fig. 5.4, assuming that E = 1 p.u., R_L = 0.2 p.u., R_s = 0.01 p.u. and X_s = 0.5 p.u. As load reactance X_L changes, the change rule of the active power P_L of the load is shown in Fig. 5.6. In case of $X_{LC} = -0.5$ p.u., the transmission limit of the maximum active power of the corresponding system is 4.5351 p.u.

(3) Variational load resistance R_L and constant reactance X_L

Assuming that the Thevenin equivalent potential and impedance at the system side are constant, the load reactance X_L is constant, and only the load resistance R_L is variational, then the extremum of the active power of the load is calculated as follows:

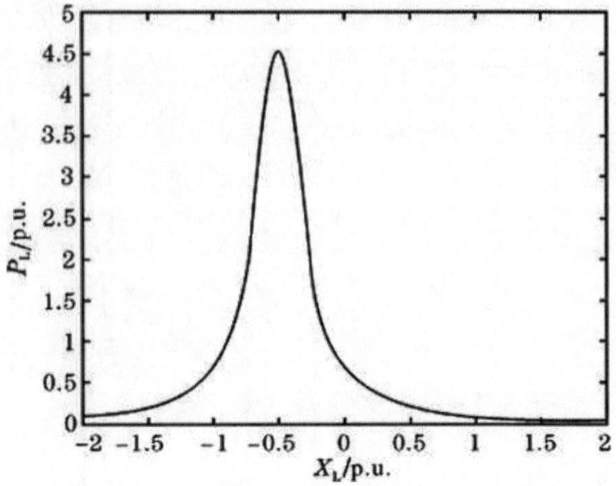

Fig. 5.6 Active power of load under constant R_L and variational X_L

$$\frac{dP_L}{dR_L} = \frac{E^2(R_s + R_L)^2 + (X_s + X_L)^2 - 2R_L(R_s + R_L)}{\left[(R_s + R_L)^2 + (X_s + X_L)^2\right]^2} = 0 \tag{5.37}$$

The load resistance under extremum is indicated by R_{LC}, and, by the monotonicity of P_L, it can be proved that the load resistance corresponds to the maximum active power of the load when meeting

$$R_{LC} = \sqrt{R_s^2 + (X_s + X_L)^2} \tag{5.38}$$

That is, $R_{LC} = \sqrt{R_s^2 + (X_s + X_L)^2}$ is the critical point of voltage stability.

For the system shown in Fig. 5.4, assume that $E = 1$ p.u., $X_L = 0.15$ p.u., $R_s = 0.01$ p.u. and $X_s = 0.2$ p.u. With the change of load resistance R_L, the change rule of the active power P_L of the load is shown in Fig. 5.7. In case of $R_{LC} = 0.35$ p.u., the active power limit that the system can supply is 1.3883 p.u.

(4) Load changes according to other rules

Assuming that the Thevenin equivalent potential and impedance at the system side are constant, the maximum load power (that is essentially conditional extremum) of the Thevenin equivalent circuit is determined according to the change rule of the load. If the changes of system conditions are not considered, the active power of the load is a binary function on load resistance R and reactance X, which is indicated as function $f\ (R, X)$

$$f = (R, X) = \frac{RE^2}{(R_s + R)^2 + (X_s + X)^2} \tag{5.39}$$

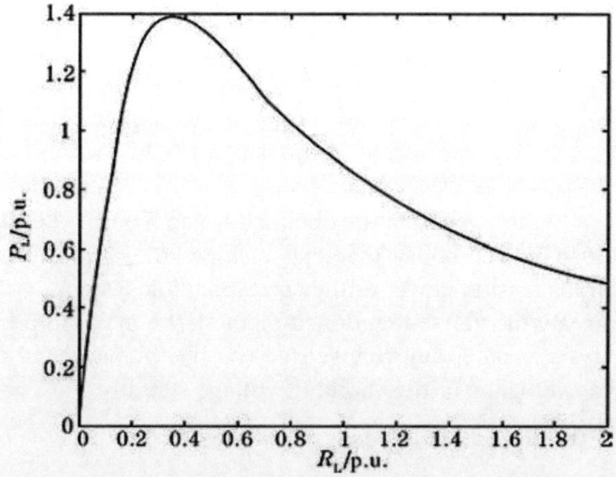

Fig. 5.7 Active power of load under variational R_L and constant X_L

The method of solving conditional extremum is used to set Lagrangian function $L\,(R_L, X_L)$, and λ refers to Lagrangian multiplier, that is,

$$L(R, X) = f(R, X) + \lambda\varphi(R, X) \tag{5.40}$$

Wherein, $\varphi(R, X)$ refers to the constraint condition corresponding to different load change rules. For example, when the load changes according to constant power factor, tanϕ is assumed to be constant power factor, and $\varphi(R, X)$ is defined as

$$\varphi(R, X) = \frac{X}{R} - \tan\phi \tag{5.41}$$

When the load change rule is that X is constant and R is variational, X_0 is assumed to be constant load reactance, and $\varphi(R, X)$ is defined as

$$\varphi(R, X) = X - X_0 \tag{5.42}$$

When the load change rule is that R is constant and X is variational, R_0 is assumed to be constant load resistance, and $\varphi(R, X)$ is defined as

$$\varphi(R, X) = R - R_0 \tag{5.43}$$

When the load change rule is that the load impedance changes according to the fixed direction angle α, R_{t0} and X_{t0} are assumed to be the load resistance and reactance at the last moment, then $\varphi(R, X)$ is defined as

$$\varphi(R, X) = \frac{X - X_{t0}}{R - R_{t0}} - \tan\alpha \tag{5.44}$$

Solution to the following equation set:

$$\begin{cases} f_R(R, X) + \lambda\varphi_{\mathrm{R}}(R, X) = 0 \\ f_X(R, X)\lambda\varphi_{\mathrm{X}}(R, X) = 0 \\ \varphi(R, X) = 0 \end{cases} \tag{5.45}$$

Let us substitute the obtained R and X into $f(R, X)$. The result is the maximum load power under the load change rule $\varphi(R, X) = 0$, and also the maximum transmission power of the corresponding system.

By the above-mentioned method, the maximum transmission power of the load system under any change rule can be obtained, and the corresponding point of the power is the critical point of voltage stability.

2. Equivalent impedance criteria

(1) Theoretical derivation

For the simple system shown in Fig. 5.4, when $|Z_{\mathrm{L}}| = |Z_{\mathrm{s}}|$ is met, the system has the unique solution of the voltage (see Sect. 4.3.5 for the specific derivation process). The point corresponds to the "waistline" position in Fig. 4.2. In a transient process, when the load impedance modulus changes from large to small, and finally it is equal to or less than the Thevenin equivalent impedance of the system, the operating point of the load will run from the upper hemisphere in Fig. 4.2 to the lower hemisphere, and the load bus voltage is in an unstable state. Therefore, the following criteria for transient voltage instability can be obtained: If the load impedance modulus $|Z_{\mathrm{L}}|$ in a transient process continuously decreases and is finally less than the Thevenin equivalent impedance modulus $|Z_{\mathrm{s}}|$ of the system, it can be regarded that transient voltage instability occurs (Note: According to the conclusion in Sect. 1, the fact that $|Z_{\mathrm{L}}|$ is equal to or less than $|Z_{\mathrm{s}}|$ is not a necessary condition for transient voltage instability, but only a sufficient condition, that is, when $|Z_{\mathrm{L}}|$ is larger than $|Z_{\mathrm{s}}|$, voltage instability may also occur, and specific conditions depend on load change trend).

(2) Tracing solution of Thevenin equivalent parameters

The premise of the application of equivalent impedance method is to obtain the change trace of the Thevenin equivalent impedance of the system in transient process. The related methods include the two-point method, the tracing algorithm for Thevenin equivalent parameters based on time-domain simulation, and the tracing algorithm for Thevenin equivalent parameters based on complete differential, etc.

(1) Two-point method

① Mathematical description

The two-point method is used to calculate the Thevenin equivalent parameters of the system by solving the power flow equations of two system operating points. In this method, the voltage magnitude U_k and load power $\dot{S}_k = P_k + \mathrm{j}Q_k$ of load bus are used as known quantities to calculate the Thevenin equivalent parameters $\dot{E}_k$ and Z_k. The algorithm is calculated as follows.

Based on the known quantities $\dot{U}_k$ and $\dot{S}_k = P_k + \mathrm{j}Q_k$, without loss of generality, let $\dot{U}_k = U_k\angle 0°$, the current phasor at k moment can be indicated as

$$\dot{I}_k = \frac{P_k - \mathrm{j}Q_k}{U_k} \tag{5.46}$$

Substitute Thevenin equivalent parameters $\dot{E}_k = E_k^{\mathrm{r}} + \mathrm{j}E_k^{\mathrm{i}}$ and $Z_k = R_k + \mathrm{j}X_k$ into Eq. (5.46) and write in the form of rectangular coordinates as follows.

$$(U_k + \mathrm{j}0) = \left(E_k^{\mathrm{r}} + jE_k^{\mathrm{i}}\right) - (R_k + \mathrm{j}X_k)\frac{P_k - \mathrm{j}Q_k}{U_k} \tag{5.47}$$

The real and imaginary parts of Eq. (5.47) are separately indicated as

$$\begin{cases} U_k E_k^{\mathrm{r}} - P_k R_k - Q_k X_k = U_k^2 \\ U_k E_k^{\mathrm{i}} - Q_k R_k - P_k X_k = 0 \end{cases} \tag{5.48}$$

There are 4 unknowns in Eqs. (5.48), namely, Thevenin equivalent parameters E_k^{r}, E_k^{i}, R_k, and X_k. The solution of the equation also requires the U_{k-1} and $\dot{S}_{k-1} = P_{k-1} + \mathrm{j}Q_{k-1}$ corresponding to another operating point, so as to list similar equation and simultaneously obtain the equations of two operating points:

$$\begin{bmatrix} U_{k-1} & 0 & -P_{k-1} & -Q_{k-1} \\ 0 & U_{k-1} & Q_{k-1} & -P_{k-1} \\ U_k & 0 & -P_k & -Q_k \\ 0 & U_k & Q_k & -P_k \end{bmatrix} \begin{bmatrix} E_k^{\mathrm{r}} \\ E_k^{\mathrm{i}} \\ R_k \\ X_k \end{bmatrix} = \begin{bmatrix} U_{k-1}^2 \\ 0 \\ U_k^2 \\ 0 \end{bmatrix} \tag{5.49}$$

Solving Eqs. (5.49) is to obtain the Thevenin equivalent parameters of the system from the load bus to the system.

② Accuracy analysis

For the two-point method, there is a basic assumption, that is, when the unknown quantities to be calculated are at the operating points $k-1$ and k, we can get

$$\begin{bmatrix} E_{k-1}^{\mathrm{r}} = E_k^{\mathrm{r}} \\ E_{k-1}^{\mathrm{i}} = E_k^{\mathrm{i}} \\ R_{k-1} = R_k \\ X_{k-1} = X_k \end{bmatrix} \tag{5.50}$$

Therefore, by this method, people can accurately calculate the parameters of Thevenin equivalent system under the condition that, at the two operating points, the internal load of Thevenin equivalent system is constant and only the load of bus side is variational; however, for the internal change of the equivalent system, because its basic assumption is not consistent with the actual situation, the parameters of Thevenin equivalent two-node system may have a large error with the actual values.

The calculation accuracy of the two-point method is analyzed in the following three cases.

Case 1: Only load parameters are variational, but the internal parameters of the system are constant, that is, assuming that (5.50) is valid, the more accurate Thevenin equivalent parameters can be obtained by Eq. (5.49).
Case 2: Load parameters are constant, but the internal parameters of the system are variational. At this moment, as it is assumed that (5.50) is no longer valid, the error of the results obtained by Eq. (5.49) is very large, and sometimes the results are obviously unreasonable.
Case 3: Load parameters and the internal parameters of the system are both variational. As it is assumed that (5.50) is no longer valid, the error of the results obtained by Eq. (5.49) is very large, sometimes the results are obviously unreasonable.

(2) Tracing algorithm for Thevenin equivalent parameters based on time-domain simulation [15]

① Mathematical description

The tracing algorithm for Thevenin equivalent parameters based on time-domain simulation can be used to calculate the Thevenin equivalent parameters of power systems at any bus in a transient process. This method uses the network algebraic equation generated in each calculation step of transient stability calculation program and the compensation method to calculate the Thevenin equivalent parameters of the system at any bus.

During time-domain simulation of power system, the following network equation must be solved at any time t to obtain the node voltage vector $\dot{U}_t$:

$$\begin{bmatrix} Y_{11} & \cdots & Y_{1i} & \cdots & Y_{1n} \\ \vdots & & \vdots & & \vdots \\ Y_{i1} & \cdots & Y_{ii} & \cdots & Y_{in} \\ \vdots & & \vdots & & \vdots \\ Y_{n1} & \cdots & Y_{in} & \cdots & Y_{nn} \end{bmatrix} \begin{bmatrix} \dot{U}_{t,1} \\ \vdots \\ \dot{U}_{t,i} \\ \vdots \\ \dot{U}_{t,n} \end{bmatrix} = \begin{bmatrix} \dot{I}_{t,1} \\ \vdots \\ \dot{I}_{t,i} \\ \vdots \\ \dot{I}_{t,n} \end{bmatrix} \tag{5.51}$$

Wherein, $Y = \begin{bmatrix} Y_{11} & \cdots & Y_{1i} & \cdots & Y_{1n} \\ \vdots & & \vdots & & \vdots \\ Y_{i1} & \cdots & Y_{ii} & \cdots & Y_{in} \\ \vdots & & \vdots & & \vdots \\ Y_{n1} & \cdots & Y_{in} & \cdots & Y_{nn} \end{bmatrix}$ refers to the admittance matrix of the system; $\widetilde{I}_t = \begin{bmatrix} \dot{I}_{t,1} \\ \vdots \\ \dot{I}_{t,i} \\ \vdots \\ \dot{I}_{t,n} \end{bmatrix}$ refers to the injection current vector of each node of the system at time t; $\widetilde{U}_t = \begin{bmatrix} \dot{U}_{t,1} \\ \vdots \\ \dot{U}_{t,i} \\ \vdots \\ \dot{U}_{t,n} \end{bmatrix}$ refers to the voltage vector of each node of the system at time t.

When the unit current is injected separately at node i, and the injection current of all other nodes is equal to 0, the following equation is solved:

$$\begin{bmatrix} Y_{11} & \cdots & Y_{1i} & \cdots & Y_{1n} \\ \vdots & & \vdots & & \vdots \\ Y_{i1} & \cdots & Y_{ii} & \cdots & Y_{in} \\ \vdots & & \vdots & & \vdots \\ Y_{n1} & \cdots & Y_{in} & \cdots & Y_{nn} \end{bmatrix} \begin{bmatrix} \dot{U}_{t,1_0} \\ \vdots \\ \dot{U}_{t,i_0} \\ \vdots \\ \dot{U}_{t,n_0} \end{bmatrix} = \begin{bmatrix} 0 \\ \vdots \\ 1 \\ \vdots \\ 0 \end{bmatrix} \tag{5.52}$$

The comprehensive impedance matrix $Z_{i\mathrm{T}}$ at equivalent node i can be obtained:

$$Z_{i\mathrm{T}} = \dot{U}_{t,i_0} \tag{5.53}$$

This process is only calculated when the network structure changes, so calculations are not done many times in the whole calculation process.

At time t, the Thevenin equivalent circuit of the system at node i is shown in Fig. 5.8.

In Fig. 5.8, $\dot{E}_{t,i\mathrm{Thev}}$ and $Z_{t,iThev}$ respectively refer to the Thevenin equivalent potential and impedance of the system found in node i at time t; $\dot{U}_{t,i}$ refers to the voltage at node i by transient stability calculation at time t; $Z_{\mathrm{ZL}i}$ refers to the impedance part of the load at node i; $i_{\mathrm{L}t,i}$ refers to the load current at node i,. $i_{\mathrm{L}t,i} = \dot{U}_{t,i}/Z_{\mathrm{ZL}i}$.

The compensation method is used to calculate open-circuit voltage $\dot{U}_{\mathrm{oc},i}$, that is, when node i is in open circuit, the load current passing through node i is equal to 0,

Fig. 5.8 Circuits of calculating Thevenin equivalent parameters of the system at node i

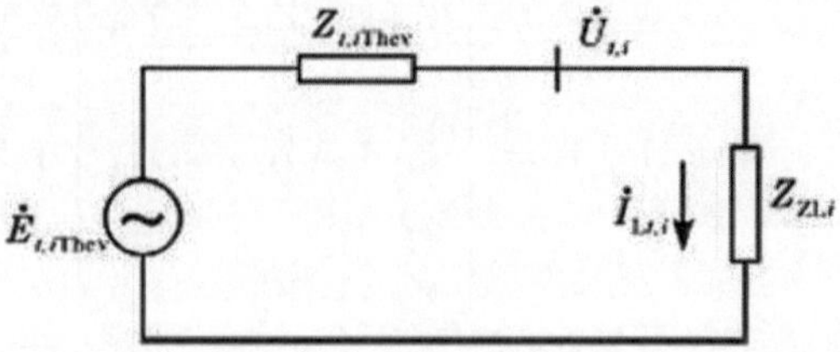

and an injection current magnitude $\Delta\dot{I}_{t,i}$ can be compensated at node i to calculate the variation of system node voltage, as shown in Fig. 5.9.

Obviously, $\Delta\dot{I}_{t,i} = -\frac{\dot{U}_{\text{oc},i}}{Z_{\text{ZL}i}}$. At this moment, if the current flowing through the impedance is equal to $\dot{I}_{\text{L}i} = -\Delta\dot{I}_{t,i}$, then

$$\dot{U}_{\text{oc},i} = -Z_{\text{ZL}i}\Delta\dot{I}_{t,i} \tag{5.54}$$

At the same time, based on the comprehensive impedance matrix $Z_{i\text{T}}$ determined above, it is found that the voltage variation at node i is

$$\Delta\dot{U}_{t,i} = -Z_{i\text{T}}\Delta\dot{I}_{t,i} \tag{5.55}$$

According to the superposition principle, the open-circuit voltage at node i is

$$\dot{U}_{\text{oc},i} = \dot{U}_{t,i} + \Delta\dot{U}_{t,i} = \dot{U}_{t,i} + Z_{i\text{T}}\Delta\dot{I}_{t,i} \tag{5.56}$$

Wherein, $\dot{U}_{t,i}$ refers to the voltage at node i obtained by transient stability calculation at time t.

Simultaneously solve Eqs. (5.52) and (5.56) to obtain

$$\begin{cases} \Delta\dot{I}_{t,i} = \frac{\dot{U}_{t,i}}{Z_{\text{ZL}i} - Z_{i\text{T}}} \\ \dot{U}_{\text{oc},i} = \frac{Z_{\text{ZL}i}}{Z_{\text{ZL}i} - Z_{i\text{T}}}\dot{U}_{t,i} \end{cases} \tag{5.57}$$

At this moment, the determined $\dot{U}_{\text{oc},i}$ refers to the Thevenin equivalent potential $\dot{E}_{t,i\text{Thev}}$ of the system at node i, as follows

Fig. 5.9 Thevenin equivalent circuit when solving open-circuit voltage at node i with compensation method

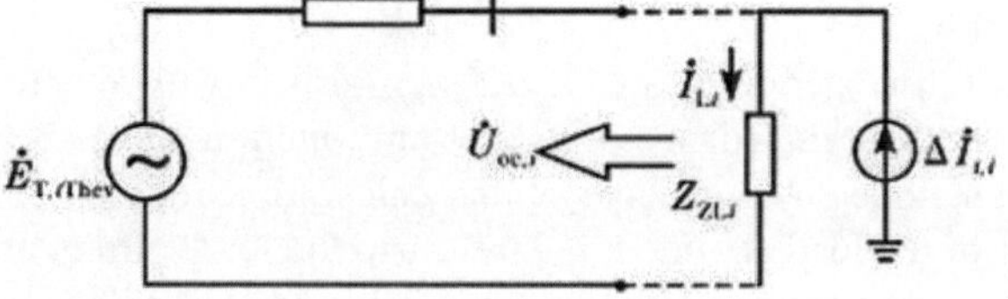

$$\dot{E}_{t,i\text{Thev}} = \dot{U}_{\text{oc},i} = \frac{Z_{\text{ZL}i}}{Z_{\text{ZL}i} - Z_{i\text{T}}}\dot{U}_{t,i} \tag{5.58}$$

Similarly, the short-circuit current $\dot{I}_{\text{sc},i}$ is also calculated according to the principle of compensation method, and the Thevenin equivalent circuit in case of short circuit of node i is shown in Fig. 5.10.

When node i is short-circuited, it is equivalent to adding an injection current magnitude $\Delta\dot{I}_{t,i'}$ at node i on the basis of the original network. According to the superposition principle, the voltage at node i in this moment is

$$\dot{U}_{t,i'} = \dot{U}_{t,i} + Z_{i\text{T}}\Delta I_{t,i'} \tag{5.59}$$

Wherein, $\dot{U}_{t,i'}$ refers to the voltage at node i after short circuit. When node i is short-circuited, let $\dot{U}_{t,i'} = 0$, then

$$\Delta\dot{I}_{t,i'} = -Z_{i\text{T}}^{-1}\dot{U}_{t,i} \tag{5.60}$$

According to the superposition principle, the short-circuit current at node i is

$$i_{\text{sc},i} = \frac{\dot{U}_{t,i}}{Z_{\text{ZL}i}} - \Delta\dot{I}_{t,i'} = \frac{\dot{U}_{t,i}}{Z_{\text{ZL}i}} + Z_{i\text{T}}^{-1}\dot{U}_{t,i} \tag{5.61}$$

According to above open-circuit voltage $\dot{U}_{\text{oc},i}$ and short-circuit current $\dot{I}_{\text{se},i}$ upon calculation, time t is obtained by solving for the ratio of the two. Thevenin equivalent impedance $Z_{t,i\text{Thev}}$ of the system at node i is shown as follows:

$$Z_{t,i\text{Thev}} = \frac{\dot{U}_{\text{oc},i}}{\dot{I}_{\text{sc},i}} \tag{5.62}$$

Similarly, at any time after the fault occurs, for different load nodes, repeat the above steps, to obtain Thevenin equivalent potential $\dot{E}_{\text{Thev}}$ and Thevenin equivalent impedance Z_{Thev} of the system at any load bus that changes with time.

As for the above Thevenin equivalent algorithm based on time-domain simulation, the power grid admittance matrix that is formed during the calculation of transient stability can be used to quickly and accurately obtain Thevenin equivalent parameters. It is properly operable and is suitable for the large disturbance inside the Thevenin

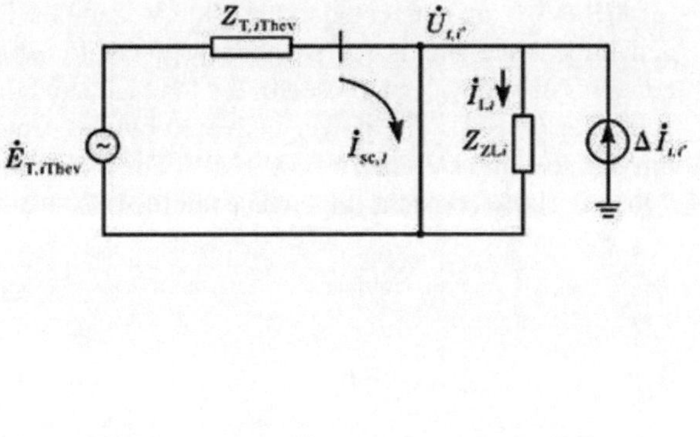

Fig. 5.10 Thevenin equivalent circuit when solving short-circuit current at node i with compensation method

equivalent system. The accuracy of Thevenin equivalent parameters upon calculation is very high.

However, in Thevenin equivalent parameters obtained using the above method, Thevenin equivalent impedance is constant when there is no switching operation (because the system admittance matrix generated by the transient stability program is constant when the network structure does not change), all changes of the system are reflected in the change of Thevenin equivalent potential. This is inconsistent with the physical system and it may be disadvantageous to further application of Thevenin equivalent parameters upon calculation. So, in practice, the following method is used to modify the system admittance matrix generated by the transient stability program at each calculation time before the above calculation: other loads except equivalent node i are regarded as impedance and merged into the system admittance matrix Y.

② Accuracy analysis

Analysis methods are as follows: the actual values of voltage, active power and reactive power of a bus (set as bus A) at any calculation step after the fault upon calculation by the transient stability simulation program are compared with the values of voltage, active power and reactive power of the bus upon calculation by the Thevenin equivalent parameters of the calculation step. If they are basically consistent, Thevenin equivalent parameters upon calculation are basically consistent with actual parameters. The equation for calculating the voltage, active power and reactive power of the bus A after the fault by using Thevenin equivalent parameters at the kth calculation step is as follows:

$$\begin{cases} U_k = \frac{E_{\text{Thev}}}{Z_{\text{Thev}}+Z_{\text{L}}} Z_{\text{L}} \\ P_k = \left(\frac{E_{\text{Thev}}}{Z_{\text{Thev}}+Z_{\text{L}}}\right)^2 R_{\text{L}} \\ Q_k = \left(\frac{E_{\text{Thev}}}{Z_{\text{Thev}}+Z_{\text{L}}}\right)^2 X_{\text{L}} \end{cases} \tag{5.63}$$

Wherein, U_k, P_k and Q_k are the voltages, active power, and reactive power of bus A upon calculation with Thevenin equivalent parameters of the kth calculation step; E_{Thev} and Z_{Thev} are the kth calculation step of Thevenin equivalent parameters; R_l and X_l are the bus load resistance and reactance of the kth calculation steep.

Taking 3-unit 10-node system as an example, the system topology is shown in Fig. 5.11. The introduction to equivalent system is as follows: 500 kV bus (Bus 6) supplies power to two loads in load area, industrial load (Bus 7) is connected to 500 kV load bus through OLTC transformer. But, the residential load and commercial load (Bus 10) are connected to the 500 kV load bus through two OLTC transformers and impedance stands for secondary power transmission systems. Lots of shunt compensation devices are used in the load area and there is a 1,600 MVA equivalent generator (Bus 3). The power delivered by two remote generators to the load area through five 500 kV lines is 5,000 MW. The main model used in the simulation: Bus 7 load is 100% constant impedance, and Bus 10 load is 100% motor load.

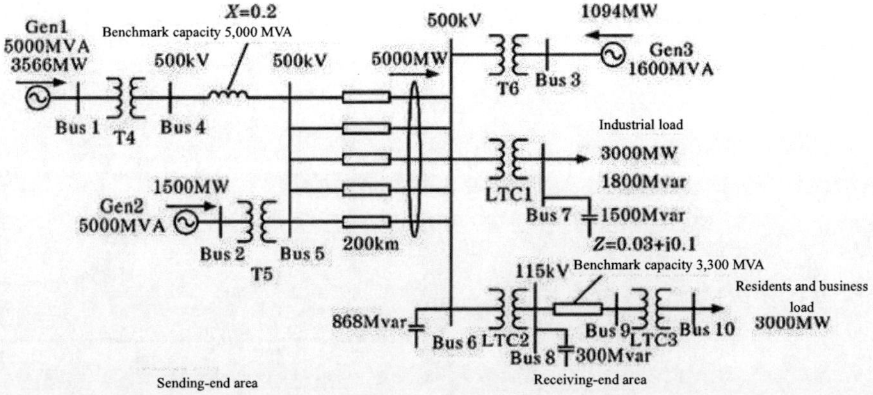

Fig. 5.11 Wiring diagram of 3-unit 10-node system

Select the Thevenin equivalent of the system from Bus 10. The disturbance is the three-phase N-1 fault on Bus 6 side of the 5th circuit between Bus 5 and Bus 6, the fault start time is 1.0 s, and the line fault clearing time is 1.0056 s.

The calculation results are shown in Figs. 5.12, 5.13 and 5.14 for bus voltage curve, line active power curve, and line reactive power curve respectively. In each figure, curve 1 is the real one upon transient stability simulation, and curve 2 is the one upon calculation by Thevenin equivalent system. According to FigS. 5.12, 5.13 and 5.14, the calculation method of Thevenin equivalent parameters based on time-domain simulation has very high calculation accuracy.

(3) Tracing algorithm for Thevenin equivalent parameters based on complete differential

The tracing algorithm for Thevenin equivalent parameters based on time-domain simulation can be used to calculate the accurate Thevenin equivalent parameters. But the limitation of this method is that the parameters of the entire network must

Fig. 5.12 Bus voltage curve

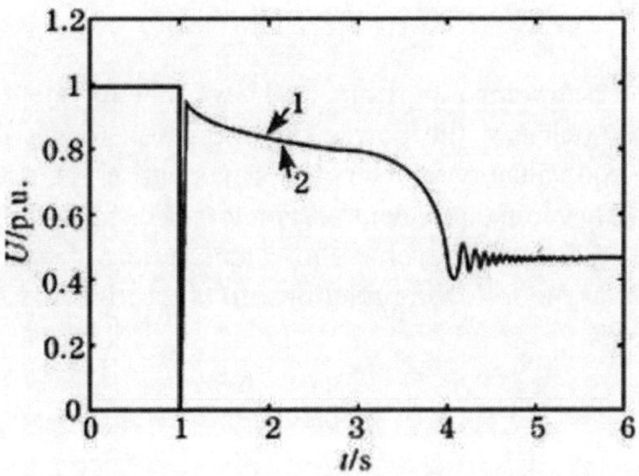

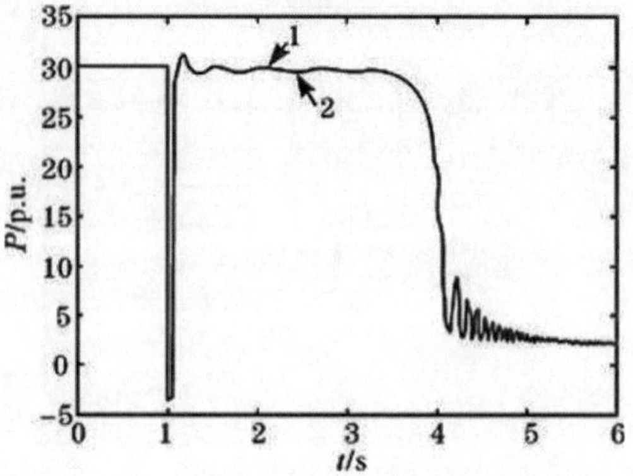

Fig. 5.13 Line active power curve

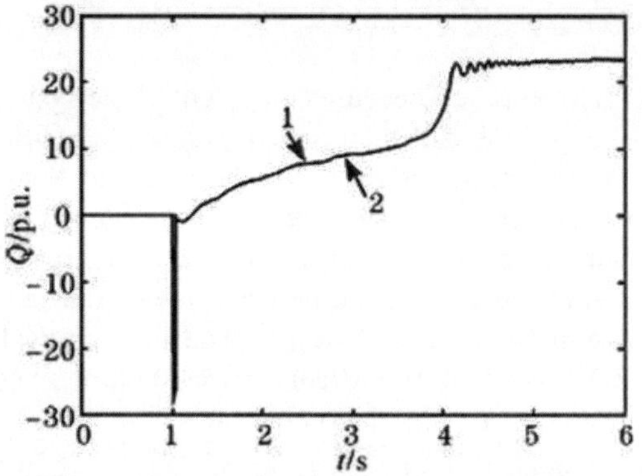

Fig. 5.14 Line reactive power curve

be known before the calculation, it is not suitable for equipment that only collects local electrical quantities. The tracing algorithm for Thevenin equivalent parameters based on complete differential can be free from the above limitation, and just rely on local electrical quantities and a small amount of external information to accurately calculate the Thevenin equivalent parameters of the system.

① Mathematical description

The tracing algorithm for Thevenin equivalent parameters based on complete differential uses the power flow equation to get the total differential of the Thevenin equivalent parameters to form equation set, and then solve the equation set to get the Thevenin equivalent parameters. The specific principle of this method is as follows

For the Thevenin equivalent system shown in Fig. 5.4, each electrical parameter has the following relationship (subscripts are omitted for each variable):

$$\dot{E} = \frac{P - \mathrm{j}Q}{\overset{*}{U}}(R + \mathrm{j}X) + \dot{U} \tag{5.64}$$

Wherein, $\dot{E}$ is Thevenin equivalent potential; $\dot{U}$ is the load bus terminal voltage; $P + jQ$ are the load active power and reactive power; $P + jX$ are Thevenin equivalent impedance; $\overset{*}{U}$ is the conjugate of $\dot{U}$. Without loss of generality, $let \dot{U} = U\angle 0°$. Separate the real part of Eq. (5.64) from the imaginary part. Then we can get

$$U^2 - UE_r + PR + QX = 0 \tag{5.65}$$

$$UE_i - PX + QR = 0 \tag{5.66}$$

Wherein, E_r and E_i are the real part and imaginary part of Thevenin equivalent potential.

The expressions of active power and reactive power can be obtained from Eqs. (5.65) and (5.66):

$$P = \frac{UE_r R + UE_i X - U^2 R}{R^2 + X^2} \tag{5.67}$$

$$Q = \frac{UE_r X + UE_i R - U^2 X}{R^2 + X^2} \tag{5.68}$$

Wherein, total differential of the Thevenin equivalent parameters and the load bus voltage magnitude (E_r, E_i, R, X and U) is calculated as follows with P and Q in the formula

$$\begin{aligned} dP &= \frac{\partial P}{\partial E_r} dE_r + \frac{\partial P}{\partial E_i} dE_i + \frac{\partial P}{\partial U} dU + \frac{\partial P}{\partial R} dR + \frac{\partial P}{\partial X} dX \\ &= \frac{UR}{R^2 + X^2} dE_r + \frac{UX}{R^2 + X^2} dE_i + \frac{E_r R + E_i X - 2RU}{R^2 + X^2} dU \\ &+ \frac{(UE_r - U^2)(R^2 + X^2) - 2R(UE_r R + UE_i X - U^2 R)}{(R^2 + X^2)^2} dR \\ &+ \frac{UE_i(R^2 + X^2) - 2X(UE_r R + UE_i X - U^2 R)}{(R^2 + X^2)^2} dX \end{aligned} \tag{5.69}$$

$$\begin{aligned} dQ &= \frac{\partial Q}{\partial E_r} dE_r + \frac{\partial Q}{\partial E_i} dE_i + \frac{\partial Q}{\partial U} dU + \frac{\partial Q}{\partial R} dR + \frac{\partial Q}{\partial X} dX \\ &= \frac{UX}{R^2 + X^2} dE_r + \frac{UR}{R^2 + X^2} dE_i + \frac{E_r X + E_i R - 2XU}{R^2 + X^2} dU \\ &+ \frac{(-UE_i)(R^2 + X^2) - 2R(UE_r X + UE_i R - U^2 X)}{(R^2 + X^2)^2} dR \\ &+ \frac{(UE_r - U^2)(R^2 + X^2) - 2X(UE_r X + UE_i R - U^2 X)}{(R^2 + X^2)^2} dX \end{aligned} \tag{5.70}$$

Replace differential equation with difference equation. Equations (5.69) and (5.70) can be written as

$$
\begin{aligned}
P_{k+1} - P_k = {} & \frac{U_k R_k}{R_k^2 + X_k^2}\left(E_{\mathrm{r}}^{k+1} - E_{\mathrm{r}}^k\right) + \frac{U_k X_k}{R_k^2 + X_k^2}\left(E_{\mathrm{i}}^{k+1} - E_{\mathrm{i}}^k\right) \\
& + \frac{E_{\mathrm{r}}^k R_k + E_{\mathrm{i}}^k X_k - 2R_k U_k}{R_k^2 + X_k^2}(U_{k+1} - U_k) \\
& + \frac{\left(U_k E_{\mathrm{r}}^k - U_k^2\right)\left(R_k^2 + X_k^2\right) - 2R_k\left(U_k E_{\mathrm{r}}^k R_k + U_k E_{\mathrm{i}}^k X_k - U_k^2 R_k\right)}{\left(R_k^2 + X_k^2\right)^2}(R_{k+1} - R_k) \\
& + \frac{U_k E_{\mathrm{i}}^k\left(R_k^2 + X_k^2\right) - 2X_k\left(U_k E_{\mathrm{r}}^k R_k + U_k E_{\mathrm{i}}^k X_k - U_k^2 R_k\right)}{\left(R_k^2 + X_k^2\right)^2}(X_{k+1} - X_k) \\
= {} & f_P\left(U_k, U_{k+1}, R_k, X_k, R_{k+1}, X_{k+1}, E_{\mathrm{r}}^k, E_{\mathrm{i}}^k, E_{\mathrm{r}}^{k+1}, E_{\mathrm{i}}^{k+1}\right)
\end{aligned}
\tag{5.71}
$$

$$
\begin{aligned}
Q_{k+1} - Q_k = {} & \frac{U_k X_k}{R_k^2 + X_k^2}\left(E_{\mathrm{r}}^{k+1} - E_{\mathrm{r}}^k\right) + \frac{U_k R_k}{R_k^2 + X_k^2}\left(E_{\mathrm{i}}^{k+1} - E_{\mathrm{i}}^k\right) \\
& + \frac{E_{\mathrm{r}}^k X_k + E_{\mathrm{i}}^k R_k - 2X_k U_k}{R_k^2 + X_k^2}(U_{k+1} - U_k) \\
& + \frac{\left(-U_k E_{\mathrm{i}}^k\right)\left(R_k^2 + X_k^2\right) - 2R_k\left(U_k E_{\mathrm{r}}^k X_k - U_k E_{\mathrm{i}}^k R_k - U_k^2 X_k\right)}{\left(R_k^2 + X_k^2\right)^2}(R_{k+1} - R_k) \\
& + \frac{\left(U_k E_{\mathrm{r}}^k - U_k^2\right)\left(R_k^2 + X_k^2\right) - 2X_k\left(U_k E_{\mathrm{r}}^k X_k - U_k E_{\mathrm{i}}^k R_k - U_k^2 X_k\right)}{\left(R_k^2 + X_k^2\right)^2}(X_{k+1} - X_k) \\
= {} & f_P\left(U_k, U_{k+1}, R_k, X_k, R_{k+1}, X_{k+1}, E_{\mathrm{r}}^k, E_{\mathrm{i}}^k, E_{\mathrm{r}}^{k+1}, E_{\mathrm{i}}^{k+1}\right)
\end{aligned}
\tag{5.72}
$$

Thus, according to Eqs. (5.65), (5.66), (5.71) and (5.72), the following equation set containing 6 equations can be obtained:

$$
\begin{cases}
U_k^2 - U_k E_{\mathrm{r}}^k + P_k R_k + Q_k X_k = 0 \\
U_k E_{\mathrm{i}}^k - P_k X_k + Q_k R_k = 0 \\
U_{k+1}^2 - U_{k+1} E_{\mathrm{r}}^{k+1} + P_{k+1} R_{k+1} + Q_{k+1} X_{k+1} = 0 \\
U_{k+1} E_{\mathrm{i}}^{k+1} - P_{k+1} X_{k+1} + Q_{k+1} R_{k+1} = 0 \\
P_{k+1} - P_k = f_p\left(U_k, U_{k+1}, R_k, X_k, R_{k+1}, X_{k+1}, E_{\mathrm{r}}^k, E_i^k, E_{\mathrm{r}}^{k+1}, E_{\mathrm{i}}^{k+1}\right) \\
Q_{k+1} - Q_k = f_q\left(U_k, U_{k+1}, R_k, X_k, R_{k+1}, X_{k+1}, E_{\mathrm{r}}^k, E_i^k, E_{\mathrm{r}}^{k+1}, E_{\mathrm{i}}^{k+1}\right)
\end{cases}
\tag{5.73}
$$

In the Eq. (5.73), there are 8 unknowns, namely, $R_k, R_{k+1}, X_k, X_{k+1}, E_{\mathrm{r}}^k, E_i^k, E_{\mathrm{r}}^{k+1}$, *and* E_{i}^{k+1}. Therefore, you must first give the value of any 2 unknowns, or give 2 equations, and then you can get the solutions of all unknowns. In practice, first, the initial value $R_0 + \mathrm{j}X_0$ of Thevenin equivalent impedance upon calculation by the dispatch center computer with the information of the entire network is sent to local equipment through communication facilities. Or the existing Thevenin equivalent method is used, as mentioned above, the initial value $R_0 + \mathrm{j}X_0$ of the system Thevenin equivalent impedance can be determined by the Eq. (5.49) under the condition of constant system potential and impedance constant load disturbance. And then, through Eq. (5.73), you can get Thevenin equivalent parameters at the next moment. After that, according to Thevenin equivalent impedance at the next moment, get the Thevenin equivalent parameters at the next moment. And so on, get a series of Thevenin equivalent parameters.

② Accuracy analysis

Still take the 3-unit 10-node system as an example. Select the Thevenin equivalent of the system from Bus 10. a disturbance with a sinusoidal power volatility of 100 MW and a frequency of 0.5 Hz is applied at Bus 7, that is, the potential and impedance of the Thevenin equivalent system change.

The tracing algorithm for Thevenin equivalent parameters based on complete differential equation needs to first give any 2 unknown initial values. During this calculation, the initial value of Thevenin equivalent impedance is $R_0 + jX_0$, which is obtained using tracing algorithm for Thevenin equivalent parameters based on time-domain simulation.

In the process of tracing algorithm for Thevenin equivalent parameters based on complete differential, the voltage, active power and reactive power at every calculated time interval at Bus 10 are calculated with a transient stability simulation program, and the calculated time interval is 0.01 s. Using the Thevenin equivalent impedance initial value $R_0 + jX_0$ and the voltage magnitude, active power and reactive power collected at Bus 10 at the first moment, Thevenin equivalent parameters at the first moment are calculated by Eq. (5.73). Use the Thevenin equivalent impedance initial value at the first moment, and then use Eq. (5.73) to calculate the Thevenin equivalent parameters at the second moment. And so on, a series of Thevenin equivalent parameters that change with time are obtained.

The calculation results are shown in Figs. 5.15, 5.16, 5.17. The figure simultaneously shows the equivalent parameters obtained by the two-point method. Figure 5.15 is Thevenin equivalent potential amplitude E_{eq}. Figure 5.16 is Thevenin equivalent

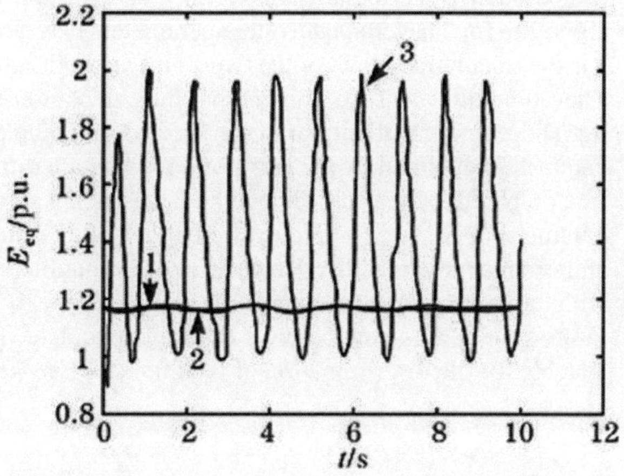

Fig. 5.15 Comparison of Thevenin equivalent potential magnitude

Fig. 5.16 Comparison of Thevenin equivalent resistance

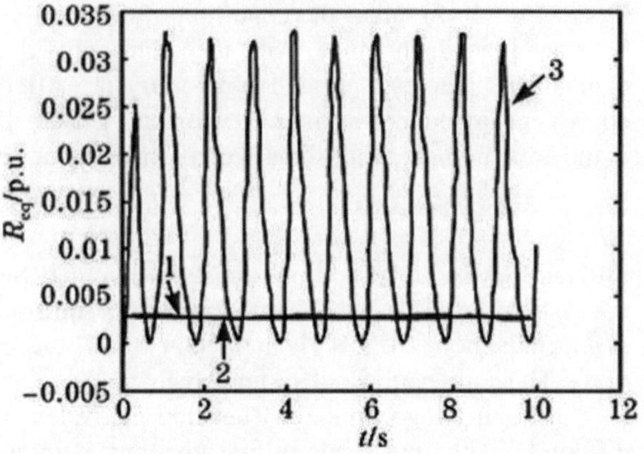

Fig. 5.17 Comparison of Thevenin equivalent reactance

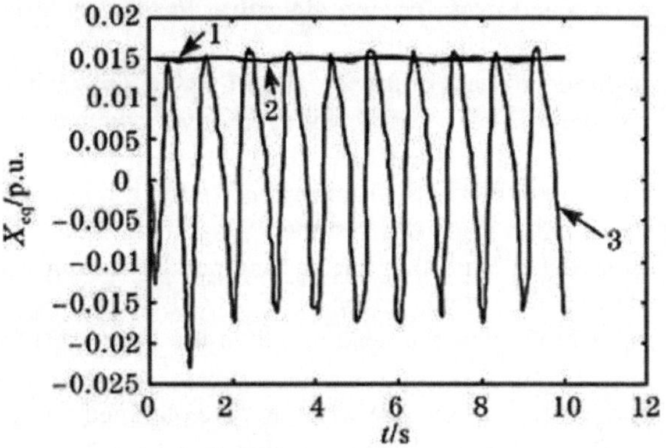

resistance R_{eq}. Figure 5.17 is Thevenin equivalent reactance X_{eq}. In each figure, curve 1 is about calculation results of tracing algorithm for Thevenin equivalent parameters based on time-domain simulation. Curve 2 is the calculation result of the tracing algorithm for Thevenin equivalent parameters based on complete differential. Curve 3 is the calculation result of the two-point method. From the results, it is close to real Thevenin equivalent parameters that the calculation result of the tracing algorithm for Thevenin equivalent parameters based on complete differential, but equivalent parameters obtained by the two-point method are different from the actual values.

Compared to the two-point method, the tracing algorithm for Thevenin equivalent parameters based on complete differential features much higher calculation accuracy and stronger practicability. However, the calculation error of this method will increase with the increase of disturbance. At the same time, if there are switching operations in the system, this method must restart to calculate. Therefore, this method is more suitable for the discrimination of medium and long-term voltage stability.

(3) Example simulation

The effectiveness of the equivalent impedance method is analyzed with the following two simulation examples.

(1) Example 1.

Data source: 3-unit 10-node system data, in which, Bus 7 is 100% constant impedance load and Bus 10 is 100% motor load.
Fault mode: The 5th circuit between Bus 5 and Bus 6 has three-phase N–1 fault on the Bus 6 side.
Fault start time: 1.0 s.
Line fault clearing time: 1.0056 s.

The calculation results are shown in Figs. 5.18 and 5.19.

Compared with Figs. 5.18 and 5.19, the voltage of Bus 10 has been decreasing in the process of the load impedance modulus changing from large to small. At the

Fig. 5.18 Bus 10 voltage variation curve

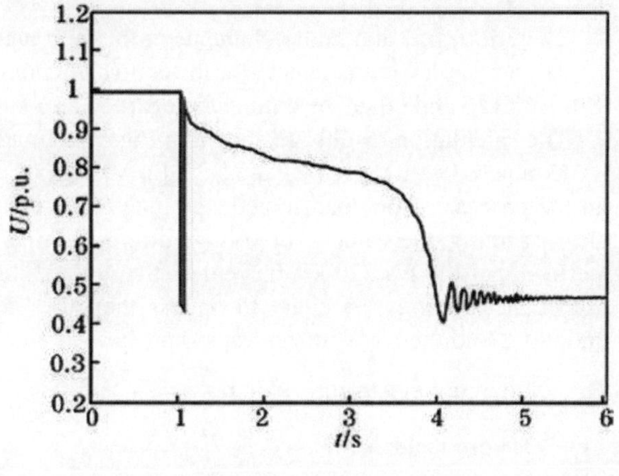

Fig. 5.19 Variation curve of load impedance and Thevenin equivalent impedance of the system at bus 10

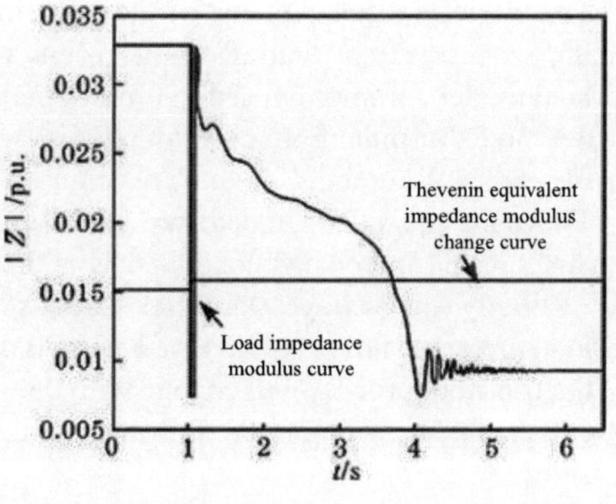

beginning, the velocity of voltage drop is relatively slow, but the voltage of Bus 10 is at the voltage instability region from the voltage stability region when the load impedance modulus is equal to the Thevenin equivalent impedance modulus, and the bus voltage drops sharply.

(2) Example 2

Data source: Data of a regional power grid in China (the load in the fault area is 40% motor load + 60% constant impedance load).

Fault mode: cascading fault mode 1. The sequence of faults is as follows:

① Transformer ground fault between B22–B1 at 0 s;
② Fault clearing of #1 transformer between B22–B1 at 0.1 s;
③ Fault clearing of #2 transformer between B22–B1 at 1.1 s;
④ Fault clearing of 220 kV single circuit between B1–B8 at 1.1 s;
⑤ Fault clearing of 220 kV double circuits between B1–B24 at 1.1 s;
⑥ Fault clearing of 220 kV single circuit between B10–B11 at 2.1 s;
⑦ Fault clearing of #1 unit of F2 power plant at 3.1 s;
⑧ Fault clearing of #1 unit of F1 power plant at 6.1 s.

The grid frame and faulty elements in the fault area are shown in Fig. 5.20.

The B6 220 kV bus is selected as the equivalent point, i.e., the whole area including B4, B6, B26 and other substations is regarded as a single load.

The calculation results are shown in Figs. 5.21 and 5.22.

Compared with Figs. 5.21 and 5.22, the voltage of B6 220 kV has been decreasing in the process of the load impedance magnitude changing from large to small. At the beginning, the velocity of voltage drop is relatively low but rises again in some periods, but the B6 220 kV bus enters the voltage instability region when the load impedance modulus is equal to or less than the Thevenin equivalent impedance modulus, and the bus voltage drops sharply.

3. Maximum electromagnetic power

(1) Theoretical basis

The induction motor is one of the most important components of load and also an important element that affects transient voltage stability. At present, the dynamic load model commonly used in time-domain simulation analysis of power systems is the model of induction motor in parallel with constant impedance load, as shown in Fig. 5.23. Wherein, $\dot{E}$ is the Thevenin equivalent potential of the system, Z_s is the Thevenin equivalent impedance of the system, $Z_R(R_R + jX_R)$ is the impedance of the constant impedance load, and M is the induction motor.

Firstly, the voltage stabilities of constant impedance load and induction motor load are analyzed respectively. The load demand P_Z of constant impedance load in the transient process will change with the voltage change at the load side, and can be calculated according to the following equation, wherein U is the voltage at the load side.

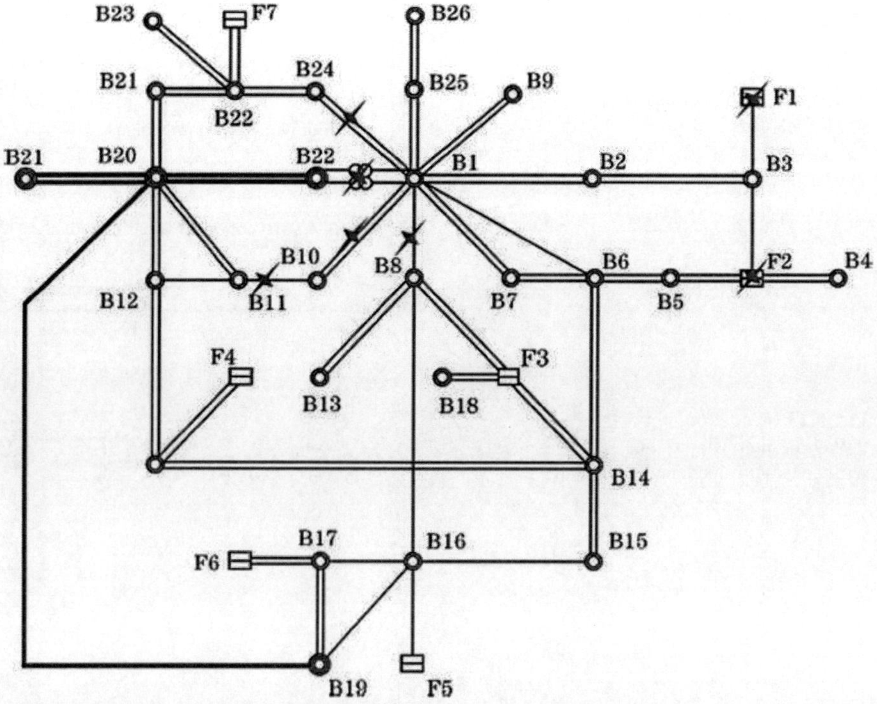

Fig. 5.20 Schematic diagram of grid frame and faulty elements in fault area

Fig. 5.21 Voltage variation curve of B6 220 kV Bus

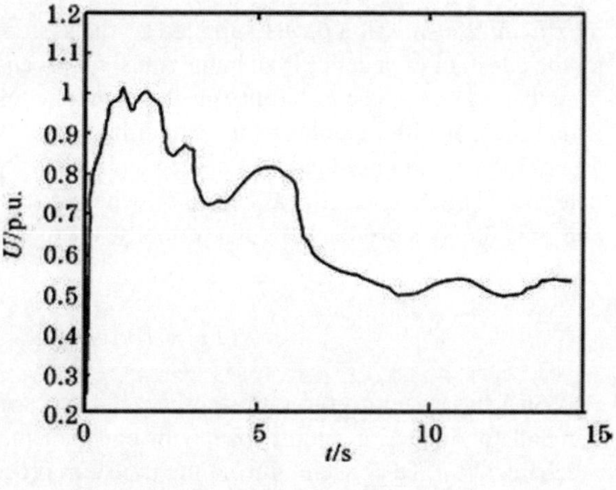

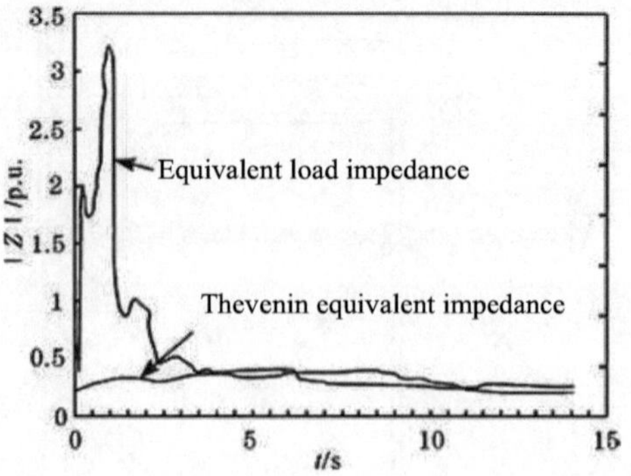

Fig. 5.22 Variation curve of equivalent load impedance and Thevenin equivalent impedance of the system at B6 220 kV Bus

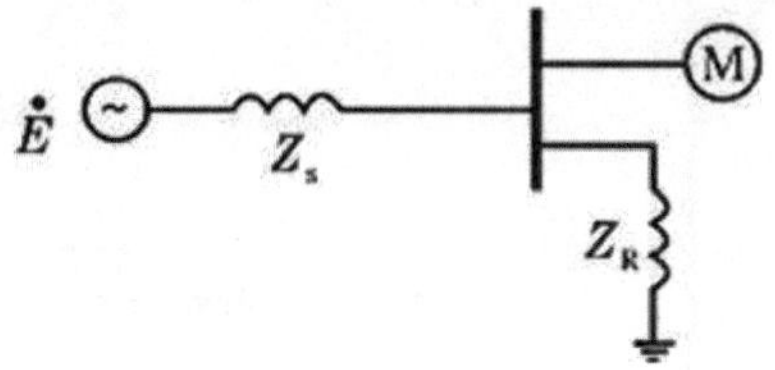

Fig. 5.23 Model of induction motor in parallel with constant impedance load

$$P_Z = \frac{U^2}{|Z_R|^2} R_R \tag{5.74}$$

The maximum transmission power of the system changes with the change of Thevenin equivalent parameters on the system side. For constant impedance load, the maximum transmission power supplied by the system is the active power consumed by the constant impedance load in the transient process, which is equal to P_Z. Therefore, there is no voltage instability mechanism for constant impedance load, and there is no problem with regard to voltage stability.

For induction motor load, the mechanical power P_T in the transient process can be determined by the mechanical torque T_t, wherein s is the slip of the induction motor, and ω is the synchronous angular velocity, all of which are calculated by per-unit value.

$$P_T = T_t(1 - s)\omega \tag{5.75}$$

When the system returns to stability, $1\text{–}s \approx 1$, and $\omega \approx 1$. Therefore, the load demand for induction motor load with constant mechanical torque is to keep its mechanical torque constant during the transient process, i.e., the mechanical power under the initial steady-state condition.

The equivalent circuit of the system is shown in Fig. 5.24. The induction motor is designed with first-order model. R_1 and X_1 are stator resistance and reactance, R_2 and X_2 are rotor resistance and reactance, s is slip, X_μ is excitation reactance,

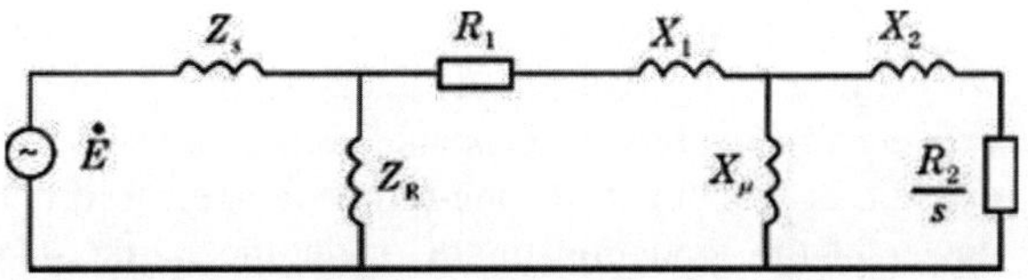

Fig. 5.24 Equivalent circuit

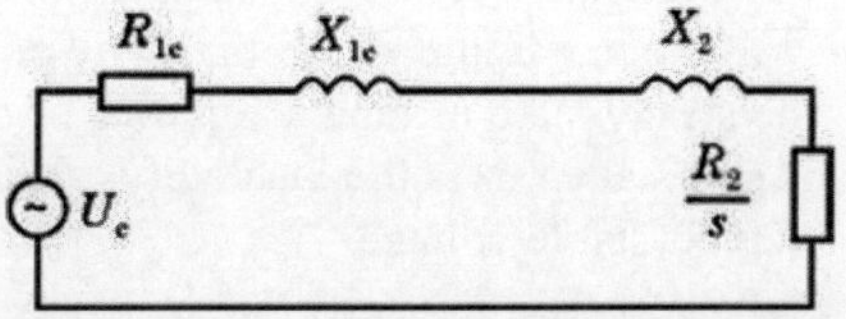

Fig. 5.25 Further equivalent circuit

Z_R is equivalent impedance of constant impedance load, $\dot{E}$ and Z_s are the Thevenin equivalent potential and impedance of the system.

After further simplification of the equivalent circuit shown in Fig. 5.24, the equivalent circuit shown in Fig. 5.25 can be obtained. In Fig. 5.25:

$$U_e = \frac{j\dot{E} Z_R X_\mu}{(Z_s + Z_R)\left(R_1 + jX_1 + jX_\mu + Z_s//Z_R\right)} \tag{5.76}$$

$$R_{1e} = \mathrm{Re}\left[(Z_s//Z_L + R_1 + jX_1)//jX_\mu\right. \tag{5.77}$$

$$X_{1e} = \mathrm{Im}\left[(Z_s//Z_L + R_1 + jX_1)//jX_\mu\right. \tag{5.78}$$

The electromagnetic power supplied by the system to the induction motor can be calculated with the following equation:

$$P_m = \frac{R_2 U_e^2}{s\left(R_{1e} + \frac{R_2}{s}\right) + (X_{1e} + X_2)^2} \tag{5.79}$$

When slip s changes, the maximum electromagnetic power can be obtained through calculation with Eq. (5.80). The maximum load power (i.e., the maximum electromagnetic power) that the system can supply to the induction motor will change with the changes of Thevenin equivalent parameters of the system in the transient process.

$$P_{m\max} = \frac{U_e^2 \sqrt{R_{1e}^2 + (X_{1e} + X_2)^2}}{\left(R_{1e} + \sqrt{R_{1e}^2 + (X_{1e} + X_2)^2}\right)^2 + (X_{1e} + X_2)^2} \tag{5.80}$$

Based on the above, the transient as well as medium and long-term voltage stability criteria (namely, the maximum electromagnetic power method) can be determined:

In the load model of induction motor with constant mechanical torque in parallel with constant impedance, the voltage stability can be determined by the relationship between the maximum electromagnetic power supplied by the system in the transient as well as medium and long-term voltage stability processes and the mechanical power of the induction motor under the initial steady-state condition. When the mechanical power of the motor is higher than the maximum electromagnetic power that the system can supply in steady state, the system voltage becomes unstable.

If the maximum electromagnetic power that the system can supply is higher than the steady-state mechanical power of the induction motor, and the induction motor is still stalled, it is the instability of the induction motor itself, and the system still outputs stable voltage.

(2) Example simulation

(1) Example 1

The power flow of a 3-unit 10-node system is shown in Fig. 5.26. Assuming Bus 7 is constant impedance load, Bus 10 is 80% induction motor + 20% constant impedance load. Parameters of induction motor are: stator resistance $R_1 = 0.5$ p.u., stator reactance $X_1 = 0.02$ p.u., rotor resistance $R_2 = 0.02$ p.u., rotor reactance $X_2 = 0.145$ p.u., and excitation reactance $X_\mu = 3.3$ p.u. A circuit of connection line between Bus 5 and Bus 6 will be tripped without fault at 0 s. The rotor angle and voltage curves of the system after the fault are shown in Figs. 5.27 and 5.28. The simulation results show that the rotor angle of the system is stable after the fault, the voltage is unstable, and the induction motor is stalled at 2.8 s. In order to verify the correctness of the voltage stability criterion based on the maximum electromagnetic power method, the Thevenin equivalent parameter curve and the maximum electromagnetic power before stalling of the induction motor and the mechanical power curve under initial

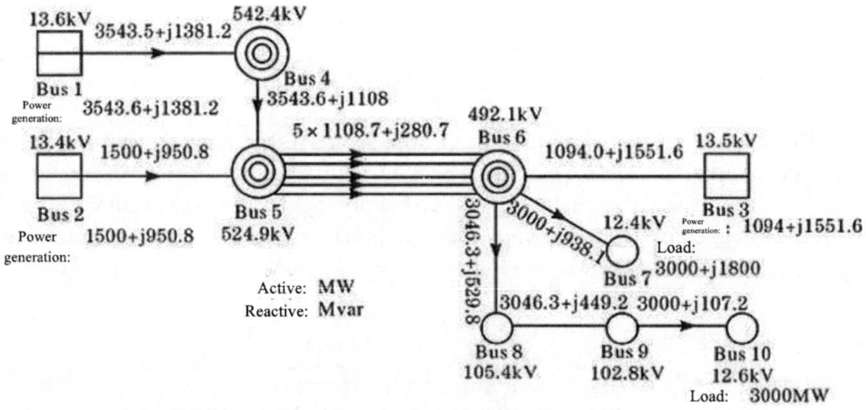

Fig. 5.26 Steady-state power flow 1 of 3-unit 10-node

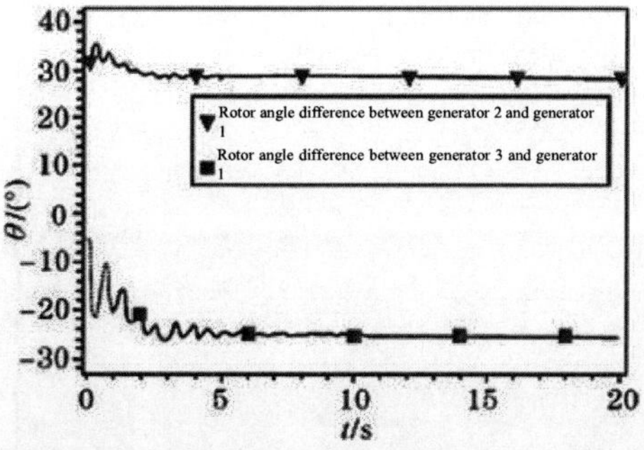

Fig. 5.27 System rotor angle curve after fault

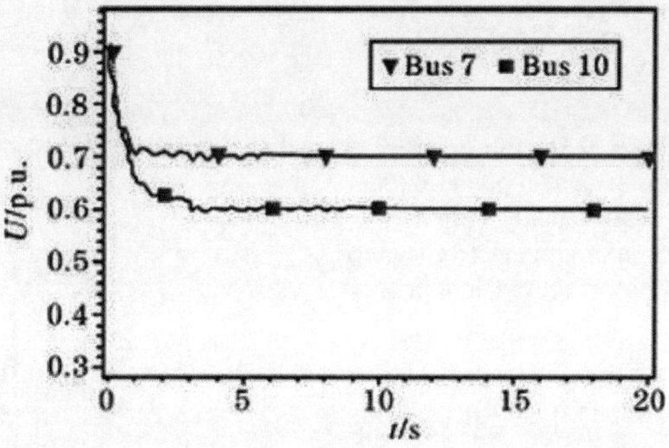

Fig. 5.28 System voltage curve after fault

steady state in the transient process after the fault occurs, as shown in Figs. 5.29, 5.30 and 5.31.

According to the criterion of the maximum electromagnetic power method, the mechanical power of the system is higher than the maximum electromagnetic power after the fault and before the stalling of the induction motor in Fig. 5.31, thus it can be judged that the voltage of the system is unstable.

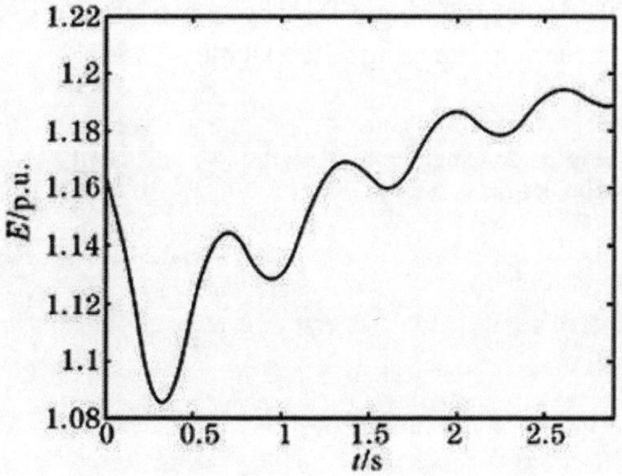

Fig. 5.29 Thevenin equivalent potential after fault

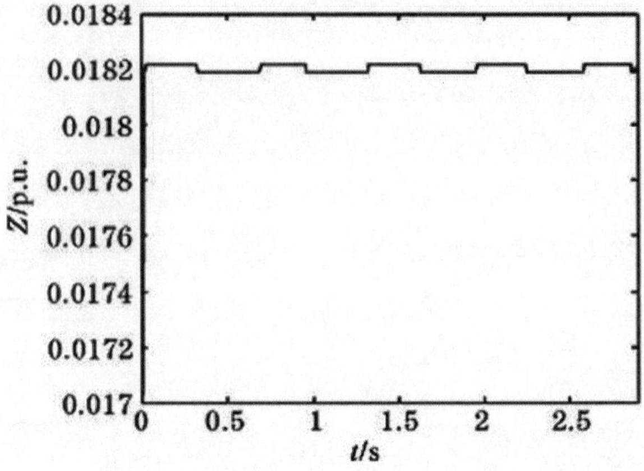

Fig. 5.30 Thevenin equivalent impedance after fault

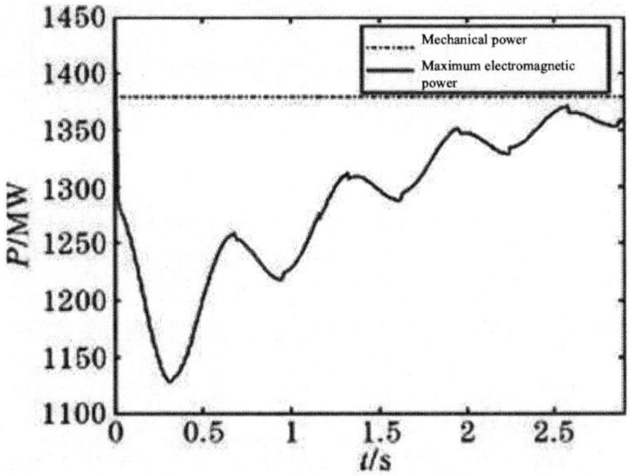

Fig. 5.31 Curves of mechanical power under initial steady state and maximum electromagnetic power in transient process

(2) Example 2

Take a single-unit single-load system as an example. The power flow diagram of a single-unit single-load system is shown in Fig. 5.32. The generator is an infinite generator, keeping E' constant.

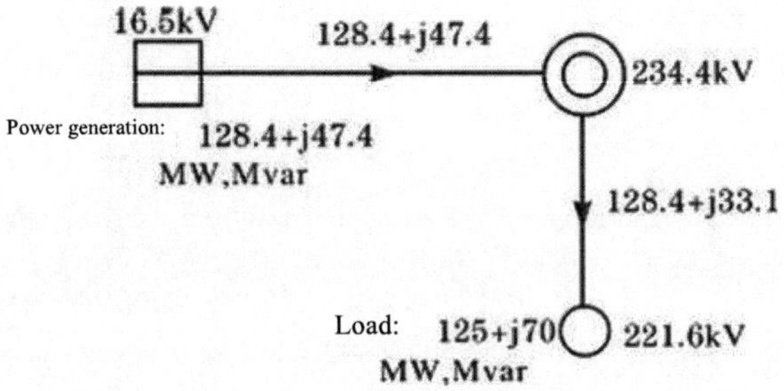

Fig. 5.32 Single-unit single-load system

A three-phase short-circuit fault occurred at the load bus at 0 s, and the fault was cleared after 0.1 s. After the fault, the load bus voltage dropped sharply, and the slip of the induction motor increased continuously until the motor is stalled at 1.84 s, as shown in Figs. 5.33 and 5.34.

The maximum electromagnetic power of the system and the mechanical power of the induction motor after the fault are shown in Fig. 5.35.

As can be seen from Fig. 5.35, after the fault is cleared, the maximum electromagnetic power of the induction motor is higher than the mechanical power, and the system voltage is stable. However, the instability of the induction motor causes the slip of the system to increase until the motor is stalled. From the foregoing description, it can be seen that this situation should be distinguished from stalling of induction motor caused by voltage instability.

4. Voltage instability judging method based on power and current variation relation

(1) Theoretical basis

Fig. 5.33 Load bus voltage after fault

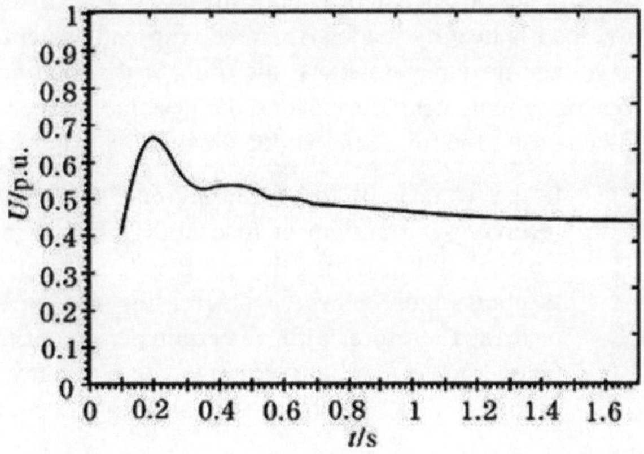

Fig. 5.34 Slip of induction motor after fault

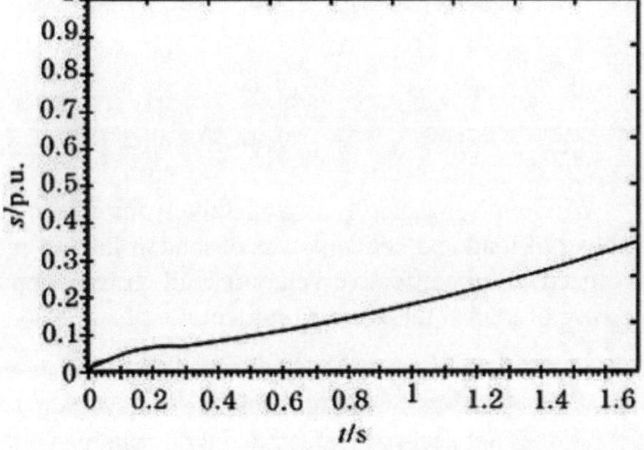

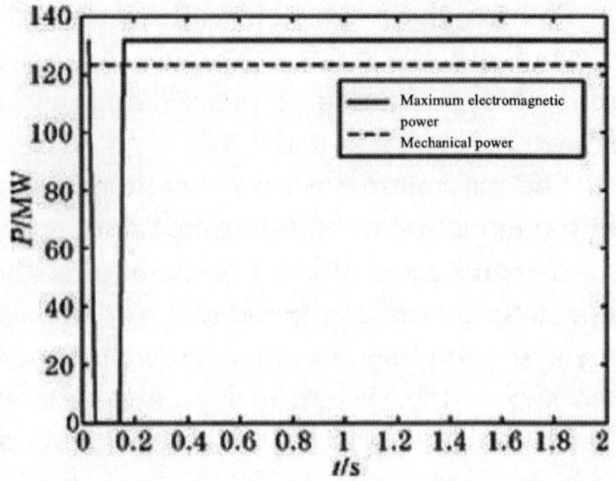

Fig. 5.35 Maximum electromagnetic power and mechanical power of induction motor after fault

Under steady state, the equivalent impedance $|Z_L|$ of the load is higher than the equivalent impedance $|Z_s|$. When the load needs more power, it always achieves its goal by reducing its equivalent impedance and thus increasing the current. However, when a serious accident occurs in the system or the channel supplying power to the load is heavily loaded, the receiving-end system or the load node may present a voltage unstable state. At this time, if the load continues to try to obtain more power by increasing the current, the system voltage will drop rapidly. Based on this mechanism, the following voltage instability criteria can be obtained.

(1) If the voltage of the receiving-end system or load node is unstable, the receiving-end system or load node must be in the unstable voltage area for a period of time during the whole voltage drop. That is, when the load attempts to obtain more power by increasing the current, the load power decreases instead. Therefore, within a certain period of time (set to t), the apparent power (active power) and the effective value of current of the receiving-end system or load node at which voltage instability occurs must meet

$$I_{k+1} > I_k \text{ and } S_{k+1} < S_k \tag{5.81}$$

or

$$I_{k+1} > I_k \text{ and } P_{k+1} < P_k \tag{5.82}$$

Wherein I_k, S_k and P_k are respectively the effective value of load current, apparent power of load and active power of load at the kth moment; I_{k+1}, S_{k+1} and P_{k+1} are respectively the effective value of load current, apparent power of load and active power of load at the k + 1st moment.

(2) When the receiving-end system or load node enters the unstable voltage area from the stable voltage area, the load voltage will drop. However, this process does not necessarily lead to uncontrollable voltage reduction, i.e., the load may

re-enter the stable voltage area from the unstable voltage area (related to the load and the dynamic characteristics of the system), at which time the voltage will remain at a certain level or rise again (this phenomenon is described in detail in the examples later in this section). Therefore, a voltage threshold value (e.g., 0.75p.u.) must be added to the criteria of Eqs. (5.81) and (5.82). When the load enters the unstable voltage area from the stable voltage area and the voltage in the unstable area continues to drop below the threshold value, it is regarded that voltage instability occurs. That is, the complete voltage instability criterion should consist of "judging the entering of the unstable voltage area" + "consequences of entering the unstable voltage area".

Two voltage instability criteria are summarized as follows.

(1) Criterion 1:

① $I_{k+1} - I_k > \varepsilon$ & $S_k - S_{k+1} > \varepsilon$ with a duration longer than or equal to 0.05 s;
② The voltage determined by ① is lower than 0.75 p.u. after continuous reduction of the voltage.

In which the variable ε is a very small positive number, which can be taken as 0.00001, the same below.

(2) Criterion 2:

① $I_{k+1} - I_k > \varepsilon$ and $P_k - P_{k+1} > \varepsilon$ with a duration longer than or equal to 0.05 s;
② The voltage determined by ① is lower than 0.75 p.u. after continuous reduction of the voltage.

Setting a duration longer than or equal to 0.05 s in criteria 1 and 2 can filter out possible interference points (which may be caused by switching operation or data acquisition errors); $I_{k+1} - I_k > \varepsilon$ & $S_k - S_{k+1} > \varepsilon$ and $I_{k+1} - I_k > \varepsilon$ & $P_k - P_{k+1} > \varepsilon$ can avoid miscalculation caused by calculation errors. In addition, the fault duration should be avoided when using criteria 1 and 2.

Criteria 1 and 2 are applicable to the voltage stability judgment of single-load bus and the voltage stability judgment of the receiving-end system when there are single or multiple lines connected between the sending and receiving-end systems.

(2) Example simulation

(1) Example 1

Data source: Data of a regional power grid in peak load mode in the wet season of 2008 in China (the load in the fault area is 40% motor load + 60% constant impedance load).

Fault mode: Cascading fault mode I. The sequence of faults is as follows:

① Transformer ground fault between B22–B1 at 0 s;
② Interruption of # 1 transformer between B22–B1 at 0.1 s;
③ Interruption of # 2 transformer between B22–B1 at 1.1 s;

④ Interruption of B1–B8 220 kV single circuit at 1.1 s;
⑤ Interruption of B1–B24 220 kV double circuit at 1.1 s;
⑥ Interruption of B10–B11 220 kV single circuit at 2.1 s;
⑦ Interruption of #1 unit of F2 power plant at 3.1 s;
⑧ Interruption of #1 unit of F1 power plant at 6.1 s;

The grid frame and faulty elements in fault area are shown in Fig. 5.36.

The B6 220 kV bus is selected as the monitoring point, i.e., the whole area including B4, B6, B26 and other substations is regarded as a single load.

Calculation results are shown below:

① If only part ① of criterion 1 or criterion 2 is used to judge voltage instability, three time points, namely, 2.26 s, 3.1 s and 6.1 s, can be found (Note: several time points can be found in the time range after 6.1 s, but the voltages corresponding to those time points are so low that they are not considered any more). This is because after the load (refers to the whole receiving end including B4, B6 and B26, the same below) enters the unstable voltage area at 2.26 s and 3.1 s, the voltage drops greatly. However, after a short time, the load and the system work together to make the load return to the stable voltage area from the unstable voltage area again very quickly, so the voltage no longer continues to drop

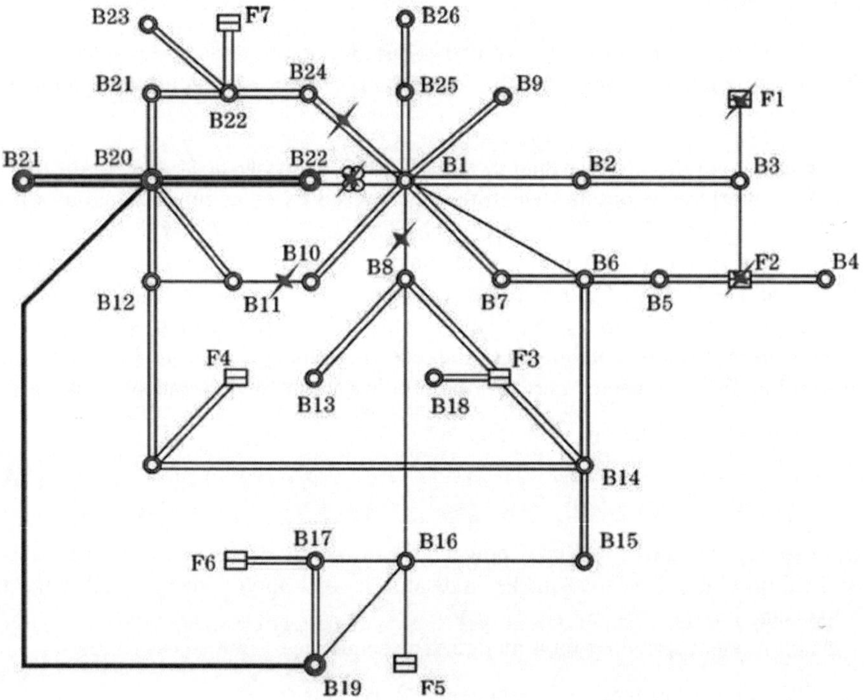

Fig. 5.36 Schematic diagram of grid frame and faulty elements in fault area

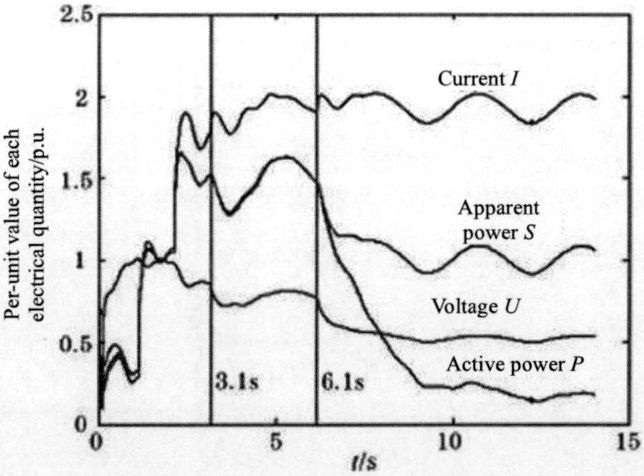

Fig. 5.37 Variation curve of electrical quantity at B6 220 kV

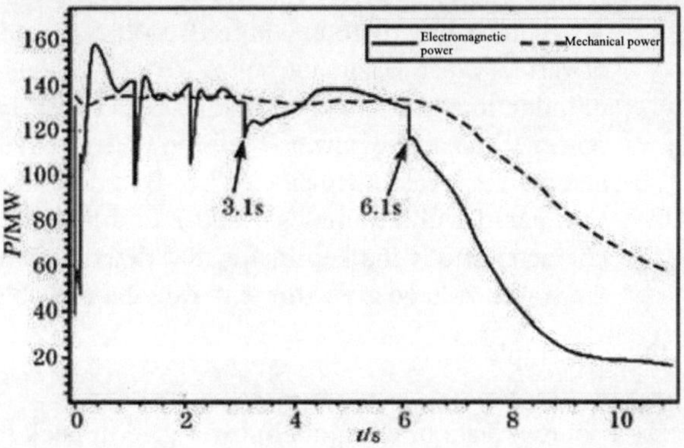

Fig. 5.38 Variation curve of power of induction motor load on B4 110 kV bus

but rises slightly. Figures 5.37, 5.38 and 5.39 show the variation curves of various electrical quantities of B6 220 kV bus and the power variation curves of induction motor load on B4 and B5 110 kV bus, wherein current I, apparent power S and active power P refer to the effective current value, apparent power and active power at B6 side of B6–B14 double-circuit line.

② If the complete criterion (5.81) or (5.82) is adopted, the voltage collapse point of 6.1 s can be directly found.

(2) Example 2

Data description: Based on the data used in Example 1 of this section, the load on 110-kV bus of B7 and B25 substations is changed to 100% induction motor.

The fault mode and voltage stability monitoring points are the same as in Example 1 of this section.

Calculation results are shown below:

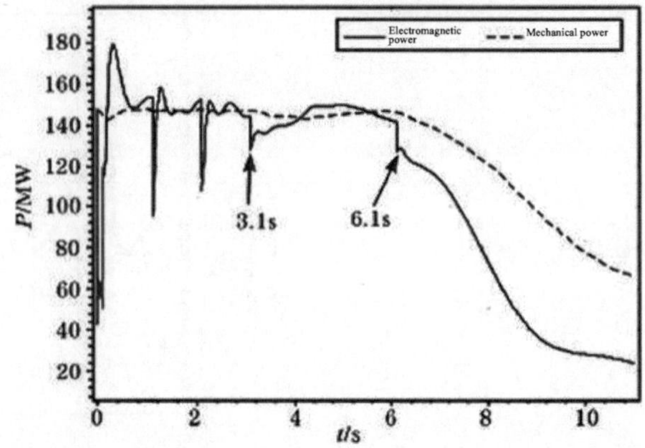

Fig. 5.39 Variation curve of power of induction motor load on B5 110 kV Bus

① If the complete criterion (5.81) or (5.82) is adopted, the voltage collapse point of 3.1 s can be directly found. Figures 5.40, 5.41 and 5.42 show the variation curves of various electrical quantities of B6 220-kV bus and the power variation curves of induction motor load on B4 and B5 110-kV bus, wherein current I, apparent power S and active power P refer to the effective current value, apparent power and active power at B6 side of B6–B14 double-circuit line.

② Compared with Examples 1 and 2 in this section, it can be seen that the load characteristic is the key factor that determines whether the load can return to the stable voltage area after entering the unstable voltage area.

(3) Example 3

Data source: Data of a regional power grid in peak load mode in the wet season of 2008 in China (the load in the fault area is 40% induction motor load + 60% constant impedance load).

Fault mode: Cascading fault mode II. The sequence of faults is as follows:

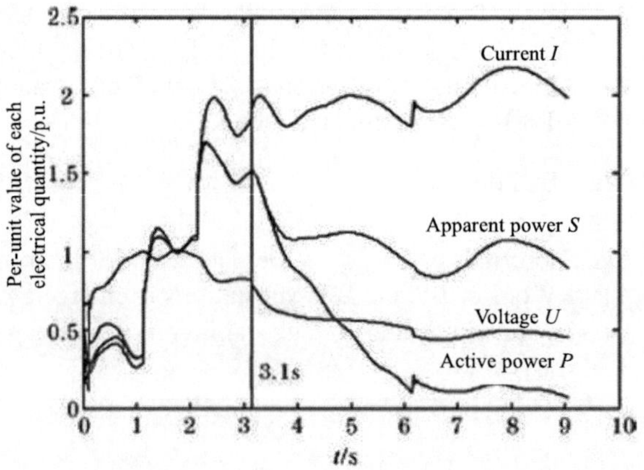

Fig. 5.40 Variation curve of electrical quantity at B6 220 kV

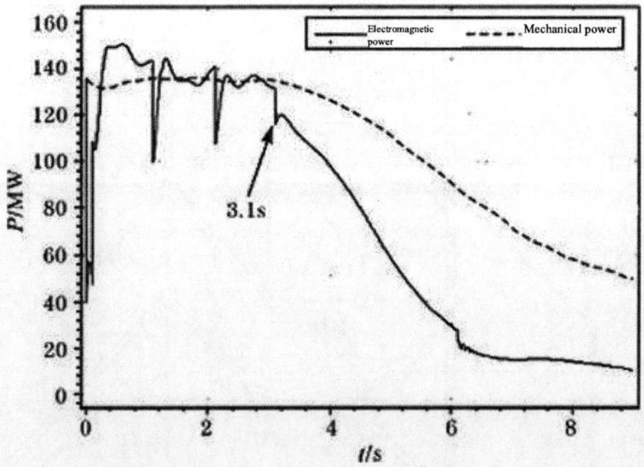

Fig. 5.41 Variation curve of power of induction motor load on B4 110 kV Bus

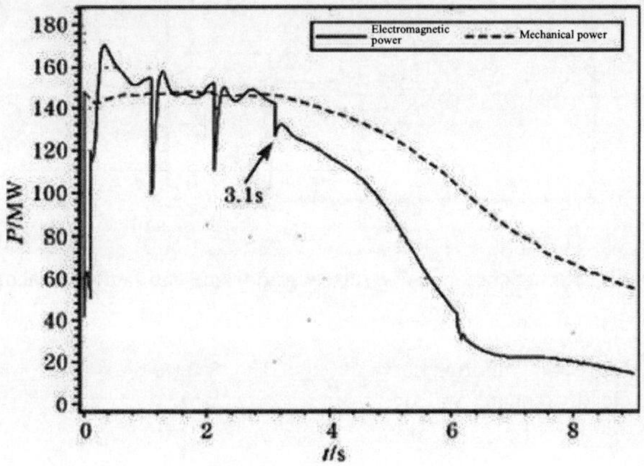

Fig. 5.42 Variation curve of power of induction motor load on B5 110 kV Bus

① Ground fault of B1–B2 220-kV primary circuit at 0 s;
② Interruption of B1–B2 220-kV primary circuit at 0.12 s;
③ Interruption of B1–B6 220-kV single circuit at 2.12 s;
④ Interruption of second-circuit B1–B2 220-kV line at 3.12 s;
⑤ Interruption of B1–B7 220-kV primary circuit at 4.12 s;
⑥ Interruption of second-circuit B1–B7 220-kV line at 5.12 s;
⑦ Interruption of #1 unit of F2 power plant at 6.12 s;
⑧ Interruption of #2 unit of F2 power plant at 7.12 s;
⑨ Interruption of #1 unit of F1 power plant at 8.12 s;

The grid frame and faulty elements in fault area are shown in Fig. 5.43.

The B6 220 kV bus is selected as the monitoring point, i.e., the whole area including B4, B5, B6 and other substations is regarded as a single load.

The calculation results show that, if the complete criterion (5.81) or (5.82) is adopted, the voltage collapse point of 8.12 s can be directly found. Figures 5.44,

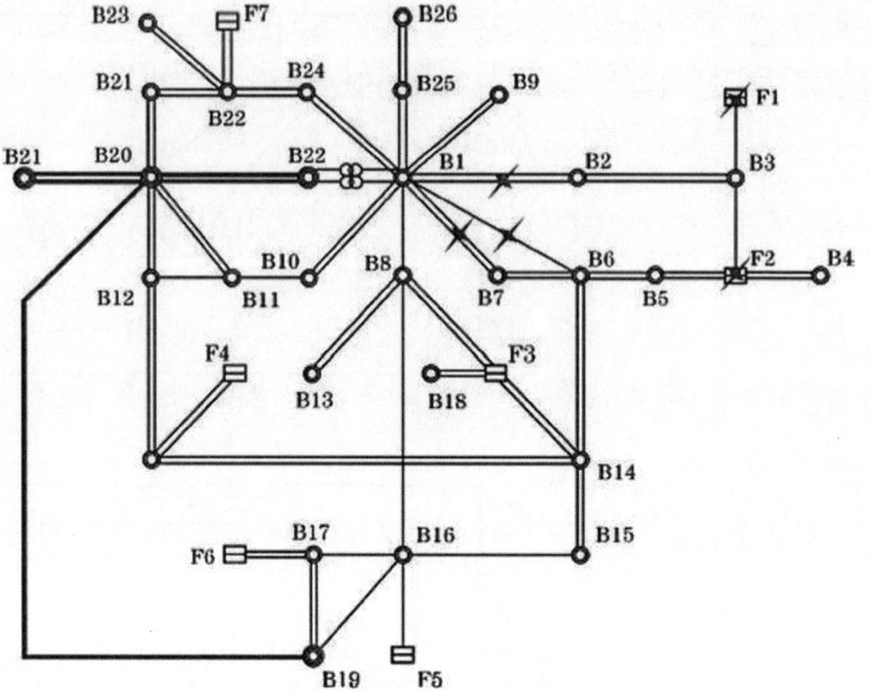

Fig. 5.43 Schematic diagram of grid frame and faulty elements in fault area

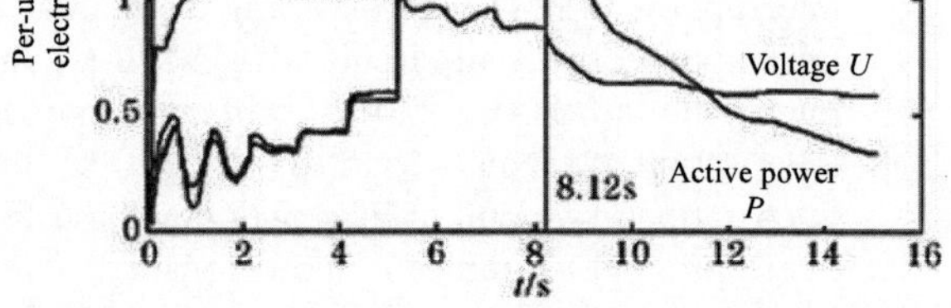

Fig. 5.44 Variation curve of electrical quantity at B6 220 kV

5.45 and 5.46 show the variation curves of various electrical quantities of B6 220 kV bus and the power variation curves of induction motor load on B4 and B5 110 kV bus, wherein current I, apparent power S, and active power P refer to the effective current value, apparent power, and active power at B6 side of B6–B14 double-circuit line.

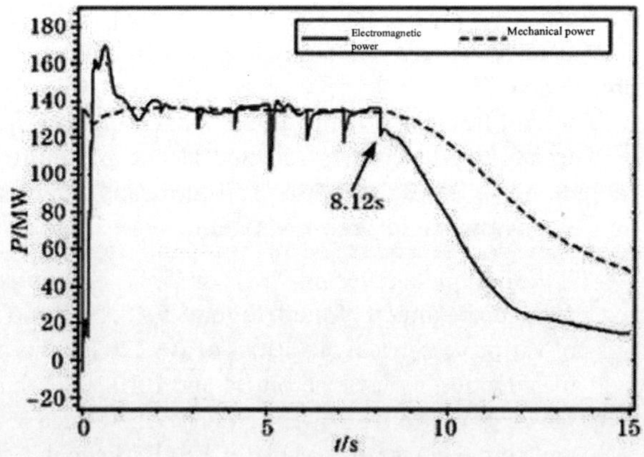

Fig. 5.45 Variation curve of power of induction motor load on B4 110 kV bus

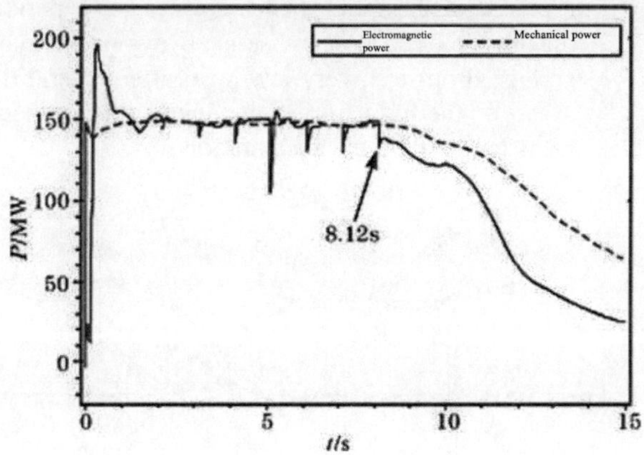

Fig. 5.46 Variation curve of power of induction motor load on B5 110 kV bus

(4) Example 4

Data source: Data of a regional power grid in peak load mode in the wet season of 2008 in China (the load in the fault area is 40% induction motor load + 60% constant impedance load).

Fault mode: Cascading fault mode III. The sequence of faults is as follows:

① Transformer ground fault between B22–B1 at 0 s;
② Interruption of # 1 transformer between B22–B1 at 0.1 s;
③ Interruption of # 2 transformer between B22–B1 at 1.1 s;
④ Interruption of B10–B11 220-kV single circuit at 2.1 s;
⑤ Interruption of #1 unit of F2 power plant at 3.1 s;
⑥ Interruption of #2 unit of F2 power plant at 4.1 s;
⑦ Interruption of first-circuit B6–B14 220-kV line at 5.1 s;
⑧ Interruption of #1 unit of F1 power plant at 6.1 s;

Note: After the last fault of the cascading fault mode 3 occurs, there are still many lines connected between the sending end (main network) and receiving end (B2, B5, and B6, etc.).

The grid frame and faulty elements in fault area are shown in Fig. 5.47.

The B6 220 kV bus is selected as the monitoring point, i.e., the whole area including B4, B5, B6 and other substations is regarded as a single load.

Calculation results are shown below:

① If the complete criterion (5.81) or (5.82) is adopted, the voltage collapse point of 6.1 s can be directly found. Figures 5.48, 5.49 and 5.50 show the variation curves of various electrical quantities of B6 220 kV bus and the power variation curves of induction motor load on B4 and B5 110 kV bus, wherein current *I*, apparent power *S* and active power *P* refer to the effective current value, apparent power and active power at B6 side of B6–B14 second-circuit line.

② Different from the previous example, the voltage in this example decreases for a period of time and then increases for a period of time after 6.1 s, but the magnitude of each decrease is always larger than that of the increase, and the voltage drops to a very low level after several times. During this process, the speed of the induction motor increases and decreases, but each deceleration takes longer than the acceleration.

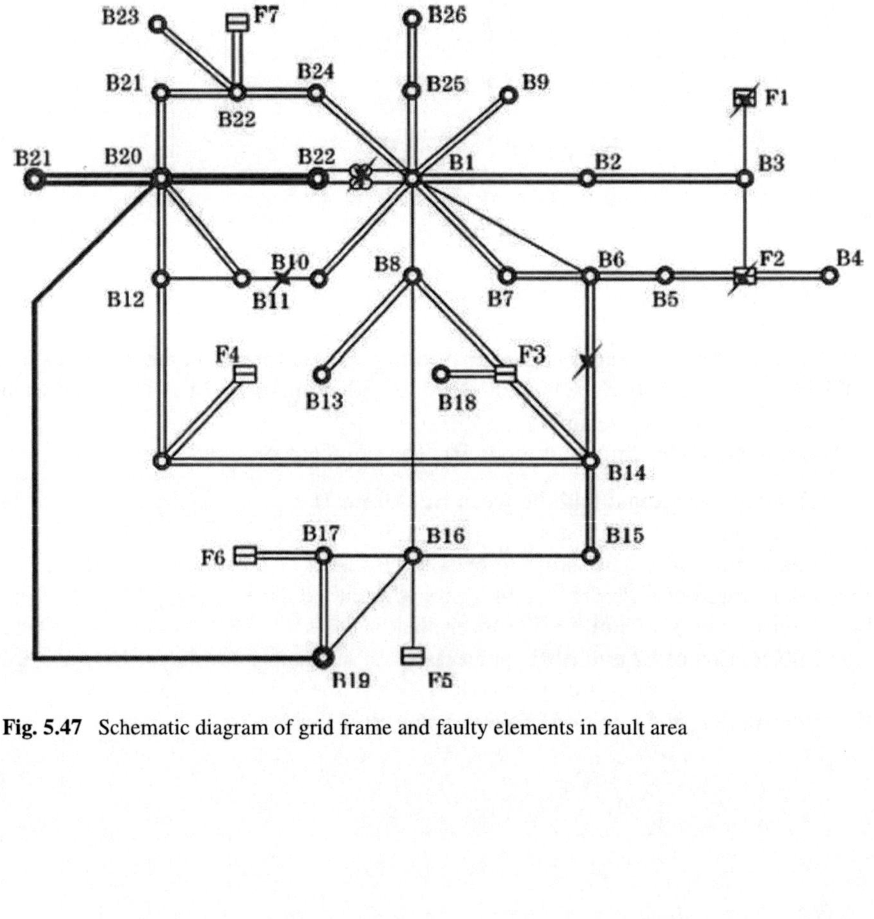

Fig. 5.47 Schematic diagram of grid frame and faulty elements in fault area

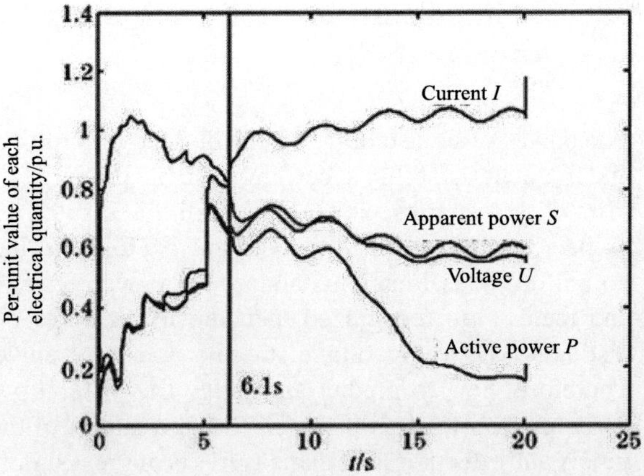

Fig. 5.48 Variation curve of electrical quantity at B6 220 kV

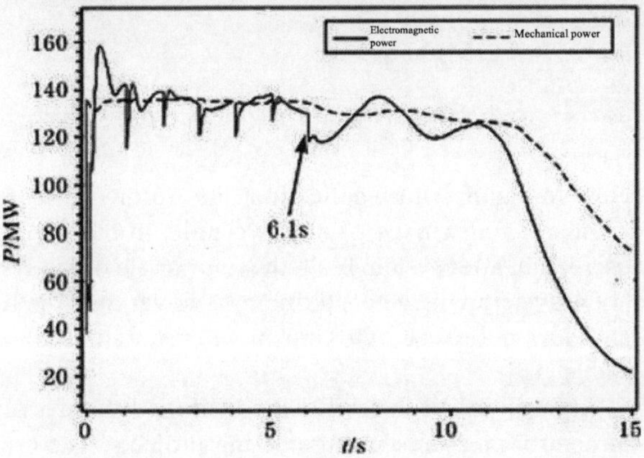

Fig. 5.49 Variation curve of power of induction motor load on B4 110 kV bus

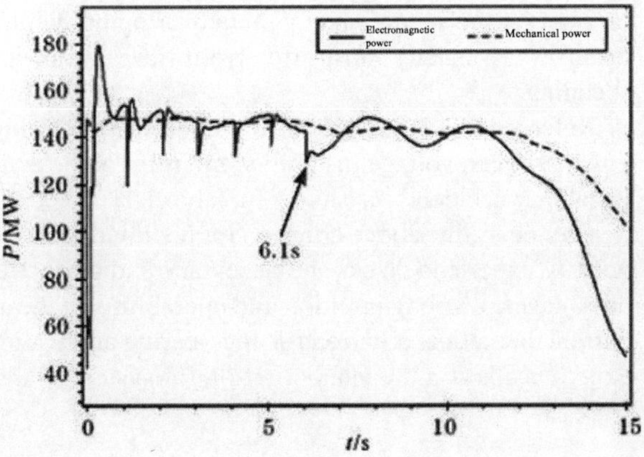

Fig. 5.50 Variation curve of power of induction motor load on B5 110 kV bus

5.3 Distinguishment Between Voltage Stability and Rotor Angle Stability

The power system stability can be divided into rotor angle stability, voltage stability and frequency stability according to stable properties [16–18]. In many power system instability scenarios, voltage instability and rotor angle instability are often intertwined, which is difficult to distinguish. However, due to historical reasons, there is no uniform classification standard for power system stability at home and abroad, and there is no recognized mechanism for voltage stability. The research on the distinction between voltage stability and rotor angle stability is not deep enough. Therefore, how to predict and judge the instability mode of the system from the variation information of the electrical quantity of the system so as to take correct, timely and effective control measures is of great significance for improving the power system stability theory and improving the operation control of the power system.

5.3.1 Engineering Empirical Approach

Now in engineering application, the following criteria are adopted: if the relative angular swing curves of any two units in the system show synchronous damping oscillation after system fault, the rotor angle of the system is stable; if the bus voltage at the system voltage center point is lower than the limit value (generally 0.75 p.u.) and does not exceed the limit time (generally 1 s), the system voltage is regarded stable [16].

In addition, the instability mode of the system is often judged by the effectiveness of control measures in engineering projects. The criterion is based on the fact that when the system rotor angle instability occurs, it is usually most effective to take the measure of the sending end generator tripping, while the system voltage instability occurs, it is usually most effective to take the measure of the receiving end load shedding.

References [19] and [20] also proposed engineering-applicable criterion to distinguish between voltage instability and rotor angle instability from a large number of instability accidents, as shown in Table 5.1.

Obviously, the above criterion for distinguishing instability modes based on engineering experience lacks the necessary and perfect theoretical basis, but it has the advantages of strong intuition and operability. In the absence of generalization of the current theoretical criterion for engineering applications, the criterion of engineering experience is still the main criterion for system analysts.

Table 5.1 Criterion for voltage instability and rotor angle instability

Bus voltage characteristics	Voltage instability	Rotor angle instability
1	The observed bus is at or near the load center	The observed bus is on the tie-line
2	The voltage drops monotonously	The voltage and the power of the tie-line periodically change in a reverse manner, that is, drop first and then rise (if the tie-line is not interrupted)
3	The voltage drop process takes a long time, and causes the OLTC to operate, for example, ten seconds to several minutes	The voltage drop process takes a short time and cannot cause OLTC to operate, for example, 1–2 s

5.3.2 Small Disturbance Analysis Method

The small disturbance analysis method mainly derives the matrix or eigenvalues related to the rotor angle instability and voltage instability mode through the Jacobian matrix, so as to judge the system instability mode [21–24]. Reference [21] studies the Jacobian matrix $|J| = 0$, and the results show that the system may suffer from rotor angle instability or voltage instability. Reference [22] derives matrices related to rotor angle instability and voltage instability through the extended Jacobian matrix and reduced-order Jacobian matrix respectively, and judges the instability mode of the system with the singularity of the matrices. Define the extended power flow Jacobian matrix:

$$J_E = \begin{bmatrix} A_{11} & A_{12} \\ A_{21} & A_{22} \end{bmatrix} \tag{5.83}$$

The matrix J_{E} has the following properties:

(1) When A_{22} is non-singular, the singularity of J_{E} is equivalent to the singularity of $A_{11} - A_{12}A_{22}^{-1}A_{21}$;
(2) When A_{11} is non-singular, the singularity of J_{E} is equivalent to the singularity of $A_{22} - A_{12}A_{11}^{-1}A_{21}$.

It can be proved as follows:

$$\operatorname{rank}(J_E) = \operatorname{rank}\begin{bmatrix} A_{11} & A_{12} \\ A_{21} & A_{22} \end{bmatrix} = \operatorname{rank}\begin{bmatrix} A_{11} - A_{12}A_{22}^{-1}A_{21} & 0 \\ A_{21} & A_{22} \end{bmatrix} \tag{5.84}$$

Therefore, the necessary and sufficient condition for singularity of J_{E} is the singularity of $A_{11} - A_{12}A_{22}^{-1}A_{21}$. Property (1) is proved. Similarly, the property (2) is proved. Let $A_g = A_{11} - A_{12}A_{22}^{-1}A_{21}$, called the generator matrix, and $A_r = A_{21} - A_{12}A_{11}^{-1}A_{12}$, called the load matrix. The positive and negative of A_g

characteristic root indicates the stability of system rotor angle, while A_r indicates the stability of system voltage. If the generator matrix A_g and the load matrix A_r tend to be singular at the same time in a given operation mode, voltage instability and rotor angle instability occur simultaneously in the power system.

The essence of the generator matrix is to retain the generator rotor angle variable, eliminate the network variable, and only take into account the influence of the network on the generator as a correction, so that make it a pure rotor angle matrix. Similarly, the load matrix only retains the voltage and phase angle of the load node, taking into account the influence of the generator on the network [22].

References [23] and [24] obtain simplified voltage and rotor angle Jacobian matrices by decoupling voltage and rotor angle respectively.

(1) Voltage Jacobian matrix

The power flow Jacobian matrix:

$$\begin{bmatrix} \Delta P \\ \Delta Q \end{bmatrix} = \begin{bmatrix} J_1' & J_2' \\ J_3' & J_4' \end{bmatrix} \begin{bmatrix} \Delta \delta \\ \Delta U \end{bmatrix} = J \begin{bmatrix} \Delta \delta \\ \Delta U \end{bmatrix} \tag{5.85}$$

Assume that the active power of the system remains unchanged, that is, $\Delta P = 0$. Substitute it into Eq. (5.85) to obtain

$$\Delta Q = \left(J_4' - J_3' J_1'^{-1} J_2'\right) \Delta U \tag{5.86}$$

Let $J_{RQ} = J_4' - J_3' J_1'^{-1} J_2'$, then J_{RQ} matrix is a simplified voltage Jacobian matrix.

(2) Rotor angle Jacobian matrix

Assume that the system only changes the active power. Let $\Delta Q = 0$, and substitute it into Eq. (5.85) to obtain

$$\Delta P = \left(J_1' - J_2' J_4'^{-1} J_3'\right) \Delta \delta \tag{5.87}$$

$$J_{RP} = J_1' - J_2' J_4'^{-1} J_3' \tag{5.88}$$

Similarly, the J_{RP} matrix is defined as a simplified rotor angle Jacobian matrix.

Calculate the minimum eigenvalues $\lambda_{Q\min}$ and $\lambda_{P\min}$ of both matrices separately, and then compare the minimum eigenvalues to distinguish the instability mode of the system under small disturbances. If $|\lambda_{Q\min}| \gg |\lambda_{P\min}|$, the system first reaches the rotor angle stability limit; if $|\lambda_{P\min}| \gg |\lambda_{Q\min}|$, the system first reaches the voltage stability limit.

On the basis of small disturbance analysis, reference [25] uses the normal form theory of vector field to obtain high-order analytical solution of system response. The degree of interaction between the stable modes is analyzed by the second-order participation factor in the high-order analytical solution. The normal form theory of vector field is a theory to recognize and understand the dynamic characteristics of strongly nonlinear systems by analyzing the nonlinear correlation between internal modes of the system, which deepens the results of linearization analysis. The high-order analytical solution contains more dynamic information of the power system, which is convenient for analyzing and understanding the dynamic essence of the system. Moreover, the second-order participation factors in the high-order analytical solution are analyzed to understand the degree of interaction between various stable modes in the power system, so as to gain a deeper understanding of the essence of dynamic characteristics of the system, which is not revealed by the traditional small disturbance method.

5.3.3 Bifurcation Analysis

In reference [26], the critical conditions for voltage bifurcation instability and rotor bifurcation instability in power systems are derived respectively by bifurcation theory.

The DAEs of the power system is expressed as

$$x = f(x, y, \mu) \tag{5.89}$$

$$0 = g(x, y, \mu) \tag{5.90}$$

Wherein f is defined as the dynamic characteristics of generator, exciter and load; g indicates the power flow model of the network and is expressed as $g = (g_1, g_2)^{\mathrm{T}}$, wherein g_1 is the active power model and g_2 is the reactive power model.

Without loss of generality, select the system differential state variable $x = (\delta)$, the algebraic state variable $y = (\theta, v)$, the system control parameter $\mu = (P_{\mathrm{L}}, Q_{\mathrm{L}})$, wherein δ is the generator rotor angle, θ and v are the voltage phase angle and magnitude of PV and PQ nodes respectively, and P_{L} and Q_{L} are load active power and reactive power respectively.

The equilibrium point described by Eqs. (5.89) and (5.90) can be determined by the following Equation:

$$F(\delta, \theta, \upsilon, \mu) = \begin{bmatrix} f(\delta, \theta, \upsilon, \mu) \\ g(\delta, \theta, \upsilon, \mu) \end{bmatrix} = 0 \tag{5.91}$$

Expand Eq. (5.91) near the equilibrium point:

$$\begin{pmatrix} D_\delta f & D_\theta f & D_\upsilon f \\ D_\delta g_1 & D_\theta g_1 & D_\upsilon g_1 \\ D_\delta g_2 & D_\theta g_2 & D_\upsilon g_2 \end{pmatrix} \begin{bmatrix} \mathrm{d}\delta \\ \mathrm{d}\theta \\ \mathrm{d}\upsilon \end{bmatrix} + \begin{bmatrix} D_\mu f \\ D_\mu g_1 \\ D_\mu g_2 \end{bmatrix} \mathrm{d}\mu = 0 \tag{5.92}$$

Take

$$J_1 = \begin{bmatrix} D_\delta f & D_\theta f \\ D_\delta g_1 & D_\theta g_1 \end{bmatrix}, J_2 = \begin{bmatrix} D_\upsilon f \\ D_\upsilon g_1 \end{bmatrix}, J_3 = [D_\delta g_2, D_\theta g_2]$$

$$J_4 = [D_\upsilon g_2], J_5 = \begin{bmatrix} D_\mu f \\ D_\mu g_1 \end{bmatrix}, J_6 = [D_\mu g_2]$$

$$J_1' = [D_g f], J_2' = [D_\theta f, D_\theta f], J_3' = \begin{bmatrix} D_\delta g_1 \\ D_\delta g_2 \end{bmatrix}$$

$$J_4' = \begin{bmatrix} D_\theta g_1 & D_\upsilon g_1 \\ D_\theta g_2 & D_\upsilon g_2 \end{bmatrix}, J_5' = [D_\mu f], J_6' = \begin{bmatrix} D_\mu g_1 \\ D_\mu g_2 \end{bmatrix}$$

According to Eq. (5.92), it is obtained that

$$[J_4 - J_3 J_1^{-1} J_2][\mathrm{d}\upsilon] = [J_3 J_1^{-1} J_5 - J_6]\mathrm{d}\mu \tag{5.93}$$

The following conclusions can be proved.

(1) The conditions for voltage static bifurcation instability of the system at the equilibrium point: $\det[J_1] \neq 0$ and $\det[J_4 - J_3 J_1^{-1} J_2] = 0$;
(2) The conditions for synchronization static bifurcation instability of the system at the equilibrium point: $\det[J_4'] \neq 0$ and $\det[J_1' - J_2' J_4'^{-1} J_3'] = 0$;

Based on bifurcation analysis method, in references [26] and [27], we study the position of system load, load power factor, load side generator output, generator reactance, and the correlation between power distribution, voltage instability, and rotor angle instability in the network.

5.3.4 Energy Function Method

According to the classical theory of the energy function method, the stable equilibrium point is located at the lowest point of the potential energy well, and each related unstable equilibrium point (UEP) is on the well wall of the potential energy well, wherein the Class I UEP corresponds to the minimum potential energy on the well wall. When critical stability is achieved, the system will escape the potential energy well near the Class I UEP. Therefore, after grasping the type and distribution of equilibrium points, possible instability mode of the system is roughly understood.

References [28] and [29] propose a unified energy function framework for studying the relationship between voltage stability and rotor angle stability. In a 2-unit 3-node system using a constant power load model, it is found that Class I UEP may have the characteristics of large rotor angle, high voltage, small rotor angle and low voltage, and these two different types of Class I UEP are associated with rotor angle stability and voltage stability respectively. Meanwhile, it is pointed out that the transient instability of the system is manifested by rotor angle instability or transient voltage instability depending on system conditions.

On this basis, the reference [30] studies in detail the impact of the load magnitude, generator output, and load model on the system equilibrium point. The research results show that, under the idea of the energy function method, both the rotor angle stability and the voltage stability are essentially regarded as a UEP mode, and the both are distinguished by different UEP modes. Under different power distribution and power change modes, they both may occur simultaneously or separately, or may transform into each other.

Analyzing the stability of the power system from the perspective of changes in the distribution of equilibrium points not only indicates the static instability mode of the system, but also provides a basis for qualitative assessment of the transient stability of the system. That is, through the bifurcation of the normal operating point of the system, the static instability mode of the system is understood; at the non-critical point, through the number, type, and distribution of UEP, the possible transient instability modes of the system and the characteristics of the system variable changes during instability.

Under the same power distribution of the same system, different load models have different numbers and distribution of UEP. If the classical basic theory of energy function method is followed and the critical point of system instability is believed to be near the dominant UEP, different load models will lead to different transient instability modes of the system, thus, the rotor angle stability problem will be highlighted under one load model (such as constant impedance load model) and the voltage stability problem will be highlighted under another load model (such as constant power load model).

5.3.5 Thevenin Equivalent Parameter Tracking Algorithm

Voltage instability and rotor angle instability are essentially different in mechanism and stability measures. However, rotor angle instability may cause voltage collapse, and voltage collapse may also cause rotor angle instability during system instability. The transient rotor angle stability and transient voltage problems are often intertwined. It is not easy to distinguish the instability mode of the system after the fault only from the form of rotor angle and voltage. However, system instability is inevitably dominated by one instability mode, and different instability modes occur one after another due to changes in system conditions [31–36].

1. Theoretical basis

(1) Rotor angle instability

Take the dual-unit single-load system shown in Fig. 5.51 as an example to illustrate the relationship between Thevenin equivalent parameter and rotor angle instability.

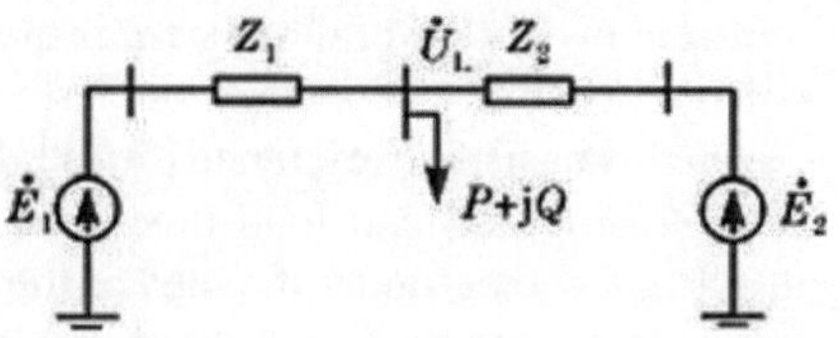

Fig. 5.51 Dual-unit single-load system model

Wherein $\dot{E}_1$ and Z_1, $\dot{E}_2$ and Z_2 are the electric potential and impedance of the power supply 1 side and 2 side respectively, and P and Q are the active power and reactive power of the load respectively.

The Thevenin equivalent parameters at the load point in Fig. 5.51 are calculated with the following equation, wherein $\dot{E}_{\text{Thev}}$ is Thevenin equivalent potential and Z_{Thev} is Thevenin equivalent impedance.

$$\dot{E}_{\text{Thev}} = \dot{E}_1 - \frac{\dot{E}_1 - \dot{E}_2}{Z_1 + Z_2} Z_1 = \frac{Z_2 \dot{E}_1 + Z_1 \dot{E}_2}{Z_1 + Z_2} \tag{5.94}$$

$$Z_{\text{Thev}} = \frac{Z_1 Z_2}{Z_1 + Z_2} \tag{5.95}$$

Due to the excitation system and the automatic voltage control device of the generator, the potentials E_1 and E_2 on the power supply side remain unchanged before and after the fault. Therefore, assume that $E_1 = E_2 = E$. Taking the electrical midpoint of the load point between the power supplies on both sides, namely, $Z_1 = Z_2 = Z$ as an example, it can be obtained that

$$\dot{E}_{\text{Thev}} = \frac{\dot{E}_1 + \dot{E}_2}{2} = \frac{E\angle\delta_1 + E\angle\delta_2}{2} \tag{5.96}$$

$$Z_{\text{Thev}} = \frac{Z}{2} \tag{5.97}$$

Figure 5.52 shows the correlation between Thevenin equivalent potential and rotor angle difference under the above assumptions. When the rotor angle difference is 0, $E_{\text{Thev}} = E$; when the rotor angle difference reaches 90°, $E_{\text{Thev}} = \frac{\sqrt{2}}{2}E$; when the rotor angle difference reaches 180°, $E_{\text{Thev}} = 0$.

Therefore, if the voltage collapse at the load point is caused by the rotor angle instability, the voltage at the load point and the Thevenin potential will decrease with the increase of the rotor angle difference.

(2) Voltage instability

Based on the basic principle of judging voltage stability by Thevenin equivalent tracking parameters, when the load equivalent impedance at the equivalent point is equal to Thevenin equivalent impedance, that is, when the system meets $|Z_L| = |Z_s|$,

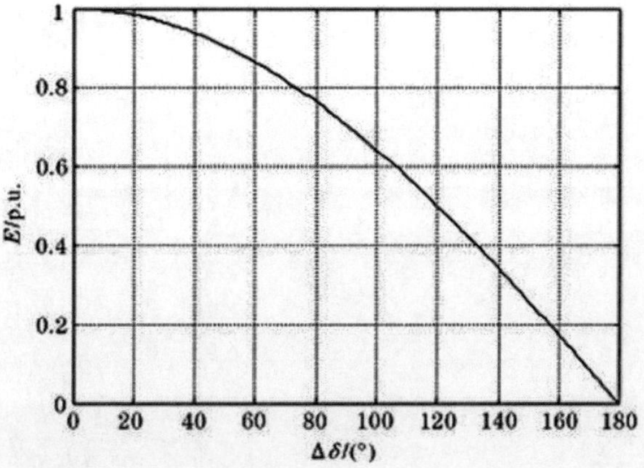

Fig. 5.52 Thevenin equivalent potential and rotor angle difference

the system has a unique voltage solution, which corresponds to a sufficient condition for voltage instability.

In summary, through the tracking of the change of Thevenin equivalent parameters during the transient process, it is possible to distinguish the dominant instability mode of voltage instability and rotor angle instability of the system during voltage collapse at the load point.

2. Identification of simple power system instability modes

Take the simple double-unit single-load system shown in Fig. 5.51 as an example. Assuming that the voltages of power supplies on both sides are equal to $E_1 \angle 0°$ and $E_2 \angle \delta$ respectively when the internal resistance is ignored, and $E_1 = E_2 = E$, then:

$$\left|\frac{E_{\text{Thev}}}{E}\right|^2 = \frac{Z_1^2 + Z_2^2 + 2Z_1Z_2\cos\delta}{(Z_1 + Z_2)^2} \tag{5.98}$$

Let $Z_1^2 + Z_2^2 = A$, $2Z_1Z_2 = B$. Then we can get

$$\frac{Z_1^2 + Z_2^2}{Z_1^2 + Z_2^2 + 2Z_1Z_2} = \frac{A}{A + B} \tag{5.99}$$

Assuming $\frac{A}{A+B} = x$, from formula (5.99), it can be obtained that

$$\frac{1}{2} \le x \le 1 \tag{5.100}$$

Let $\left|\frac{E_{\text{Thev}}}{E}\right| = y$. From formula (5.98), it can be obtained that

$$y^2 = x + (1 - x)\cos\delta \tag{5.101}$$

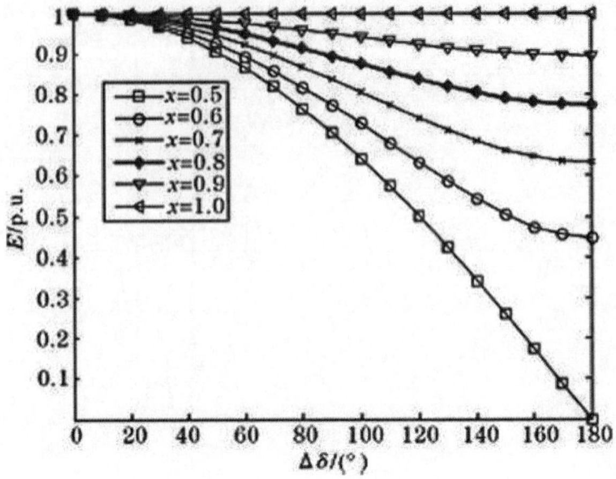

Fig. 5.53 Correlation between Thevenin equivalent potential and rotor angle difference at different Thevenin equivalent points

When $E = 1$ p.u., the correlation between Thevenin equivalent potential and rotor angle difference and the position of load point is shown in Fig. 5.53.

As can be seen from Fig. 5.53, the magnitude of Thevenin equivalent potential at different load points of the system decreases monotonically when the rotor angle increases. When the load point is located at the oscillation center after the instability of rotor angle of the system, Thevenin equivalent potential reaches the minimum, as the curveshown in Fig. 5.52.

Through the above analysis, when the voltage at the load point collapses in the simple double-unit single-load system, the criteria distinguishing rotor angle instability from voltage instability are as follows:

(1) If Thevenin equivalent potential at the load point continuously declines during the instability, it is rotor angle instability;
(2) If load impedance modulus at the load point is less than Thevenin equivalent impedance modulus during the instability, it is voltage instability.

3. Identification of complex power system instability modes

Regardless of the complexity of the power system, the system can be equivalent to a single-unit system, namely, Thevenin equivalent potential source from the perspective of a certain load point, which supplies power to the node load via equivalent impedance at any instant. Therefore, when the voltage at the load point collapses, rotor angle instability and voltage instability can still be identified using the above method.

In case of rotor angle instability, the voltage at the node near the oscillation center will decline sharply, and Thevenin equivalent potential will also decline sharply. Moreover, load equivalent impedance modulus at the load point is larger than Thevenin equivalent impedance modulus. Therefore, this way can be applied to identify rotor angle instability mode.

When the voltage at the load node collapses, the voltage will decline sharply, but Thevenin equivalent potential changes by a little, and load equivalent impedance modulus at the load point is smaller than Thevenin equivalent impedance modulus. Therefore, this way can identify the voltage instability mode.

When the voltage of the local receiving-end system collapses, the voltages at multiple load nodes will decline sharply. Under the influence of voltage drop at nearby nodes, Thevenin equivalent potentials at some load nodes also decline. In this case, it is difficult to identify the instability mode through the change of Thevenin equivalent parameter of one node. However, there is still a point at which the load equivalent impedance is less than Thevenin equivalent impedance in the system in which the voltage has collapsed, which can also be used to identify the voltage instability mode.

In terms of the receiving-end system connected with the sending-end system by a single line, the whole receiving-end system can be equivalent to a single load in according to the Thevenin's theorem. Then, the instability mode can be identified with the above method.

In terms of the receiving-end power grid or looped network connected with the sending-end power grid by multiple lines, since the sending-end power grid cannot be equivalent to a single port according to the Thevenin's theorem, it is difficult to identify the instability mode with the above method, and further in-depth research is needed.

5.3.6 Combination of Singularity Induced Bifurcation (SIB) Theory and Energy Function Method

1. Algorithm flow

Time-domain simulation method and transient energy function method [37–42] have their own advantages and disadvantages. In addition, the SIB may cause the transient voltage instability. Therefore, based on the time-domain simulation method, combined with transient energy function method and SIB theory, a transient stability analysis method is proposed in this section. The calculation time of the method is the same as that of the time-domain simulation method. Its advantages are mainly in the following aspects: The stability margin is provided. The unstable conditions can be accurately and reliably identified. The problems on transient rotor angle stability and transient voltage stability can be distinguished effectively.

The mechanism of transient rotor angle stability problems is relatively clear, while the mechanism of transient voltage stability problems has been being researched. However, it is a universally recognized fact that any load model has an important impact on the transient voltage stability. The existing load model can generally be represented by ZIP and induction motor models. For the constant impedance model, there is no voltage stability problem in the system. For the constant current and constant power models, there may be a singular surface in the time-domain simulation, and the voltage of the system may collapse. For the induction motor model, the SIB phenomenon in the DAE will be eliminated, and the time-domain simulation can be conducted smoothly. However, the induction motor instability may still cause local voltage instability or collapse. In this circumstance, the induction motor model can be approximated with the following method.

Assuming that the positive constant ε is a very small number, the system of which the load model is induction motor model can be represented by a singularity disturbance model:

$$\begin{cases} \dot{x} = f(x, y) \\ \varepsilon\dot{y} = g(x, y) \end{cases} \tag{5.102}$$

Obviously, the system is well-conditioned and solvable in any state in the form of singularity disturbance model. However, since the constant ε is very small, when g_y approaches singularity, a small change in x will lead to a large change in y, which exactly corresponds to the situation that the rotor angle changes are little, but the voltage declines sharply in physics.

In the actual simulation calculation, when the lowest voltage of the system is lower than a certain threshold (such as 0.8 p.u.), an adjoint matrix is constructed by replacing the induction motor model with a constant power load model and calculated at each step of the simulation. If the calculation results do not converge, it means that the system trajectory meets a singular surface, that is, the voltage of the system collapses; if the calculation results converge, it means that the system trajectory does not meet a singular surface, and the calculation can be continued until the system trajectory meets a singular surface or the time set in the simulation is reached. It can be seen that if the voltage at the load node is the algebraic variable y, in which the transient process is ignored, it is easy to know that the disturbance system features a typical transient voltage collapse near the DAE singular point, thus correlating the singularity with the transient voltage stability. The schematic diagram of the method is shown in Fig. 5.54.

The transient stability domain of power system can be deemed as consisting of the stable manifold of UEP and the singular surface. In the time-domain simulation, it is necessary to check whether the system state has exceeded the potential energy peak state in addition to monitoring the algebraic equation singularity of the system.

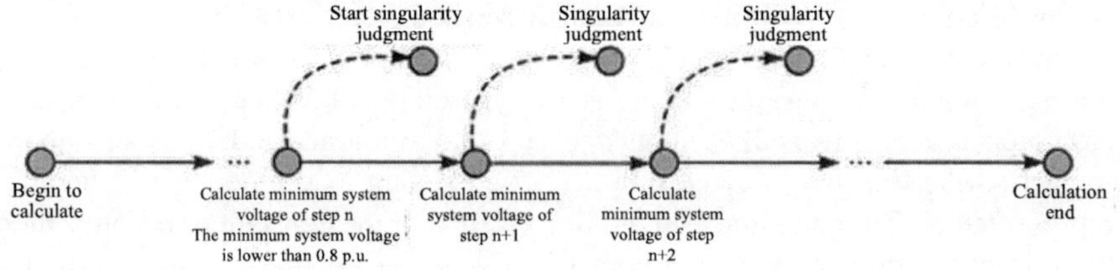

Fig. 5.54 Schematic diagram of calculation method

If the system transient trajectory intersects with the singular surface, the transient voltage instability will occur. Taking the energy of the system at the instability point as the critical energy can define the energy margin corresponding to the transient voltage stability. If the potential energy at the system transient trajectory exceeds the critical energy, the transient rotor angle instability will occur. Taking the energy of the system at the instability point as the critical energy can define the energy margin corresponding to the transient rotor angle stability. If the system transient trajectory does not intersect with the singular surface, the potential energy does not exceed the critical energy, it means that the transient system is stable.

The calculation steps of the method are as follows:

(1) Input original data and calculate the power flow.
(2) Perform the transient stability simulation.
(3) Determine the approximate controlling UEP $\hat{x}$ according to the continuous fault trajectory $x_F(t)$ based on the PEBS method. The critical energy V_{cr} is the value of the energy function $V(\bullet)$ at the controlling UEP, namely, $V_{cr} = V(\hat{x})$.
(4) Calculate the energy function value $V(\bullet)$ at the fault clearing time t_{cl}, namely, $V_{cl} = V(x_F(t_{cl}))$.
(5) If $V_{cl} < V_{cr}$, the transient rotor angle of the system is stable after a fault. At this time, it is necessary to perform a stable time-domain simulation after fault clearing in order to judge whether the system trajectory meets a singular surface. If so, the transient voltage instability occurs, otherwise, the system transient is stable.
(6) If $V_{cl} \geq V_{cr}$, it is necessary to perform a stable time-domain simulation after fault clearing and calculate the transient energy at each step in order to judge whether the system trajectory meets a singular surface or the system energy reaches the critical energy. If the system transient trajectory intersects with the singular surface, the instability mode is the transient voltage instability; if the system energy reaches the critical energy, the instability mode is the transient rotor angle instability.

The corresponding algorithm flow chart is shown in Fig. 5.55.

2. Analysis of calculation examples

The WSCC 4-unit 11-node system is selected for simulation study. The system wiring diagram and parameters are shown in Fig. 5.56 and Tables 5.2, 5.3 and 5.4. Bus 4 is set as an infinite bus. Other generators adopt classical second-order model. The generator damping coefficient is 1 s/rad. The system frequency is 50 Hz. All branch impedance is ignored. The faults are the three-phase ground short circuits on the Bus 6 side in the primary line between Bus 5 and Bus 6. The faulty line is cut off after the fault occurs for a period of time. Since the load model has a significant influence on the transient voltage stability of the system, 2 scenarios (see Table 5.5) are constructed in the calculating example for different load models and simulated respectively.

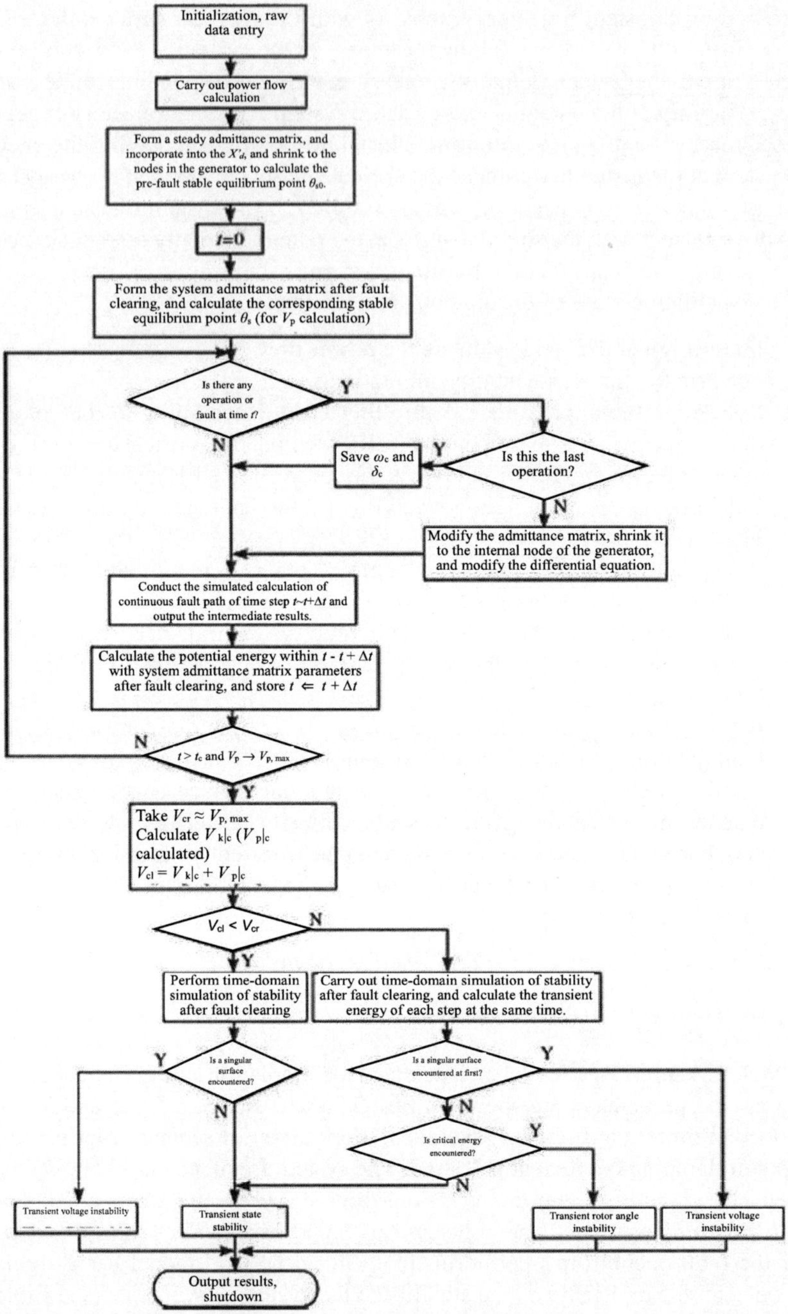

Fig. 5.55 Algorithm flow chart

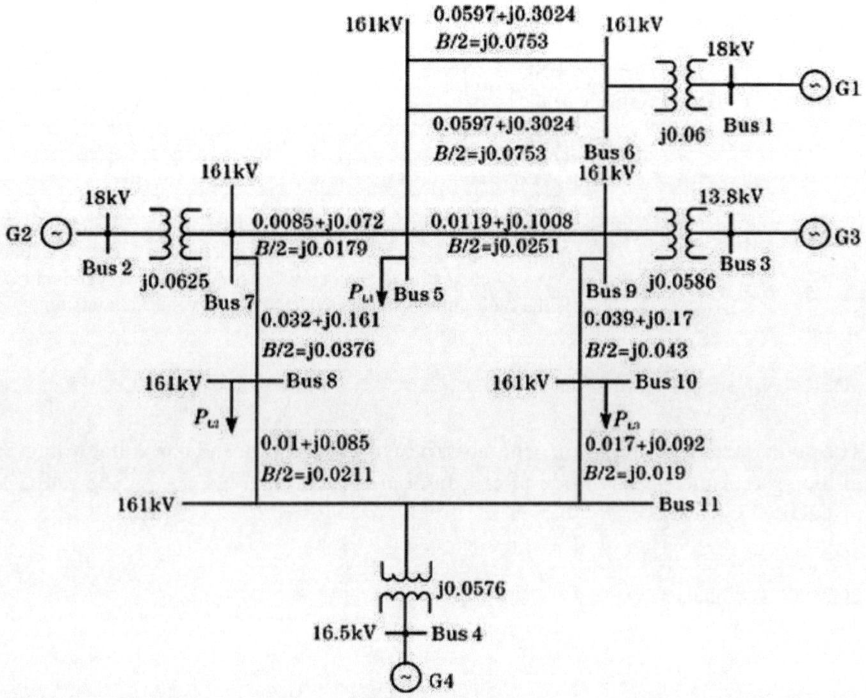

Fig. 5.56 Wiring diagram of WSCC 4-unit 11-node system

Table 5.2 Generator parameters of WSCC 4-unit 11-node system

Generator number	G1	G2	G3	G4
X'_d	0.1198	0.1196	0.1813	0.0608
H (s)	6.40	6.40	3.01	23.64

Table 5.3 Load parameters of WSCC 4-unit 11-node system

Bus 1		Bus 2		Bus 3		Bus 4		Bus 5		Bus 8		Bus 10	
P	U	P	U	P	U	U	θ	P_L	Q_L	P_L	Q_L	P_L	Q_L
1.4	1.03	1.4	1.04	0.8	1.04	1.03	0	2.0	0.6	1.8	0.4	1.8	0.4

Table 5.4 Induction motor parameters

Busbar number	5	8	10
s_0/(rad·s^{-1})	4.3982	4.7124	5.6549
M/(s^2·rad^{-1})	0.0048	0.0057	0.0064

Table 5.5 Table of calculating examples of different load models of WSCC 4-unit 11-node system

Load model	Scene 1	Scene 2
Z/%	100	0
I/%	0	100
P/%	0	0
Motor/%	0	0
T_{cl}/s	0.24	0.22

Note Z, I and P are the percentages of constant impedance loads, constant current loads and constant power loads respectively; T_{cl} is the fault duration when system critical values are unstable

(1) Example 1

The potential energy diagram, generator rotor angle diagram and bus voltage diagram in the system instability mode after a fault are shown in Figs. 5.57, 5.58 and 5.59 respectively.

Fig. 5.57 Potential energy diagram in system instability mode

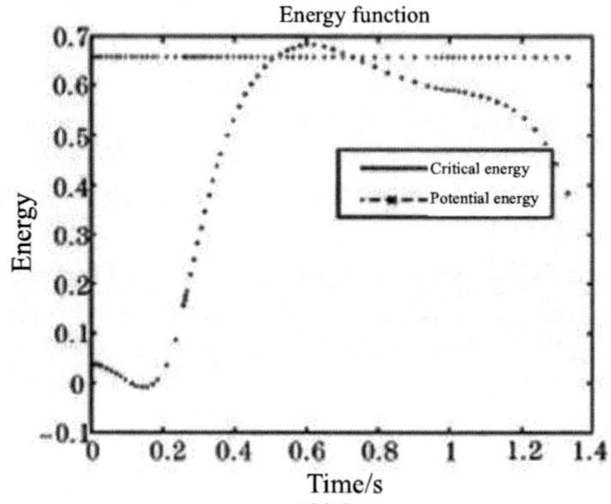

Fig. 5.58 Generator rotor angle diagram in system instability mode

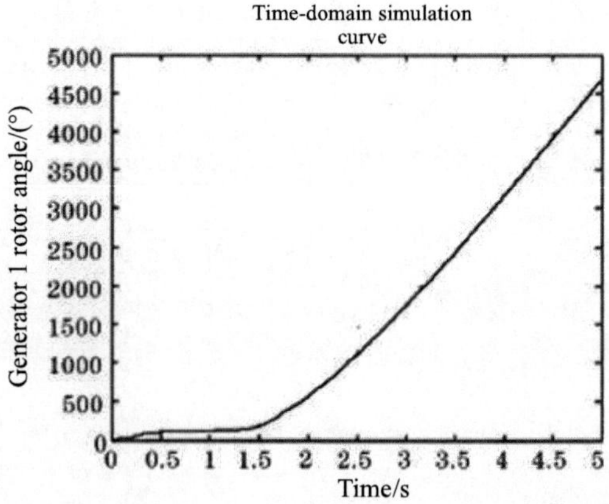

Fig. 5.59 Bus voltage diagram in system instability mode

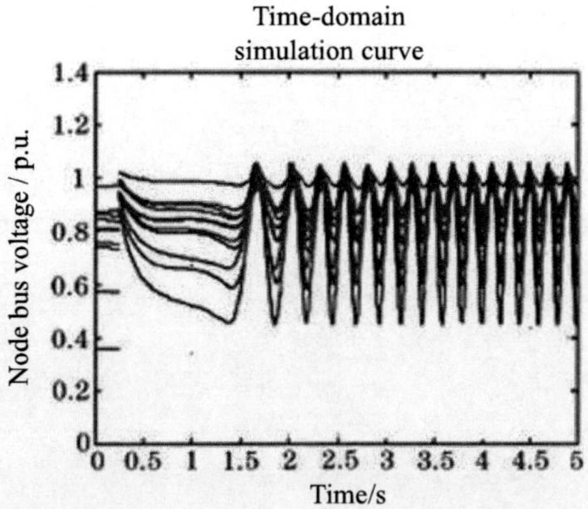

Simulation results in calculating Example 1 show that when the system adopts constant impedance load models, there is no SIB problem; the SIB phenomenon in the DAE will disappear; the time-domain simulation can be performed smoothly in the whole simulation period; the potential energy at the trajectory in the simulation exceeds the critical energy; it can be judged that the instability type is the transient rotor angle instability through the further analysis of the instability curve.

(2) Example 2

Potential energy diagram, generator rotor angle diagram and bus voltage diagram in the system instability mode after a fault are shown in Figs. 5.60, 5.61 and 5.62 respectively.

Simulation results in calculating Example 2 show that when the system adopts the constant power load model, its trajectory first meets the singular surface, and the instability type is the transient voltage instability. It can be seen from the calculating example that the constant power load model easily causes SIB in the multi-unit system. There is often no power supply around the load nodes in the multi-unit

Fig. 5.60 Potential energy diagram in system instability mode

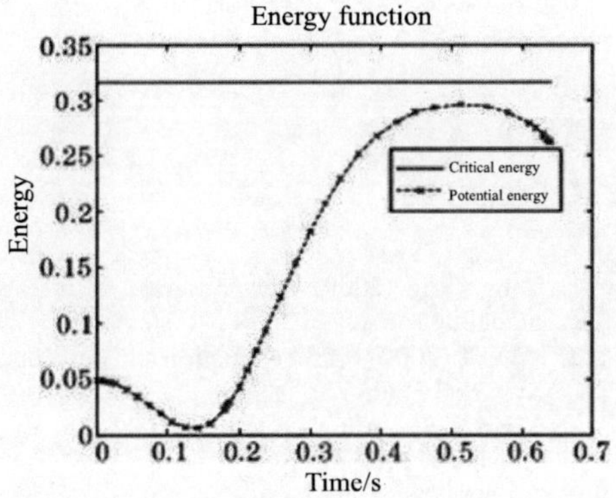

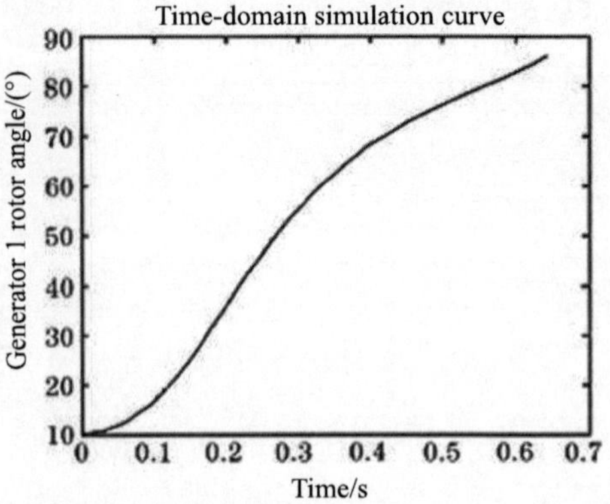

Fig. 5.61 Generator rotor angle diagram in system instability mode

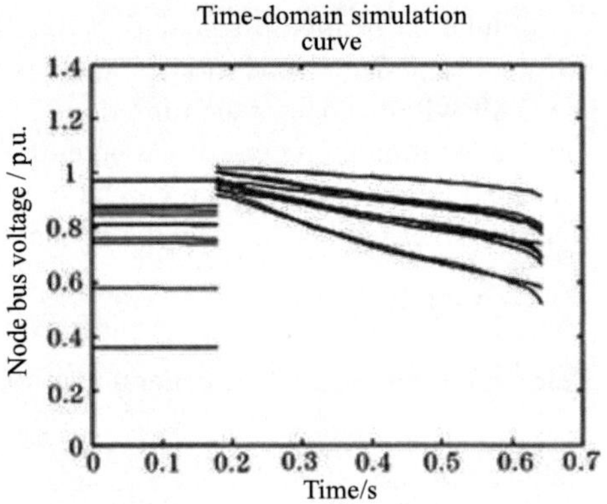

Fig. 5.62 Bus voltage diagram in system instability mode

system. When the voltage of the load and those at its surrounding nodes are very low, the system may not meet the power requirement of the load, so that the conditions for constructing the algebraic equation to describe the system are not met. In addition, the static load has no state variable, so this situation can only be caused by the change of generator state variables. In a word, the change of generator rotor angle causes the SIB.

References

1. Yong Tang (2009) New progress in research on multi-time scale unified simulation and modeling for AC/DC power systems. Power Syst Technol 33(16):1–8
2. Yong T (2002) Present situation and development of power systems. Automation Electr Power Syst 26(17):66–70
3. Yixin N, Shousun C, Baolin Z (2002) Theory and analysis of dynamic power system. Tsinghua University Press, Beijing

4. Fouad AA, Stanton SE (1981) Transient stability of a multimachine power system. Part I and II. IEEE Trans Power Appar Syst 100(7):3408–3424
5. Chiang HD, Wu FF, Varaiya PP (1988) Foundation of the potential energy boundary surface method for power system transient stability analysis. IEEE Trans Circ Syst 35(6):712–728
6. Xue Y, van Custem T, Ribbens-Pavella M (1989) Extended equal area criterion justifications, generalizations, applications. IEEE Trans Power Syst 4(1):44–52
7. Sheng L, Jing W (1996) Analysis of power system transient stability energy function. Shanghai Jiao Tong University Press, Shanghai
8. Venkatasubramanian V, Schattler H, Zaborszky J (1992) Voltage dynamics: study of a generator with voltage control, transmission, and matched MW load. IEEE Trans Autom Control 37(11):1717–1733
9. Yixin Yu (1999) Review on voltage stability studies. Automation Electr Power Syst 23(21):1–8
10. Yixin Y, Chengshan W (1999) Power system stability theory and method. Science Press, Beijing
11. Venkatasubramanian V (1994) Singularity induced bifurcation and the van de pol oscillator. IEEE Trans Circ Syst I 41(11):765–769
12. Venkatasubramanian V, Schattler H, Zaborszky J (1995) Local bifurcations and feasibility regions in differential-algebraic systems. IEEE Trans Automatic Control 40(12):1992–2013
13. Beardmore RE (1999) A study of bifurcation in singular differential equations motivated by electrical power system. Ph.D. Dissertation. Brunnel University, London
14. Yong T, Weifang L, Huadong S et al (2010) Analysis of voltage stability criterion considering load characteristics. Proc CSEE 30(16):12–18
15. Yong T, Huadong S, Jun Y et al (2009) Tracing algorithm for Thevenin equivalent parameters based on complete differential equation. Proc CSEE 29(13):48–53
16. Power Industry Standard of the People's Republic of China (2001) Guide on security and stability of power systems (DL755-2001). China Electric Power Press [CEPP], Beijing
17. IEEE/CIGRE Joint Task Force on Stability Terms and Definitions (2004) Definition and classification of power system stability. IEEE Trans Power Syst 19(2):1387–1401
18. Huadong S, Yong T, Shiying M (2006) A commentary on definition and classification of power system stability. Power Syst Technol 30(17):31–35
19. Liu CQ (1998) A discussion of July 2, 1996 outage in WSCC. Power Eng Rev 18(10):60–61
20. Liu Q (2007) Power system stability and generator excitation control. China Electric Power Press, Beijing
21. Zhihong F (1990) Research on voltage stability of power systems [Doctoral Dissertation]. Tsinghua University, Beijing
22. Wen H, Zhenxiang H (1997) A new method for judging voltage stability and generator rotor angle stability of power systems. Proc CSEE 17(6):367–368
23. Srivastava SC, Varma RK, Tyagi AK (1997) Singularity of reduced Jacobian matrixs for identifying voltage and angle instability. In: Proceedings of the 4th international conference advances in power system control, operation and management, Hong Kong, vol 2, pp 493–498
24. Zhuowei G, Guoqing L (2004) A new method for identifying instability model of electrical power systems. North China Electr Power 8:6–9
25. Padiyar KR, Rao SS (1996) Dynamic analysis of small signal voltage instability decoupled from angle instability. Electric Power Energy Syst 18(7):445–452
26. Peng Zhiwei H, Guogen HZ (2000) Analysis of voltage stability and synchronism stability in power systems. Proc CSU-EPSA 12(1):1–5
27. Wen H, Zhenxiang H (1996) Relationship between voltage collapse and angle instability. Autom Electric Power Syst 20(12):16–19
28. Overby TJ, Pai MA, Sauer PW (1992) Some aspects of the energy function approach to angle and voltage stability in power systems. In: Proceedings of the 31st conference on decision and control, Tucson, pp 2941–2946
29. Overbye TJ, Pai MA, Sauer PW (1992) A composite framework for synchronous and voltage stability in power systems. Int Symp Circ Syst San Diego 5:2541–2544
30. Jing Z, Jinyu W (2006) Analysis of power system voltage stability with equilibrium points. Power Syst Technol 30(10):13–17

31. Tianyu A, Suquan Z (2007) Research on a weak voltage node group's Thevenin equivalents tracing method. Relay 35(24):21–25
32. Zhiming W, Yi W, Jingyou X et al. (2002) Prediction of Thevenin's equivalent parameters on electrical power network side for the muster of key load nodes. Proc CSEE 22(2):16–20
33. Jundong D, Zhizhong G (2006) A method for on-line determination of the operation range of voltage stability. Proc CSEE 26(4):113–117
34. Guodong L, Xiaoru W (2008) Uncertain models for identification of electric power system Thevenin equivalent parameters. Proc CSEE 28(28):74–79
35. Li Laifu Y, Jilai LZ (2005) Research on parameters drift problem in tracking Thevenin equivalent. Proc CSEE 25(20):1–5
36. Yong T, Weifang L, Huadong S et al (2009) Method identifying voltage instability and angle instability based on tracking Thevenin equivalent parameters. Proc CSEE 25(12):1–6
37. Moon YH, Cho BH, Lee YH et al (1999) Energy conservation law and its application for the direct energy method of power system stability//IEEE Power Engineering Society 1999 Winter Meeting. New Jersey: IEEE Press 1:695–700
38. Yong M, Lei C (2007) Transient energy function of power systems including induction motor model. Sci China Ser E: Technol Sci 37(9):1117–1125
39. Chiang HD, Tong JZ, Miu KN (1993) Predicting unstable modes in power systems: theory and computations. IEEE Trans PWRS 8(4):1429–1437
40. Praprost KL, Loparo KA (1994) An energy function method for determining voltage collapse during a power system transient. IEEE Trans CAS, Part I. Fund Theory Appl 41(10):635–651
41. Wuzhi Z, Yong T (2010) Transient voltage stability analysis of differential-algebra equation in power systems. Proc CSEE 30(25):10–16
42. Wuzhi Z (2008) Mechanism research of transient voltage stability in receiving-end grid. Ph.D Thesis. China Electric Power Research Institute, Beijing

Chapter 6
Medium and Long-Term Voltage Stability

Medium and long-term voltage stability is the stability of a power system that is influenced by slow-changing elements (such as: OEL, OLTC action, power recovery characteristics of loads and others) and the slow changing process of the power system (such as AGC, load growth, operation mode adjustment and others) after the transient stability is guaranteed. The object of medium and long-term voltage stability analysis is the voltage phenomenon after transient time frame, the time range can be tens of seconds to tens of minutes even several hours, it studies the ability that the system reaches the acceptable operating equilibrium point state after disturbance.

The increasing long-distance power transmission capacity of power systems results in increasingly prominent heavy load problem of power transmission network, and the medium and long-term voltage stability of powers system after transient stability will be one of the important problems for safe and stable operation of a power system step by step, threatening the safe and stable operation of the power system. It will certainly become one of the important items for power system computation and analysis to analyze the long-term voltage stability problems of power system for avoiding large scope blackout due to voltage collapse and study effective measures for preventing accident expansion.

6.1 Multi-time Scale Full Dynamic Simulation of Power System

6.1.1 Study Range

The full dynamic simulation of power system is a long-term dynamic simulation of the disturbed power system, i.e., the general long-term dynamic stability computation of power system should take into account the long-term and slow dynamic characteristics of power system that are not considered in the general transient stability

© Science Press 2021

Y. Tang, *Voltage Stability Analysis of Power System*, Power Systems,
https://doi.org/10.1007/978-981-16-1071-4_6

simulation, including the dynamic responses of relay protection system, automatic control system, thermal and hydraulic systems of power plant and nuclear reaction system.

The full dynamic stability computation of power system is mainly used to analyze the dynamics of power system from electromechanical transient (tens of seconds) to medium and long-term process (tens of minutes or even hours). The scope of long-term dynamic simulation and research is mainly to—(as follows).

(1) Carry out medium and long-term voltage stability analysis to study the mechanism for voltage stability of power systems and effective measures for preventing voltage collapse;
(2) Simulation study of complex, serious and interlock accidents to understand the dynamic process and essential reasons of the accident, so as to study the right anti-accident measures;
(3) Study accident development processes and provide training to operators on their emergency handling capacity;
(4) Study and arrange load shedding strategies;
(5) Study the effectiveness for emergency reactive power support;
(6) Study the allocation of spinning reserve and distribution of spinning reserve units;
(7) Study auto generation control strategies;
(8) Study the impact of response from boiler control system (including reactor) and auxiliary equipment of power plant after large disturbance on the operation characteristics of power plants and coordinate the control and protection system of power plants;
(9) Assess the system capacity to endure extremely serious faults (severe faults beyond the normal design standard) in the planning and design stages to study measures for reducing the frequency of those severe faults and preventing fatalities.

6.1.2 Simulation Model

Power system is actually a complex large-scale nonlinear dynamic system. It contains a large number of variables based on time constants. Some variables are characterized by rapid change while others are characterized by slow change. The dynamic response related to power system stability can be divided into three processes: rapid change (electromagnetic transient), normal rate (electromechanical transient) and slow change (medium and long-term dynamics, therefore, the power system is at least a three-time scale dynamic system.

Due to the multi-time scale characteristic of the power system, when conducting power system simulation modeling, in order to increase the computational efficiency, electromagnetic transient process is generally considered to be over, electromagnetic transient variable has attenuated, and the rapid dynamic of electromagnetic process is ignored in electromechanical transient and medium and long-term dynamic simulation.

Dynamic behavior of the power system can be fully described by DAE (DAE) with continuous variables and discrete variables.

$$\dot{X} = f(X, Y, Z_c, Z_d) \tag{6.1}$$

$$0 = g(X, Y, Z_c, Z_d) \tag{6.2}$$

$$\dot{Z}_c = h_c(X, Y, Z_c, Z_d) \tag{6.3}$$

$$Z_d(k+1) = h_d(X, Y, Z_c, Z_d(k)) \tag{6.4}$$

In the equation, X indicates all transient state variables; Y indicates the algebraic variables in network equation, including the magnitude and phase of node voltage; Z_c indicates the long (slow) state variables with continuous changes; Z_d indicates the long (slow) state variables with discrete changes. Equation (6.1) describes the transient behavior of the power system, Eq. (6.2) describes network characteristics, and Eqs. (6.3) and (6.4) describe the medium and long-term dynamic characteristics in the system, such as OEL, OLTC, secondary voltage control and others.

In electromechanical transient simulations, it is considered that medium and long-term dynamic processes have not started to change yet, medium and long-term dynamic variables stay constant, that is, with transient voltage stability time-domain simulation uses Eqs. (6.1) and (6.2) to convert Z_c and Z_d variables as constants. Under the medium and long-term voltage stability time frame, the system dynamics will be determined by the action of low-dynamic elements, thus the medium and long-term voltage stability time-domain simulation can directly use Eqs. (6.1)–(6.4); without sacrificing the accuracy of medium and long-term analysis, time frame decomposition technology can be adopted to ignore the rapid process of the system, in order to replace the differential Eq. (6.1) describing rapid process with the algebraic Eq. (6.5):

$$0 = f(X, Y, Z_c, Z_d) \tag{6.5}$$

It means that adopting quasi steady state (QSS) approximate model.

6.1.3 Simulation Method

With the solving technology of DAE, the calculation method based on fixed step size is a widely-used numerical method with the advantage of simple implementation, while the direct use in medium and long-term time-domain simulation can result in excessive calculated amount of simulation. As the models (6.1)–(6.4) describing general power system dynamics include the rapid transient process, in order to accurately simulate this dynamic, smaller step sizes must be used, such as 0.01 s, while the time span in medium and long-term dynamic process is often tens of minutes that require at least a hundred thousand steps for simulation which will waste a lot of time. Due to the large time span of medium and long-term voltage stability study, more efficient numerical integration methods with automatic variable step sizes must be adopted in algorithm; meanwhile, time constants of the model in simulation vary greatly, that is the typical stiffness and nonlinear dynamic system in which the simulation needs to adopt high numerical stability algorithm. Besides, if there are requirements of calculation time and speed, measures adopted to improve computational efficiency shall be considered.

Up to now, the calculation methods of medium and long-term voltage stability simulation mainly include: ① Consider the full dynamic simulation methods of transient process and medium and long-term dynamic processes at the same time; ② QSS simulation method replacing the transient process with its QSS model.

1. Full dynamic simulation

The full dynamic simulation of a power system refers to that unifying the electromechanical transient process, the medium-term dynamic process and the long-term dynamic process of power system organically for simulation, which mainly used in studying the long electromechanical transient process of the power system after large disturbances.

Gear algorithm is recognized as one of the most effective methods in solving stiff problems, especially for the system that requires low calculation accuracy, the numerical integration method of the existing FDS mostly adopts Gear variable step size method (also known as BDF method, backward integration method) such as SIMPOW program [1] of Sweden, EUROSTAG program [2, 3] of French and Belgium, and ETMSP program [4] of EPRI, USA. Its advantages are uniform model and numerical integration method can be used for transient processes and medium and long-term dynamic processes, and large step method can be used in the medium and long-term dynamic processes for simulation. For variable step size Gear integration method, its calculation steps mainly include four parts: prediction, repeated correction, truncation error calculation, and automatic variable order and variable step size control. Gear method is self-starting and starts with the 1st order, as only the Gear method stability domain of 2d order and below can cover the left half plane, thus the maximum order used is the 2nd order. Repeated correction adopts the quasi-Newton method, so that Gear method can meet numerical stability and convergence requirements of stiff system solution. While the main disadvantages of this method in application practice are: ① The calculation step size in electromechanical transient process is so small that results in low overall simulation efficiency; ② Algorithm is difficult to deal with the discontinuous links in the model.

China Electric Power Research Institute started the study of key technology in transient state and medium and long-term dynamic process simulation since 1998 and developed full dynamic simulation program (FDS) [5–15]. FDS adopts a combined numerical integration algorithm which is suitable for unified electromechanical transient and medium and long-term dynamic simulation of a power system, in order to overcome the main disadvantages of Gear variable step size method. The algorithm organically combines the advantages of fixed step size trapezoidal integration method and variable step size Gear method, so as to make them switch automatically in simulation through a certain switching strategy. The detailed descriptions are as follows:

(1) In the electromechanical transient process of full simulation of a power system, fixed step size trapezoidal integration method is used for iterative solution to differential equations of dynamic elements and algebraic equation of power network. In this way, people can not only avoid the smaller step size of Gear method in transient process, but also avoid programming complexity caused by discontinuous processing for the reason that the discontinuous links in control system often occur in the transient stability process.
(2) Variable step size Gear integration method is adopted in medium and long-term dynamic process, so as to combine the differential equation of dynamic elements and algebraic equation of power network for solution. For medium and long-term dynamic process simulation, larger step size calculation can be used to calculate trapezoidal integration methods whose step size is faster than fixed step sizes.
(3) Both fixed step size and variable step size methods can switch automatically based on a certain strategy in simulation, so as to greatly shorten the simulation time and improve computational efficiency of the program on the premise of guaranteeing the numerical stability and simulation precision.

Therefore, the combined numerical integration method for full dynamic simulation of the power system is mainly composed of three parts: fixed step size trapezoidal integration method, variable step size Gear integration method and automatic switching strategy of both methods.

In terms of numerical stability, the numerical stability domain of both single fixed step size trapezoidal integration method and single variable step size Gear integration method (less than or equal to the 2nd order) include the left half of the complex plane to meet the requirements of full dynamic simulation of the power system, and now they are respectively applied in mature power system simulation programs. As a result, the new numerical integration algorithm method can greatly shorten the simulation time and improve computational efficiency of the program on the premise of guaranteeing the numerical stability and simulation precision, overcome the disadvantages that the single fixed step size trapezoidal integration method cannot meet the requirements of medium and long-term dynamic simulation, and the integration step of single variable step size Gear integration method in transient stability stage is too small, which results in too slow simulation.

The switching strategy of the new numerical integration method is based on the different characteristics in electromechanical transient process and medium and long-term dynamic processes: ① Electromechanical transient process belongs to the rapid change stage of the system, the minimum bus voltage in the grid is often less than

0.75 p.u., the variable values of solution change dramatically. At this time, if Gear integration method is adopted, the step size will be shorter with the value less than 0.01 s and even less than 0.001 s that results in slow solution; if fixed step size trapezoidal integration method is adopted, although the number of iterations will be more than 2–3 times, fixed step size integration method solves with DAE without the simultaneous solution of large equation sets, so the speed is faster. ② Medium and long-term dynamic process belongs to the slow change stage of the system. The minimum bus voltage in the grid is often higher than 0.75 p.u. and the variable values of solution change slowly. At this time, if Gear integration method is adopted, the step size will be larger with the value of longer than 0.05 s even longer than several seconds; if fixed step size trapezoidal integration method is adopted, although the number of iterations are generally 2–3 times, fixed step sizes are often about 0.01 s, the simulation speed is slower than that of Gear integration method. Seen from ① and ②, it can be judged that the system is in electromechanical transient or in medium and long-term dynamic process according to the number of iterations of bus voltage, step size and fixed step size integration methods. Besides, in order to reliably judge the dynamic process of the system, one algorithm should continuously calculate certain steps.

All in all, full dynamic simulation of the power system adopting the combined numerical integration method could strengthen the advantages and avoid the disadvantages, effectively solving the slow computation and complex handing in discontinuous links of the existing variable step size Gear integration method in electromechanical transient stage of power systems, so as to greatly improve the simulation efficiency and practicability of FDS of the power system. The full dynamic simulation method has the following main advantages: record carefully the dynamic characteristics of elements, with higher simulation precision; better reflect the full process of voltage instability and provide reliable information for analyzing the voltage collapse mechanism; obtain the preventive and corrective measures for preventing voltage instability.

The time-domain simulation method has made undeniable contributions in analyzing the large disturbance voltage stability of power systems and revealing the voltage instability mechanism, but characteristics of power system present difficulties for the time-domain simulation method in the following two aspects:

(1) Power system is a high-dimensional strongly nonlinear system covering multiple elements with different time constant and dynamic characteristics. The differential equations of the power system model is highly stiff, and in consideration of the numerical stability, convergence of iteration, cumulative errors and other factors of integration by parts, too large step sizes of integration are not allowed. Therefore, it involves heavy computation and needs much time.
(2) With many operating states and imaginary faults in the power system, another integration is required if the network structure, control parameters or fault settings change. Furthermore, time-domain simulation only can draw the conclusion that whether the system is stable or not, any information about

system stability margin is not available, which makes time-domain simulation difficult to realize real-time and on-line stability assessment and control of the power system.

The above disadvantages limit the application of full dynamic simulation in voltage stability analysis, now the method can simulate the long (tens of seconds to several hours) dynamic process of the power system which is mainly used to analyze the influence of dynamic elements on voltage stability, understand the characteristics of voltage collapse, check the correctness of stability criteria, and provide the post analysis for complex and severe accidents.

2. QSS simulation method

In recent years, van Cutsem and others proposed to replace the transient process with the QSS equilibrium equation [16, 17]. This method has the advantages of fast calculation speed and clear physical concept, and has been widely concerned in medium and long-term voltage stability analysis. QSS analysis method approximately simulates the long-term response process of the system with a series of short-term equilibrium points driven by long-term dynamic processes, which is a compromise between static power flow calculation and full time-domain simulation calculation. Its core concept is to replace the transient process of the system with balance points, and only keep the dynamic equations related to the medium and long-term processes. However, the QSS approximation is based on the following assumption: there is a stable equilibrium point in short-term dynamic process; the short-term dynamic processes can reach the equilibrium point in a short enough time.

The basic process of QSS simulation calculation is as shown in Fig. 6.1. In the actual simulation calculation, the simulation time step size usually takes 1–10 s. Each point on the curve in the figure represents a transient equilibrium point, namely, the solution to Eqs. (6.2) and (6.5) when Z_c and Z_d are fixed at the current value. Points

Fig. 6.1 Schematic diagram of QSS simulation

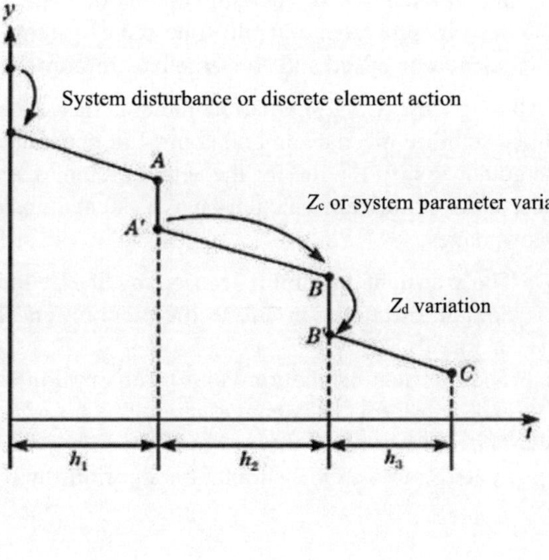

A to A' and B to B' represent the jump of equilibrium points, and the instantaneous jump of y state is derived from the line breaking, load surge, OLTC action, OEL action and others described by Eq. (6.4). Check the state of the equipment in each step of the time-domain simulation, and change the state when the action condition of the equipment is satisfied. The curves $A'B$ and $B'C$ in the figure are driven by the continuous dynamic process in the system, which is the system slow dynamic change described in Eq. (6.3), or caused by the slow change of the parameters in the system, such as the continuous increase of the system load. Since the simulation step size is generally smaller than the time constant of Eq. (6.3), the explicit numerical integration method can be used. In addition, discrete state transitions may occur in each simulation time step, so the single step size integral method is used. The medium and long-term dynamics element process is described by Eqs. (6.3) and (6.4). The action effect of discrete dynamic element can be reflected in the jump of AA' and BB', while the action effect of continuous dynamic element can be reflected in the change processes of point A' to point B and Point B' to point C.

Compared with the full dynamic simulation method, the QSS simulation method has the advantages of fast speed and good convergence in the study of the medium and long-term voltage stability, as shown in the following aspects:

(1) The transient process is replaced by balance equations. In the full simulation calculation, all transient and medium and long-term dynamic processes need to be solved, and the differential equation has a higher order. QSS simulation method approximately replaces the transient process with its balance equations, greatly reducing the number of differential equations to be solved and the amount of computation.

(2) Time scale decomposition can be used to solve transient and medium and long-term processes separately. In the full dynamic simulation, due to the limitation of transient process, the step size cannot be too large, and the increase of time steps may lead to error accumulation. In the case of stiff problems of differential equation system, large step size may affect the stability, convergence and efficiency of numerical calculation. The QSS simulation method avoids the stiff problems of multi-time scale system, the step size is larger, and the calculation speed and convergence are improved.

Although the QSS simulation method has obtained many conclusions about voltage stability mechanism and control in general sense, and obtained some valuable guidance information for the actual system operation, however, as the premise of the QSS simulation calculation is system transient stability, it also has some disadvantages:

(1) The transient instability caused by the reduction of the transient stability domain of attraction due to the medium and long-term state change cannot be found;

(2) The transient oscillation Hopf bifurcation type instability caused by medium and long-term state changes cannot be found;

(3) The transient instability caused by medium and long-term dynamics cannot be treated, that is, QSS simulation algorithm will fail when Eqs. (6.2) and (6.5)

have no solutions under given Z_c and Z_d values. This phenomenon indicates that medium and long-term state changes may lead to loss of the transient stability equilibrium point of the system.

6.2 Relationship Between Medium and Long-Term Voltage Stability and Static and Transient Voltage Stability

6.2.1 Relationship Between Medium and Long-Term Voltage Stability and Static Voltage Stability

Steady voltage stability refers to the fact that the system voltage can be maintained or restored to the allowed range after the system suffered a small disturbance, without voltage collapse; it is mainly used to define the voltage steady state stability reserve of the system under normal operation mode and post-accident operation mode. Most of the steady state voltage stability analysis methods are based on some understanding of the voltage stability mechanism in steady state, and the system operating state in network transmission power limit is usually taken as the limit state of steady state voltage stability. The basic model for analyzing the steady state voltage stability is the power flow equation or the extended power flow equation of the power system. Therefore, in essence, static voltage stability analysis is a problem to study whether there is a feasible solution to the power flow equation. The advantage of steady state voltage stability analysis is that the study of the solution of a complex differential–algebraic equation can be regarded as the study of the existence of real solutions of a simple nonlinear algebraic equation; its disadvantage is that it cannot reflect the dynamic characteristics of each component.

Medium and long-term voltage stability analysis is an extension of steady state voltage stability analysis. The capacity of medium and long-term voltage stability analysis is greatly improved at the cost of that the system model and numerical methods are complicated. Compared with steady voltage stability analysis based on power flow, medium and long-term voltage stability analysis has the following advantages:

(1) There is no assumption of PV node or $V\theta$ node. In power flow calculation, the assumption of PV nodes and $V\theta$ nodes are often set, but such electrical elements are non-existent in the actual operating system. In time-domain simulation method, when the load increases, the generator terminal voltage will drop, and the load increment will be distributed among the generators through the frequency characteristics of the generator, and the simulation is consistent with the actual situation of the power system.

(2) It can accurately calculate the limitations of various controllers. In the power flow calculation of steady voltage stability analysis, regulate the reactive power output to maintain a constant generator terminal voltage, represent the generator nodes as PV nodes, give a simple constant to the reactive power output of

generator, and give a constraint to the minimum reactive power output Q_{min} and the maximum reactive power output Q_{max}, when the generator reactive power output exceeds this constraint, this node is changed to *PQ* node, in which the reactive power constraint is roughly considered. In the medium and long-term simulation calculation, the maximum reactive power output constraint of the generator can be caused by the maximum stator current limit or the maximum excitation current limit of the generator. In the time-domain simulation, the generator excitation limit and stator current limit related to voltage stability are simulated in detail.

(3) It can reflect the law of time domain of control action and element action. In power flow calculation, although some automatic regulation behavior of some devices can be dealt with, the characteristics of regulating law in time are not considered, so it rarely reflects the actual situation of the system.

Through medium and long-term voltage stability simulation analysis, it can study the mechanism of the occurrence and development of voltage collapse, to more clearly grasp the time-varying characteristics that the system changes with the operating conditions, accurately know the order of action of each element and its effect, and the effect on system dynamic performance when the system is deviating from the steady equilibrium point in slow dynamic, whereas, the information is difficult to know in steady voltage stability analysis. It can be predicted that with the rapid development of computer hardware technology and the development of medium and long-term voltage stability simulation method, this method will become a useful supplement to steady voltage stability method, and will be gradually popularized and applied.

6.2.2 Relationship Between Medium and Long-Term Voltage Stability and Transient Voltage Stability

Transient voltage stability, namely, short-term large disturbance voltage stability, refers to the capacity of an electric power system to avoid voltage collapse after large disturbance. In power systems, there are a large number of fast-response dynamic elements, such as induction motor load, generator and excitation control system, HVDC, and SVC, and their transient action can directly lead to transient voltage instability of the system. Medium and long-term voltage stability refers to the voltage stability of the system under the action of dynamic elements and control devices with slow response, such as OLTC, generator stator current limitation, OEL, and secondary voltage control, and their slow dynamic behaviors can directly lead to medium and long-term voltage instability of the system.

The dynamic process of a power system (from electromechanical transient process to medium and long-term dynamic process) is a continuous process, not a discrete separate process. The transient process has an effect to the medium and long-term processes, and the medium and long-term process also affects the subsequent new transient processes.

A typical case of transient voltage instability caused by the medium and long-term voltage instability is: after an initial disturbance, the system continues to operate for a short time; from this moment, the system is driven by a medium and long-term dynamics described by the variables Z_c and Z_d; due to medium and long-term voltage instability, the system stability condition deteriorates, which causes either the generator to be out of synchronism due to limited excitation current, or the generator tripping due to limited stator current, or the induction motor speed loss. In such cases, medium and long-term voltage instability is the cause, while transient voltage instability is the final result. The slow system state deterioration due to medium and long-term instability eventually leads to a sudden transition process of voltage collapse.

6.3 Analysis of Medium and Long-Term Voltage Stability Scenarios

6.3.1 Overexcitation Limit of Generators

The over-excitation limiter (OEL) of a generator is also called the maximum excitation limiter or exciting current limiter. It is a device set in the AVR to prevent the over current of generator rotor winding, which limits the output of the regulator to limit the generator rotor current by calculating the calorific value of the excitation winding for reaching certain constant when the excitation current exceeds the long-term operating maximum, so as to protect the rotor of the generator.

The generator over-excitation limit model is determined by the generator overheat load capacity. It is to avoid the generator overheating caused by long time excitation overcurrent. As shown in Fig. 6.2, the rotor overcurrent time is related to excitation current peak value.

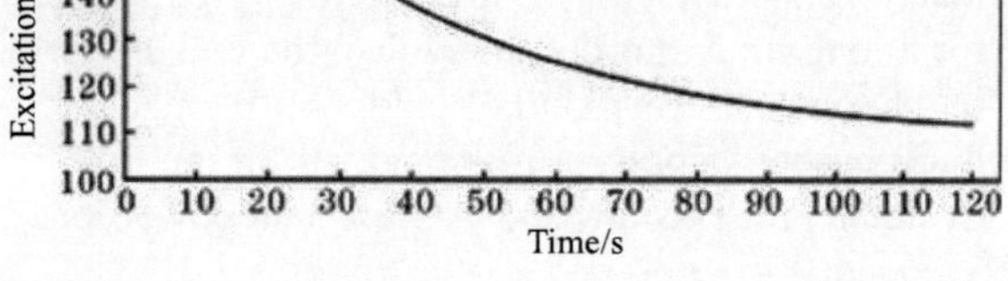

Fig. 6.2 Generator excitation winding current overload capacity

Excitation regulator is the main means of regulating voltage in the power system, but the regulating range is limited to a certain extent. Once the excitation regulation of a generator in the system reaches the maximum limit value (generally caused by long-term too low system voltage, AVR causing excitation increase or AVR fault), due to the thermal capacity limitation of the excitation winding, when the over excitation capacity of the unit reaches the set operating time, the over excitation limiter will reduce the excitation current to the rated value, resulting in a sudden reduction of reactive power in the system and a sudden drop of voltage in the system. Since other generators are generally in a state near the limit at this time, the sudden excitation reduction of the generator may cause a chain reaction of other generators, thus causing voltage instability of the system. After the OEL action, measures such as generator shutdown are generally not taken immediately, but to keep the generator in operating state as far as possible.

6.3.2 On-Load Voltage Regulating Transformer

The regulating of on load tap changing transformer (OLTC) is considered to be the leading mechanism of medium and long-term voltage instability, the main reason is: when the load side voltage drops, the OLTC action is equivalent to bring certain active and reactive power reserve to the load side, but actually OLTC did not increase the reactive power of the system; after a period of time, OLTC restored its power while restoring the load side voltage, which is unfavorable to the voltage stability of the whole system; the constant adjustment of OLTC also makes the primary side voltage lower, which could eventually leads to voltage collapse.

The OLTC tap shall be regulated step by step while monitoring the tap position and the variation of voltage and current. It will take several seconds for the tap changer to complete a gear change. However, in order to avoid frequent operation of tap changer for voltage regulation, delay time shall be set for voltage booster and drop. Generally, a delay time of 30–60 s shall be set for 110 kV voltage regulation, and 60–120 s seconds for 35 kV voltage regulation. According to the range of such delay time, the function of OLTC only affects the medium and long-term voltage stability of the system, and is independent of transient voltage stability.

6.3.3 Constant Temperature Load

During continuous low voltage of the grid, constant temperature loads and other load regulators attempt to restore load power. For example, a heating load will work for a long time in order to regulate the temperature to the temperature required by the thermostat. This kind of loads can be regarded as constant impedance loads in the transient stability time frame, but in the time frame of medium and long-term dynamic process, they show as constant power loads.

Since constant power load will seriously deteriorate the voltage stability of the system, the constant power characteristic of constant temperature load should be considered in medium and long-term voltage stability analysis.

6.3.4 *Continuous Increase of the Load*

In the process of power system operation, the power load changes constantly with time, and the slow and continuous increase of the load is normal in the operation of power systems. Since the increase of load is slow, it is not easy to detect the increase in a short time. After a period of continuous increase, the system may approach the critical state, thus causing voltage instability. Therefore, it is necessary to analyze the voltage stability of the power system under the condition of slow and continuous load increase.

For the long-term voltage stability problem caused by the continuous increase of load, it is mainly because, after the continuous increase of load, the increase of generator excitation current may lead to the action of over-excitation limiting device, resulting in the serious voltage decrease in local areas, and the voltage stability problem. The example of a typical daily load curve is shown in Fig. 6.3.

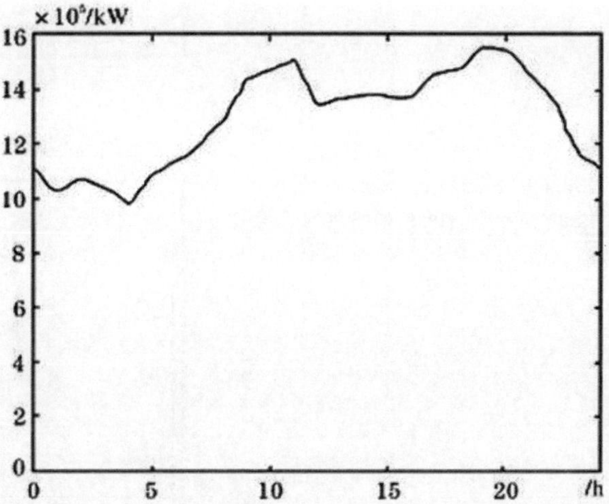

Fig. 6.3 Example of a typical daily load curve

6.4 Example of Medium and Long-Term Voltage Stability Analysis

In this section, an example of simulating calculation of medium and long-term voltage instability is given [18, 19]. The system wiring diagram is shown in Fig. 5.11. The main models used in the simulation are as follows: the transformer (Bus 9-Bus 10) is an OLTC transformer, and other taps remain unchanged; the load on Bus 7 is a constant power model, and other loads are constant impedance models; the Generator 2 and Generator 3 (Bus 2 and Bus 3) have OEL devices, and the Generator 1 (Bus 1) is an infinite generator. For other models and parameters, please refer to reference [19]. There is a 500 kV line between Bus 5 and Bus 6 and it trips at 1 s. The evolution of system status after the line trips is shown in Figs. 6.4, 6.5, 6.6 and 6.7.

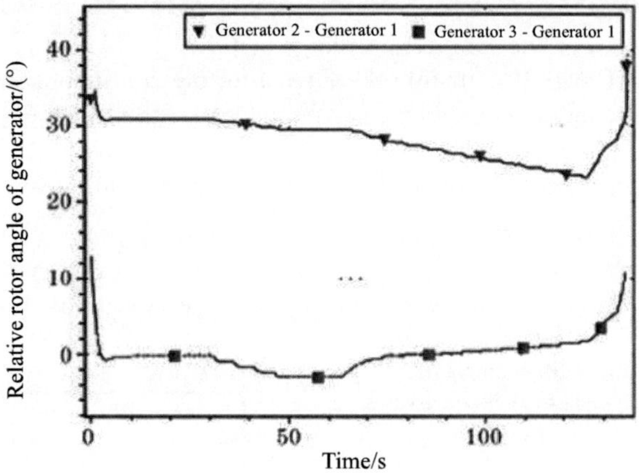

Fig. 6.4 Change of relative rotor angle of generator

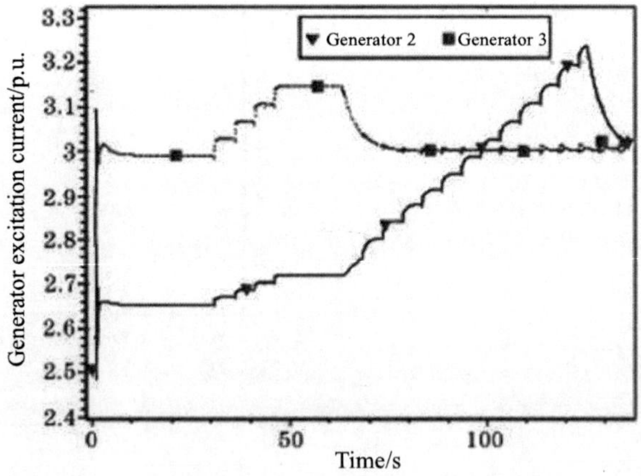

Fig. 6.5 Excitation current of generator 2 and generator 3

Fig. 6.6 Voltage of bus 9

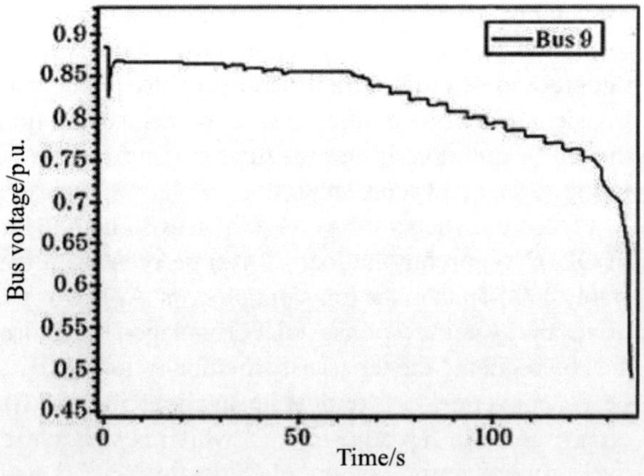

Fig. 6.7 Tap transformation ratio and load voltage

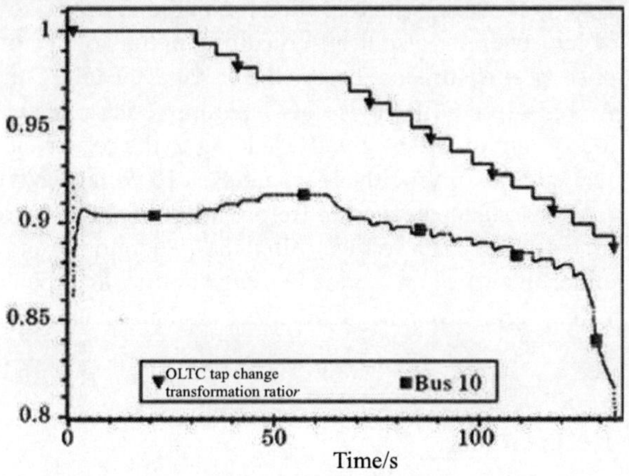

Figure 6.4 shows the swing curves of relative rotor angle between Generator 2 and Generator 1 and between Generator 3 and Generator 1 respectively. It can be seen from Fig. 6.4 that the initial fast transient process caused by this disturbance will disappear quickly, indicating that the system can keep the transient rotor angle stability; the unit can still keep synchronous operation despite the rotor angle swing in the subsequent medium and long-term processes, indicating that the system can also maintain the medium and long-term rotor angle stability.

The excitation current of the over-excitation limiter in Fig. 6.5 shows the response of the electromotive force E_q of generator 2 and generator 3, which is proportional to the excitation current. As shown in the figure, after the disturbance, the excitation current of generator 2 and generator 3 will suddenly rise. If the excitation current exceeds the rotor current limit, it will start the inverse time mechanism of over-excitation limiter. After the disturbance, the operation of OLTC imposes a very heavy reactive power demand on the generator. This demand further deteriorates the rotor overload, until the over-excitation limiter is eventually excited, causing the excitation current to return to the rated value. Note that the overexcited limiter is of integral type so that E_q is forced to Elim q. The subsequent tap change causes the transient excitation current to rise, the risen excitation current is quickly detected by the over-excitation limiter (as the maximum excitation current of generator 2 and 3 in Fig. 6.5), and is corrected.

Figure 6.6 shows the voltage of Bus 9, namely, the bus voltage on the HV side of OLTC supplying the load. It can be seen from the figure that Bus 9 can operate stably at 0.87p.u. in the transient process. After this point, it is OLTC converter which drives the system response. OLTC managed to restore the Bus 10 voltage on the load side by reducing the tap transformation ratio r. After an initial time delay of 30 s, the OLTC converter starts to operate, at about 58 s, the Bus 10 voltage rises to 0.915 p.u. after 4 times of tap adjustments, which is very close to the pre-accident level, and the excitation current output of Generator 2 and 3 also increases to meet the system's demand for reactive power (Fig. 6.5). However, at 63 s, the OEL device of Generator 3 starts to work, which limited the output current, so that the reactive power output of the generator also drops, resulting in the voltage drop of Bus 10 at the load point once more. In order to ensure the voltage, the OLTC tap continued to act for 12 times. At 125 s, the OEL device of Generator 2 starts to act and the reactive power gap of the system increases greatly, leading to the occurrence of voltage collapse.

Figure 6.7 shows the load side Bus 10 voltage and the response of transformation ratio of load tap converter. Before the generator excitation limit, each tap action will increase the load voltage; after the generator excitation limit, the tap change has almost no effect on the load voltage at first, and then, these changes reduce the load voltage and have a negative effect on the voltage stability of the system.

References

1. ABB Corporation (2005) Program operation manual of SIMPOW. ABB
2. Stubbe M, Bihain A, Deuse J (1989) STAG: a new unified software program for the study of the dynamic behavior of electrical power systems. IEEE Trans Power Syst 4(1):129–138
3. Astic JY, Bihain A, Jerosolimski M, et al (1994) The mixed Adams-BDF variable step size algorithm to simulate transient and long term phenomena in power systems. IEEE Trans Power Syst 9(2): 929–935
4. Sanchez-Gasca JJ, D′ Aquila R, Raserba JJ, et al (1993) Extended-term dynamic simulation using variable time step integration. IEEE Comput Appl Power 6(4): 23–28
5. Yong T (1999) A compensation method for simultaneous faults solution in power system stability simulation. Proc CSEE 19(7):77–79
6. Yong T (1997) An implicit integration alternating solution method for power system dynamic stability simulation. Power Syst Technol 21(2):1–3
7. Yong T, Xinli S, Wenzhuo L, Xiaoxin Z (2002) Power system full dynamic simulation Part I: numerical method. Power Syst Technol 26(9):7–12
8. Yong T, Wenzhuo L, Xinli S, Xiaoxin Z (2002) Power system full dynamic simulation Part II: fault simulations. Power Syst Technol 26(10):1–5
9. Yong T, Xinli S, Wenzhuo L, Xiaoxin Z (2002) Power system full dynamic simulation Part III: long-term dynamic models. Power Syst Technol 26(11):20–25

10. Yong T, Xinli S, Wenzhuo L, Xiaoxin Z (2002) Power system full dynamic simulation Part IV: case studies. Power Syst Technol 26(12):5–8
11. Yong T (2002) Present situation and development of power system. Autom Electr Power Syst 26(17):66–70
12. Xinli S, Zhaoxu L, Yongzhuang Li, Yong T, Zenghuang L (2008) Analysis on speed governing system model for fossil-fuel generating plant and its application in power system stability simulation. Power Syst Technol 32(23):44–49
13. Wenzhuo L, Yong T, Lei W, Xinli S (2008) Modelling and simulation technologies for large UHVDC power grid. Power Sys Technol 32(22):1–3, 7
14. Yong T (2009) New progress in research on multi-time scale unified simulation and modeling for AC/DC power systems. Power Syst Technol 33(16):1–8
15. Xinli S, Yong T, Wenzhuo L, Zhong Wuzhi Wu, Guoyang LT (2009) Mixed numerical integral algorithm for dynamic simulation of the power system. Proc CSEE 29(18):23–29
16. van Cutsem T (1996) Voltage stability analysis in transient and mid-term time scales. IEEE Trans Power Syst 11(1):146–154
17. van Cutsem T, Mailhot R (1997) Validation of a fast voltage stability analysis method on the Hydro-Quebec system. IEEE Trans Power Syst 12(1):282–292
18. Taylor CW (1992) Concepts of undervoltage load shedding for voltage stability. IEEE Trans Power Deliv 7(2):480–488
19. Taylor CW (2002) Power system voltage stability. Translated by Wang Weisheng. China Electric Power Press [CEPP], Beijing

Chapter 7
Technical Measures to Improve Voltage Stability

Power system voltage stability is related to many factors, and different accident mechanisms may lead to voltage collapse accidents with different dynamic characteristics. Therefore, there are variable measures to prevent voltage instability, with different emphasis according to different voltage instability modes.

When the power system approaches short-term voltage instability, the fast-responding dynamic reactive power regulator near the load center can improve the post-accident voltage stability level of the system and help the voltage recovery of the system. Generally, the dynamic reactive power regulator mainly includes generator, synchronous condenser and SVC. There shall be enough dynamic reactive power reserve before disturbance, but it should be noted that the overload capacity provided by generator and synchronous condenser has a time limit, they may transfer the short-term voltage stability problem to the long-term voltage stability problem, while SVC is only equivalent to ordinary capacitor in the ultimate state of regulation, and the emergency control measures shall also include the concentrated or low voltage load shedding measures to prevent voltage collapse.

The most important measure to prevent the long-term voltage instability is to restore the long-term stable operation equilibrium point and avoid the short-term voltage instability caused by the long-term dynamic process. Measures to be taken include switching of reactive power compensation, re-dispatching and load shedding. For multi-layer power transmission and distribution systems, OLTC shall be coordinated and controlled to avoid further deterioration of voltage stability level. In addition, when the generator has reactive power reserve, the generator terminal voltage can be increased through secondary voltage control.

The capacity of the system to provide dynamic reactive power support and the voltage reactive power characteristics of the load are the key factors to determine the voltage stability. Therefore, in general, increasing the dynamic reactive power reserve of the system and improving the voltage reactive power characteristics of the load are the main measures to improving the voltage stability.

© Science Press 2021

Y. Tang, *Voltage Stability Analysis of Power System*, Power Systems,
https://doi.org/10.1007/978-981-16-1071-4_7

7.1 Classification of Measures for Improving Voltage Stability

The measures to improve the power system voltage stability should be comprehensive and all-around. The various measures not only aim at specific phenomena (such as transient voltage stability, medium and long-term voltage stability) but also cooperate with each other to play a role together. At present, in the planning, design, operation and management of power system, corresponding measures shall be considered to improve the voltage stability of power system.

This section summarizes and classifies the main measures to improve the power system voltage stability.

1. Measures at the planning and design stage

(1) To optimize the network structure. Voltage instability accidents usually occur in heavy load power systems. However, after research, it is found that the so-called heavy load does not mean that most of the line load in the system are very heavy, and the actual situation is that the overall load rate of the power grid before the occurrence of voltage instability accidents is not too high, most of the line load is still relatively light, and only a few of the lines operate with heavy load. Therefore, it is necessary to plan and optimize the network structure reasonably in the planning and design stage to ensure that the system has sufficient transmission capacity, and to avoid the fault of multi-loop parallel lines or AC–DC parallel lines as far as possible, which may cause large-scale load transfer and lead to voltage collapse caused by cascading fault.

(2) To strengthen reactive power planning. The basic requirements of reactive power planning are to analyze the reactive power characteristics under the steady state and dynamic operation conditions of the power grid, reasonably configure the reactive power compensation equipment and voltage regulation means, improve, control and optimize the distribution and flow of reactive power in the whole network, so as to improve the system operating stability, suppress line voltage rise, and meet the requirements of safety operation and voltage regulation of the power grid in normal operation and after accident disturbance. Reactive power planning shall meet the principle in layers and zones, the operation standard of normal and post-accident mode voltage, and the requirements of power grid stability and dynamic reactive power reserve under accident disturbance.

(3) To provide adequate reactive power compensation equipment in the load center. From the theoretical study and the analysis of previous voltage instability accidents, one of the important reasons for voltage collapse in power systems is the lack of reactive power compensation capacity in the system. Therefore, in the power grid planning stage, it is necessary to consider enough reactive power compensation capacity (including steady state and dynamic reactive power compensation capacity), so that the load center (receiving-end power grid) has

sufficient reactive power regulation capacity and strong voltage support, with a certain margin reserved to meet the demands of future load increase.

2. Measures for dispatching operation stage

(1) To optimize the operation mode. Power system is a large nonlinear system. Under certain load conditions, there are many kinds of power flow states that meet the requirements of power generation and transmission, that is, there are many feasible equilibrium points in the system. The important measure to improve the power system voltage stability is to select a proper equilibrium point and make the system operating equilibrium point have a large voltage stability margin. In addition, in some operation modes, there are fault modes that may lead to voltage instability in the system. Adjusting the operation mode can often avoid this potential danger.

(2) To optimize reactive power compensation. The optimization of reactive power compensation can, on the basis of ensuring system voltage safety, reduce the reactive power transmission of the line, help realize the balance of reactive power in layers and zones, improve the voltage stability while reduce the line active power loss, reduce the operating cost of the system and improve the economic efficiency of power grid operation.

(3) To establish a voltage safety monitoring and control system (such as AVC) to monitor the voltage and reactive power of the power grid in real time. By using the real-time monitoring system, the dispatcher can find the weak link of voltage stability in the system in time, and take measures to eliminate the hidden danger in the system before the accident occurs.

(4) To strengthen the power factor management at user side. Strengthening the power factor management at the consumer side can reduce the amount of reactive power absorbed by the consumer from the power grid. In case the reactive power compensation capacity at the system side is fixed, it is equivalent to increasing the reactive power reserve of the system.

3. Real-time control measures

(1) Under-voltage load shedding. Under-voltage load shedding is one of the most effective measures to solve the voltage stability problem at present. It has the characteristics of simple principle and high reliability, and has been widely used in power systems at home and abroad. Under-voltage load shedding can be divided into centralized control type and distributed type. At present, the decentralized under-voltage load shedding control is mainly adopted in China.

(2) Low-voltage splitting. The function of under-voltage disconnection is to disconnect the regional power supply network or power interconnection before the accident spreads to the whole power grid to prevent the further expansion of the accident scope and avoid the accident of collapse in the whole network. There are some risks in the under-voltage disconnection, and it is difficult to set the parameters of the device. In the actual project, it is necessary to set and check according to the specific situation of the power grid to ensure the reliability of the device.

(3) Emergency control of OLTC taps. OLTC is one of the main voltage control devices in a power system. When the load suffers a large disturbance, if the load-side voltage level is low, OLTC can automatically regulate the tap to recover the load-side voltage level, so as to recover the corresponding load power. However, the tap adjustment only adjusts the distribution relationship of reactive power at different voltage levels, and cannot increase reactive power. Increasing the load-side voltage level will inevitably reduce the main-grid-side voltage level, thus, and further affect the load-side voltage, which may eventually lead to voltage collapse. In order to avoid and prevent voltage collapse, the OLTC tap shall be locked in the case of voltage drop on the power supply side, if the change of tap is not conducive to stability of the system, and then unlock the tap once the voltage is recovered.

(4) Fast control of DC transmission. Both the active power transmitted by DC and the reactive power consumed by the converter can be quickly controlled by the control system. This fast controllability can be used to improve the operation performance of an AC system. For example, according to the operation requirements of AC system, the improvement of voltage stability of AC systems can be achieved by rapidly increasing or reducing the active power or DC current setting value of DC transmission and regulating the reactive power absorbed by converter.

(5) Automatic switching of reactive power compensation devices. After a fault, quickly putting in the mechanical switching with sufficient capacity or silicon controlled capacitor can effectively prevent the further voltage drop and avoid the voltage collapse. In a system where the shunt reactor is in normal operation, fast removal of reactor can be used to replace the input of the capacitor.

4. Application of new technology

In addition to the above measures, to improve the safety and stability of power system, there are measures such as using new power electronic equipment and control of the existing element control system.

(1) To install SVC, STATCOM, TCSC and other new FACTS equipment, research and use new control strategies, so as to provide dynamic reactive power support to the system and improve the voltage stability of the system.

(2) To improve the excitation control system of generator so that the generator can change from maintaining generator terminal voltage constant to maintaining the constant of HV side bus voltage of boosting transformer. This is equivalent to reducing the impedance of the transmission and improving the transmission capacity of the system.

7.2 Generator Control

Synchronous generator is one of the most important reactive power supply and the most important dynamic reactive power reserve in a power system. The capacity of synchronous generator to provide reactive power is very important to prevent voltage instability accidents of power system.

7.2.1 Reactive Power Reserve Capacity of Generator

Figures 7.1 and 7.2 show an example of steady-state *PQ* operation limit and V-shaped curve of generator [1, 2], from which we can see how the generator reactive power output is limited by the heating of generator excitation winding or generator armature winding. When the power system approaches long-term voltage instability, this limit is one of the important reasons for the final voltage instability of the system.

It can be seen from Fig. 7.1 that if the generator terminal voltage can be mainly maintained near the rated voltage, the reactive power limit of the generator is usually limited by the heating of the excitation winding. If the generator terminal voltage drops significantly, the armature winding current becomes the limiting factor.

According to $Q = P\tan\varphi$ (φ is the power factor angle), when the generator works at a high power factor, the reactive power reserve of the generator is sufficient. As the generator is the most important dynamic reactive power supply with fast regulation characteristic in the power system, once the power system has a voltage emergency,

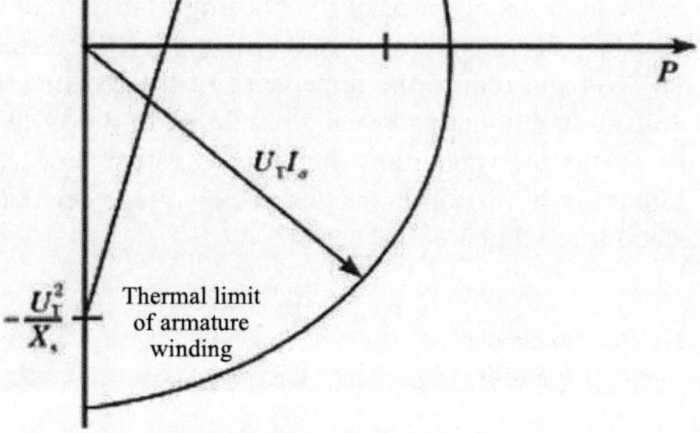

Fig. 7.1 Operating limit diagram of generator

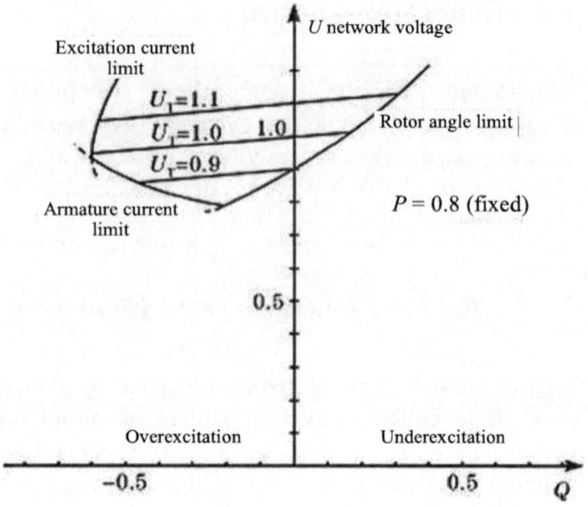

Fig. 7.2 *QV* curve of generator with constant active power output

the magnitude of reactive power reserve capacity of the generator is very important to ensure the voltage stability of the system and prevent the voltage collapse.

Therefore, a very effective method to improve the power system voltage stability is: in the normal operating state, use the shunt capacitor bank to replace the generator as the reactive power supply output, so that the power factor of the generator is close to 1.0 in the normal operation. In this way, the generator reserves a large number of reactive power reserve, which can respond quickly when the power system has voltage drops due to the reactive power shortage, so as to avoid the voltage collapse.

7.2.2 *HV Side Voltage Control of Generator*

At present, for most of the generators operating in the power system, the control target of the matching AVR is generally to maintain the generator terminal voltage constant. If the control target of AVR can be changed to maintaining the bus constant voltage at the high voltage side of the boosting transformer at the generator outlet (that is, the HV side voltage control), it will shorten the distance between the generator and the load and reduce the impedance of the transmission, which is beneficial to the voltage stability and rotor angle stability of the system.

Taking the single-unit single-load system shown in Fig. 7.3 as an example, the influence of two different generator voltage regulation characteristics on voltage stability is compared.

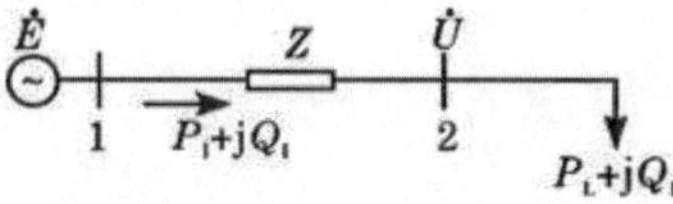

Fig. 7.3 Single-unit single-load system

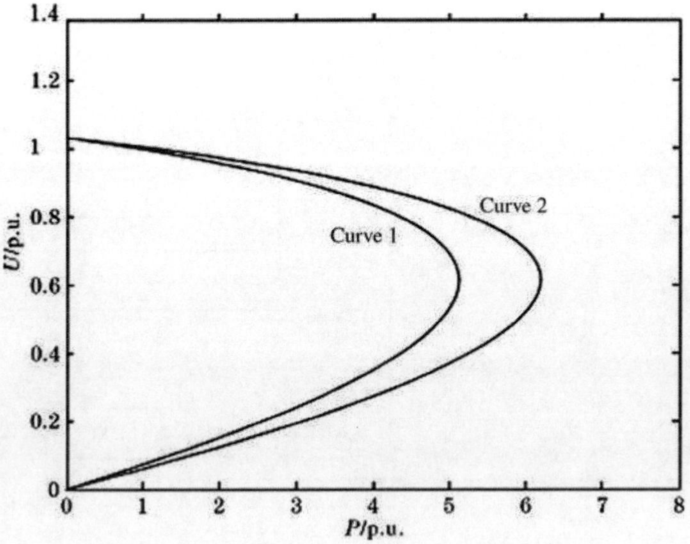

Fig. 7.4 *PV* characteristic curve of single-unit single-load system

Assume the following two situations:

(1) In Fig. 7.3, AVR of the generator can maintain the generator terminal bus voltage constant;

(2) The AVR (HV side voltage control) of the generator in Fig. 7.3 can maintain the bus voltage at the high voltage side of the boosting transformer.

The *PV* characteristic curves of these two cases can be obtained respectively, as shown in Fig. 7.4.

In Fig. 7.4, curve 1 is the *PV* Curve that maintains the generator terminal bus voltage constant, and Curve 2 is the *PV* curve that maintains the voltage of HV-side bus of the boosting transformer constant. It can be seen from Fig. 7.4 that after the HV side voltage control is adopted, the maximum transmission power allowed by the system is significantly increased. Meanwhile, unlike the increase of parallel Capacitor Compensation at the load side, with the increase of the maximum transmitted power, the critical voltage of voltage instability does not increase, so it can improve the system voltage stability better.

At present, the HV side voltage control is mainly realized by adding (transformer) voltage drop compensation to the existing excitation system. In this method, there is not need to collect the HV side electrical quantities of the boosting transformer, so it does not need to add additional equipment, which is easy to implement and has good economy. According to the current domestic and foreign application examples, it has achieved good results.

Fig. 7.5 Structure diagram of an HV side voltage controller

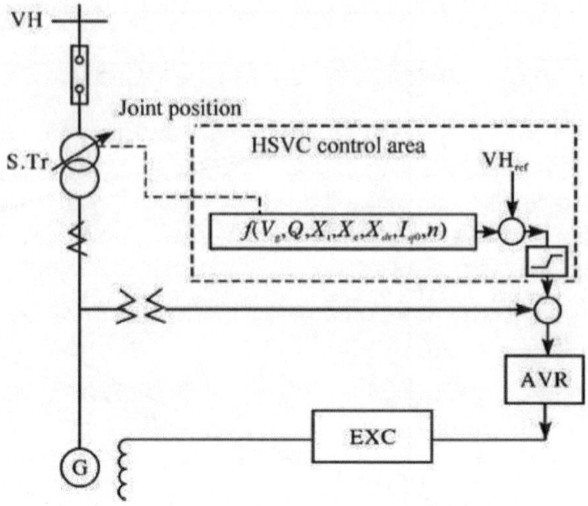

Figure 7.5 shows the structure diagram of a HV side voltage controller [3–5].

7.2.3 Generator Output Control

The occurrence of voltage collapse accidents is closely related to the load level of the power grid. Before the accident, the power system is usually under heavy load, and some key lines are overloaded, with high reactive power loss and low line terminal voltage. If the system suffers a fault disturbance or the load continues to increase, the voltage level at the end of the line gradually drops and eventually a voltage collapse occurs.

In the voltage collapse caused by cascading events, improper power flow transfer control is an important reason for the loss of system stability. Because the disconnection of lines causes a large number of power flows to transfer to the lines in normal operation, the power flow on these lines gradually increases, and the line terminal voltage is lower and lower, and finally the voltage collapse occurs due to the failure to maintain stability.

For this kind of accident, fast and reasonable power flow redistribution is an effective measure to prevent voltage collapse. In the normal operating state before the accident, by regulating the active power output of some generators in the system, and making the distribution of power flow more balanced as fast as possible, reduce the number of heavy load lines and the level of power flow, which can effectively prevent the occurrence of voltage collapse. In the emergency operating state after accident, regulating the active power output of some generators in the system, and making sure the load of overload or heavy load line is reduced to a reasonable range to prevent further line cascading trip, can avoid the expansion of the accident and prevent a large range of voltage collapse caused by cascading fault.

The output control of the generator can be executed by means of the safety automatic device or by ordering of the dispatching operator. When a line with very heavy load appears in the system, order the units at sending end to reduce output or shutdown, while order the units at the receiving end to increase output or start up, so as to reduce line load, improve voltage stability margin, and reduce the risk of voltage instability. For the medium and long-term voltage stability, because the duration of voltage drop process is relatively long, the method of manual order is feasible.

7.3 Reactive Power Compensation for Transmission Systems

From the point of view of voltage stability, reducing the transmission of reactive power on the line, achieving the balance of reactive power in layers and zones, and maintaining the bus voltage in the power grid at a higher level are good for improving the voltage stability of the system. From the point of view of voltage collapse accidents in actual systems, the lack of reactive power support capacity of the receiving-end system is often one of the reasons leading to voltage collapse.

In order to improve the reactive power support capacity of a power system, the method of increasing reactive power compensation of the power transmission system can be used. The methods include shunt compensation and series compensation. The equipment used includes shunt capacitor bank, shunt reactor, static var compensator (SVC), static var generator (SVG/STATCOM), series capacitor, condenser, and others. See the following introduction.

1. Shunt capacitor and shunt reactor

Taking the simple power system as shown in Fig. 7.6 as an example, the effect of the shunt capacitor compensation on improving voltage stability is illustrated.

For the single-unit single-load system as shown in Fig. 7.6, the nose curve as shown in Curve 1 in Fig. 7.7 can be drawn [6]. If the shunt capacitor compensations with different capacities are put into Bus 2 (Fig. 7.6), the other nose curves as shown in Fig. 7.7 can be obtained with other conditions unchanged. From Fig. 7.7, it can be seen that after adding shunt capacitor compensation at Bus 2, the maximum transmission power corresponding to the voltage collapse point of the system increases, which is equivalent to that the voltage stability margin of the system is increased (It shall be noted that the critical voltage corresponding to the voltage collapse point on the *PV*

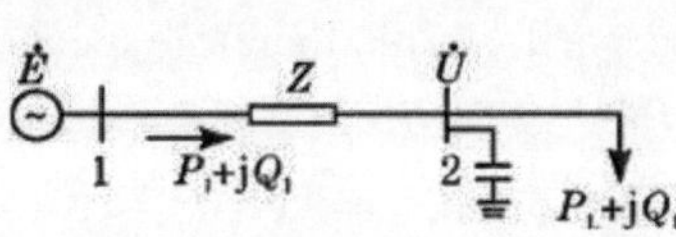

Fig. 7.6 Single-unit single-load power system (with shunt capacitor compensation)

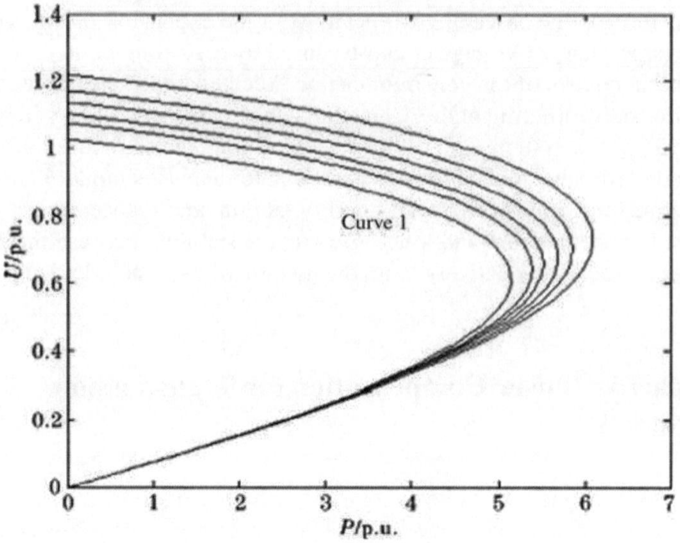

Fig. 7.7 Nose-type curves' family (with shunt compensation)

curve increases with the increase of the maximum transmission power. If the reactive power compensation capacity is too large, the critical voltage may exceed the initial operating voltage).

Compared with other reactive power compensation devices such as condenser and SVC, the greatest advantage of the shunt capacitor bank is its low cost. After being widely used in the power system, it can reduce the reactive power flow, improve the voltage level, and improve the power factor of dynamic reactive power supply such as generators, so as to improve the voltage stability of the system. One obvious disadvantage of shunt capacitor bank is that its reactive power output is proportional to the square of voltage, so when the system voltage drops and the reactive power output needs to be increased, its reactive power output will be greatly reduced; when there is overvoltage in the system and the reactive power output needs to be reduced, its reactive power output will be greatly increased.

Under the current manufacturing level, the switching speed of mechanical switching capacitor bank can reach 0.15–0.75 s. In this time range, if there is a shunt capacitor bank with adequate capacity to be put into system operation, it is possible to help a system with transient voltage instability to resume steady operation. However, due to the overvoltage and inrush current in the switching process, the shunt capacitor bank cannot be switched frequently. Therefore, a certain switching delay (the delay time can be adjusted, generally in tens of seconds or more) is added in actual use, which limits the voltage control capacity of mechanical switching shunt capacitor bank, therefore it cannot solve the transient voltage instability problem of power system.

The above discussion on shunt capacitor bank is also applicable to the shunt reactor, and switching off a shunt reactor is equivalent to putting the shunt capacitor into operation.

2. SVC

SVC overcomes the shortcomings of mechanical switching shunt capacitor and reactor, and can regulate the voltage quickly and accurately. Switching capacitor bank is not limited by transient process. SVC can regulate the voltage according to a certain slope which is related to its steady-state gain, and the control range is usually between 1 and 5% [7, 8].

Taking the single-unit single-load system shown in Fig. 7.3 as an example, if the SVCs with different capacities are added to the Bus 2 (Fig. 7.8), a nose curve as shown in Fig. 7.9 can be obtained. In the figure, Curve 1 is the *PV* curve without SVC compensation.

It can be seen from Fig. 7.9 that with the increase of load power, SVC can maintain the voltage at the load node constant within a certain change range load power. If the capacity of SVC installation is large enough, the transmission capacity of the system

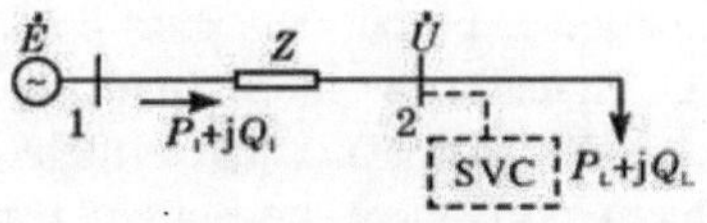

Fig. 7.8 Single-unit single-load power system (with SVC compensation)

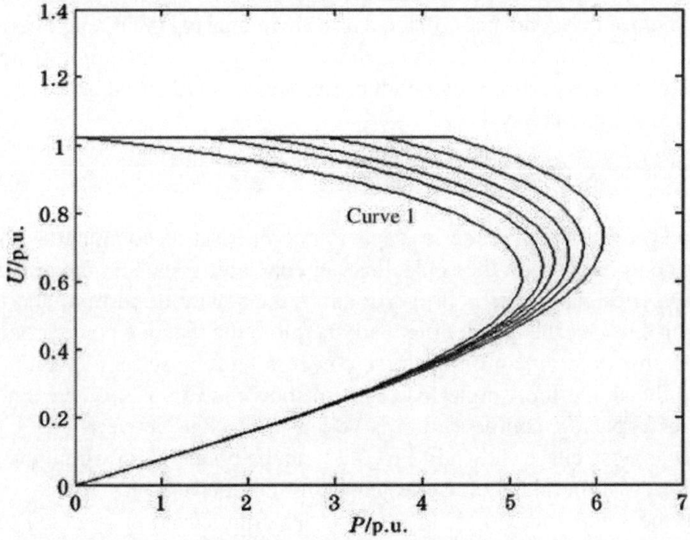

Fig. 7.9 Nose-type curves' family (with SVC compensation)

can be greatly improved while ensuring the voltage level at the load node. According to the different installed capacity, SVC has different control limits. When the control limit is reached, SVC is equivalent to the shunt capacitor.

3. Condenser

From the point of view of reactive power compensation characteristics, synchronous condenser is a good reactive power supply, its control capacity of voltage and reactive power is basically the same as synchronous generator. Compared with SVC, the advantage of the synchronous condenser is that it can maintain the rated current operation at its own limit operating point, unlike SVC, which shows capacitor characteristics—reactive power output is proportional to the square of voltage. In addition, the synchronous condenser has a certain overload operation capacity, and can output a large amount of reactive power in tens of seconds after a fault, which is very beneficial to maintaining the transient stability of the system.

The disadvantage of synchronous condenser is the high initial investment and operating cost. Compared with SVC, the investment for synchronous condenser is much higher, and its loss at full load operation is about 1.5%, and no-load loss is about 0.5%. In addition, the operation and maintenance of synchronous condenser is complex, and the start-up and operation of the unit need a lot of maintenance work. The synchronous condenser outputs the short-circuit current to the system at the moment of fault, which increases the short-circuit capacity of the system. In the current power system, due to the increasing size of the power grid and the installed capacity, there is the problem of too large short-circuit current in many power grids, and current limiting measures are often needed, so it is not suitable to adopt a large number of synchronous condensers.

At present, synchronous condensers are seldom used in China. Most of the applications of synchronous condensers in the world are related to HVDC engineering, which is used to increase the short-circuit capacity of DC inverter side. For example, three +300/−165 Mvar synchronous condensers are installed in Nelson River inverter station by Manitoba Hydro [1].

4. Fixed series capacitor compensation

The greatest advantage of the series capacitor compensation compared with the shunt capacitor compensation is the self-adjusting characteristics, i.e., the reactive power put out by series capacitors is proportional to the square of current, and the higher the line current (i.e., the heavier the load, the more the reactive power compensation required at this time), the more reactive power put out by series capacitors.

Taking the single-unit single-load system shown in Fig. 7.3 as an example, if the fixed series capacitor compensation is added to the line between bus 1 and bus 2 (Fig. 7.10), a nose curve shown in Fig. 7.11 can be obtained, in which the curve 1 is the *PV* curve without the fixed series capacitor compensation.

As can be seen from Fig. 7.11, after series compensation is added to the line, the maximum transmission power corresponding to the voltage collapse point of the system increases, while the critical voltage corresponding to the voltage collapse

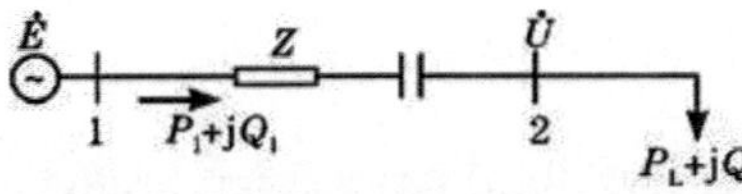

Fig. 7.10 Single-unit single-load power system (with fixed series capacitor compensation)

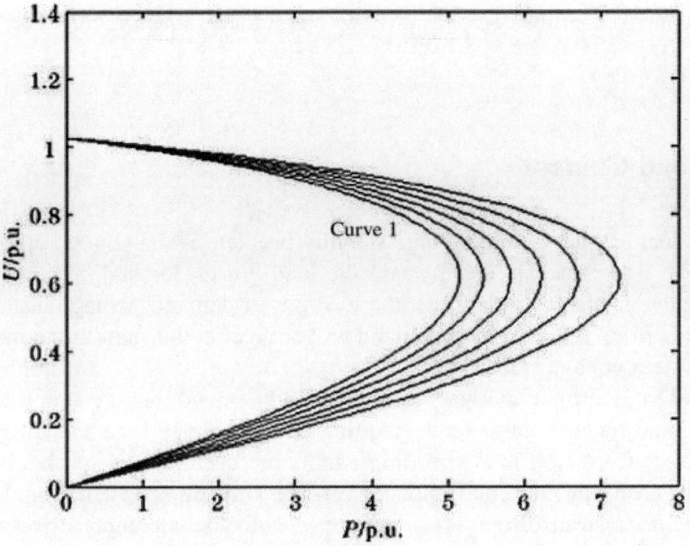

Fig. 7.11 Nose-type curves' family (adding fixed series capacitor compensation)

point of the system does not increase. This is also an advantage of series compensation compared with the shunt compensation.

In addition to regulation of the voltage and reactive power distribution, series capacitor compensation can also increase the natural power of the line. Under the ideal condition of uniform compensation along the line, adding 50% series capacitor compensation can increase the natural power of the line to $\sqrt{2}$ times to the original natural power.

The series capacitor can be installed on both sides of the line or at a point in the middle of the line. The specific installation location shall be determined after a comprehensive analysis according to the actual situation. The basic installation site selection principle is to ensure the voltage along the line as uniform as possible, and the voltage at each load point is within the allowable range. Considering the limitation of voltage magnitude and relay protection, the compensation at a compensation point on the line must be limited. For example, when the series compensation of a certain line is 70%, a 35% series compensation device is usually installed at both ends of the line respectively. The compensation for the line with series compensation currently in operation in China is generally not more than 50%.

5. Thyristor controlled series capacitor

In addition to the advantages of fixed series compensation, TCSC has more flexible regulation and higher control capacity on reactive voltage, and can also be used to suppress sub-synchronous resonance caused by series compensation. However, the investment of TCSC is large and the setting is more complicated, so in practical application, fixed series compensation with a certain capacity and TCSC are often adopted.

7.4 Load Control

The physical essence of the voltage stability problem is the contradiction between the system transmission capacity and the load power demand. When the system transmission capacity cannot meet the load power demand, voltage instability will occur. Therefore, reducing load demand will be an effective measure to maintain the power system voltage stability.

At present, the main methods to reduce load demand include concentrated load shedding and under-voltage load shedding (under-voltage load shedding). Among them, the concentrated load shedding adopts the event-driven mechanism, which is usually based on the conclusion of off-line simulation calculation. When one (several) line (s), transformer (s), generator (s) and other elements are disconnected, a trip command is issued in real time to shed the load at one time. The under-voltage load shedding will shed the load by rounds according to the collected real-time voltage and/or other electrical quantities and based on the preset working threshold and delay. The under-voltage load shedding method is described below.

7.4.1 Current Situation Analysis of Under-Voltage Load Shedding [7]

Taking under-voltage load shedding measures and automatically shedding the corresponding load by the device according to the duration of under-voltage are currently recognized as the least cost and the most effective method to solve the problem of voltage stability. The implementation scheme of low-voltage load shedding measures is related to many factors, such as voltage instability, dynamic characteristics of the load, voltage setting range of low voltage relays, and proportion for low-voltage load shedding to be arranged. The specific low-voltage load shedding scheme shall be determined by comprehensively considering various influencing factors.

There are two main types of low-voltage load shedding measures. One type is the concentrated load shedding, under which the system operation state is analyzed based on the system operating condition data from EMS or independent channels. Under certain specific conditions, if the voltage of key nodes falls below the preset threshold value, a load shedding signal will be sent out for adopting load shedding measures for the preset areas or nodes. The other type is an under-voltage load shedding device installed on nodes in a decentralized way. If the voltage of a node is lower than a preset threshold value for a certain delay, a certain amount of load is shed at the node.

The reliability of decentralized devices is relatively high, which will not have a direct or decisive impact on the operation of other elements due to the failure of one element. Moreover, it takes the local measured voltage as the load shedding criterion and does not depend on the communication device. The centralized device determines the low voltage of the whole area in the load shedding decision making through the communication links between the key nodes of the system. It can not only use the voltage level as a prediction index, but also select other indexes. The selection of the under-voltage load shedding scheme shall be determined according to the actual situation of the power system. Some of the foreign early under-voltage load shedding measures were decentralized. For example, the under-voltage load shedding device installed by TVA in 1981, which was installed in 9,161 kV substations with a delay time of 1.0–1.75 s. Other power companies such as Western Washington and Consumers Power Co., etc. in the United States also adopted decentralized devices. At present, the decentralized under-voltage load shedding control is adopted in China, but centralized control is the development trend.

Table 7.1 shows the typical configuration of under-voltage load shedding measures abroad.

At present, the characteristics of foreign under-voltage load shedding measures include:

(1) Most under-voltage load shedding measures are taken mainly for solving the problem of long-term voltage stability, which are mainly studied by static analysis methods. Therefore, the action delay setting is relatively long. For example, in the scheme of Quebec, Canada, the action delay is 6–11 s; in the scheme of Puget Sopud region of Pacific Power, the action delay is 3.5–8 s.
(2) Under-voltage load shedding measures are highly targeted, which is generally aimed at a few specific faults that can cause voltage stability problems. For example, most of the measures in Table 7.1 are aimed at the loss of main transformers or key transmission channels.
(3) The setting of the working voltage is closely related to the operating characteristics of the system. Different systems have large differences in power grid structure and operating state. The working voltage of some systems is relatively low, and the voltage drop after a disturbance is large, so their threshold values are relatively low. However, the working voltage of some systems is relatively high, and the voltage drop after a disturbance is small, so their threshold values are relatively high. Therefore, the setting of the working voltage shall be based on the actual operation of the power grid and combined with the specific analysis of its fault voltage characteristics.

Table 7.1 Typical under-voltage load shedding measures abroad

Power company	Overview
Pacific power Puget Sopud region	Non-centralized control is used to reduce voltage drop caused by 500 kV double-circuit line breaking and solve long-term voltage stability problem. If the voltage is lower than 10% and the time delay is 3.5 s, the load is shed by 5%. If the voltage is lower than 8% and the time delay is 5 s, the load is shed by 5%; if the voltage is lower than 8% and the time delay is 8 s, the load is shed by 5%
Quebec, Canada	People solve the problem of long-term voltage stability by sending control signals at two control centers installed in a 735 kV substation. If the voltage is lower than 0.94 and the time delay is 11 s, 400 MW load is shed; if the voltage is lower than 0.92 and the time delay is 9 s, 400 MW load is shed; if the voltage is lower than 0.9 and the time delay is 6 s, 700 MW load is shed
TEPCO	Monitoring units are installed in four 500 kV substations to control the under-voltage load shedding measures of 275 kV or 154 kV/66 kV substations, so as to solve the problem of long-term voltage stability. The minimum square root method is used to detect the slow change of voltage for 10 s to minute level voltage sequence, and detect the fast change of voltage for time sequence within 1 s. The voltage change rate is mainly detected to judge the voltage stability, so as to send out control signals
Entergy	The under-voltage load shedding measure based on EMS system data sends out load shedding signals when the load of the load-center substation (Franklin) exceeds 360 MW, the main transformer is breaking and the voltage of the four key 115 kV nodes (Franklin, McComb, Brookhaven and Liberty) is lower than 0.92 p.u. The shed load is divided into two areas by regions, and the shed load amount is 160 MW and 90 MW respectively
New Mexico	Concentrated load shedding control measures are divided into fast cut back and slow load shedding. Load shedding measures are issued by detecting the voltage drop of several key nodes. The fast cut back measure shed the load in 13 rounds after a time delay of 0.6 s, with a total shed load of 341 MW. The slow load shedding measure takes the condition after LTC action into account. When the voltage is lower than the threshold value (0.9 p.u.) for 10 s, 16.7 MW load is shed
BC hydro	Concentrated load shedding control measures are used to solve the voltage collapse problem of 500 kV system caused by loss of transmission lines or insufficient reactive power support, including two independent subsystems, Vancouver Island and Lower Mainland, which respectively detect the voltages of three key substations and the standby of some reactive power equipment to determine whether load shedding measures are needed or not

(continued)

Table 7.1 (continued)

Power company	Overview
Southern Sweden	Based on EMS control measures, by detecting node voltage, generator current limiter and reactive power data, start the gas turbine, switch the shunt reactance, and shed the load accordingly

(4) Rounds of load shedding are different. For example, in the scheme of Quebec, Canada, 3 rounds are set, while 9 rounds are set in the scheme of Ontario, Canada. In order to avoid load over-shedding, multiple rounds are set.
(5) The load model has a certain influence on the configuration of under-voltage load shedding measures. For areas with less induction motor load, the static load model is generally adopted, and the action delay time can be set to be longer, reaching more than ten seconds or even longer. Only for industrial areas with more induction motor load, in order to ensure the normal operation of the induction motor load, the action delay time of under-voltage load shedding measures must be considered as short as possible, which can be set at about 1.5 s.

China has long attached importance to under-voltage load shedding measures. The *Guide on Security and Stability of Power System* (2001 Revision) stipulates that in areas with concentrated load, when the working voltage is dropped, it shall be considered to shed part of the load automatically or manually (if necessary), or to disconnect the load in a planned way to prevent voltage collapse. Some domestic power grids are configured with voltage load shedding measures according to their respective conditions, and some of them are listed in Table 7.2.

At present, the characteristics of domestic under-voltage load shedding measures include:

(1) Under-voltage load shedding measures in some areas are mainly aimed at the specific fault setting of a single bus, and consider the voltage reduction problem caused by short-circuit fault. Under-voltage load shedding measures can identify short-circuit faults, and their action delay times are counted after the short-circuit fault disappears and the voltage returns to a certain level, so the action delay time can be set to be very short. However, in other areas, the fault is considered more complicated, so the transient process after the fault is more complicated. It is not enough to judge whether the voltage returns to a stable state only with the capacity of the equipment itself. It needs to increase the action delay (set to be longer) to avoid the transient process.
(2) The threshold value of working voltage is related to the operating characteristics of the power grid. The threshold value of working voltage is higher in areas with higher working voltages, for example, the threshold value of working voltage in Qinghai scheme is 0.90 p.u.; for areas with lower working voltages, the threshold value of working voltage can be set lower, for example, the threshold

Table 7.2 Configuration of under-voltage load shedding measures for domestic power grid

Power grid		First-round working voltage	Delay time (s)	Total shed load amount	Remarks
Heilongjiang		209 kV	0.5	Active load 80 MW, reactor 15 Mvar	A total of 4 rounds, and all of them are basic round
Jilin	Yanbian area	0.85 p.u.	0.3	130 MW	A total of 6 rounds, including 4 basic rounds and 2 special rounds
	Panshi area	0.85 p.u.	0.2	180 MW	A total of 7 rounds, including 4 basic rounds and 3 special rounds
	Jiaohe area	0.85 p.u.	0.3	15 MW	Only 1 round
South Hebei power grid		0.85 p.u.	0.7	695 MW	2 rounds
Shanxi		0.87 p.u.	0.7–1.3	30–40%	3-4 rounds
Zhejiang		195–200 kV	3.0–3.7	1408 MW	2 rounds
Fujian		0.83 p.u.	0.2	2649 MW	5 rounds, 1 special rounds
Hunan		0.8–0.85 p.u.	1.2–21.2	Divide by the equal capacity	5 rounds
Ningxia		0.85 p.u.	1	1756 MW	5 rounds, including 3 basic rounds, 1 special round and 1 standby round
Qinghai		0.90 p.u.	1.5	2104.6 MW	4 rounds, including 3 basic rounds and 1 special round
Guangdong		0.83 p.u.	0.3–1	11532 MW	2 rounds

value of working voltage in Hunan scheme is set to 0.8p.u. However, in general, the threshold value of the first-round working voltage is about 0.85 p.u.

(3) The shed load amount depends on the installation range of the device and the sensitivity of node voltage to load. For overall measures, a large amount of load shall be shed in case of failure. For example, in Guangdong scheme, it needs to shed 11,532 MW. For measures for single node, the shed load amount is very

small, which depends on the effect of voltage recovery after load shedding. For example, only 15 MW load needs to be shed in Jiaohe area, Jilin.

(4) Load shedding rounds are generally multiple rounds, which are related to domestic devices with low pressure load shedding measures. The domestic devices with low pressure load shedding measures are generally set at 5 rounds, including 4 basic rounds and 1 special round. As a result, there are often 4 basic rounds in the domestic allocation scheme, but 1–2 cycles are also used in a few of the devices.

7.4.2 Configuration Method of Under-Voltage Load Shedding Measures

1. Fault types and operation modes to be considered

The safety and stability standards for the capacity of power systems to withstand large disturbances is divided into three levels in the *Guide on Security and Stability of Power System* of China: The first-level standard is common single fault, requiring the system to maintain stable operation and normal power supply in case of such faults; the second-level standard is relatively serious fault, requiring the system to maintain correct work and stable operation of protection, switch and reclosing in case of such faults. When necessary, it is allowed to take stable control measures such as generator shutdown or load shedding. The third-level standard is rare and serious fault, when the system suffers from such faults and causes stability disruption, measures must be taken to prevent the system from collapsing and avoid long-term large-scale power outage and massive power outage to the most important users (including auxiliary power), so as to minimize the load loss and ensure the power system shall resume normal operation as soon as possible. The measures taken against these three levels of standards are the three lines of defense to ensure the safe and stable operation of the power system. In China, under-voltage load shedding measures are often configured in the third line of defense to prevent cascading collapse events and reduce load losses when local voltage collapses occur in the system.

If under-voltage load shedding measures are configured in the third line of defense, the rare and serious faults defined by the third-level safety and stability standard shall be mainly considered in the configuration. In actual configuration, the following fault types can be considered:

(1) Three-phase tripping fault of single circuit in double circuits on the same tower;
(2) The double shunt-transformer of the central substation is in fault state;
(3) The switch does not work in case of fault;
(4) Loss of large-capacity power plants;
(5) DC transmission line bipolar blocking;
(6) Disconnection of the power interconnection from the main power grid;
(7) Cascading failure;

(8) Other accidental factors.

Since voltage collapse accidents basically occur under heavy load conditions, the operation mode to be considered in the configuration of under-voltage load shedding measures is mainly the large load operation mode.

2. Voltage instability types to be considered

In terms of time scale, voltage stability can be divided into transient voltage stability and medium and long-term voltage stability. Because voltage instability accidents with different time spans have different influencing factors and the dynamic characteristics in the process of voltage instability are also different, the requirements for under-voltage load shedding measures are also different. In the specific configuration, it is necessary to use the full simulation program to simulate the serious faults that may cause voltage instability, and to verify the effectiveness and rationality of under-voltage load shedding measures in different types of voltage instability accidents.

3. Basic principles to be observed in configuration of under-voltage load shedding measures

Ensuring the safe and stable operation of the power system is the basic requirement for all stability control measures. In addition, economy is also one of the issues to be considered. For under-voltage load shedding measures, the configuration guideline is to minimize load shedding on the basis of keeping the system stable and not causing large-scale power outages. Based on this guideline, the configuration principles of under-voltage load shedding measures are as follows:

(1) After a single element fault disturbance corresponding to the first-level safety and stability standard occurs in the power system, if the system can remain stable and the voltage is within a reasonable range, the automatic under-voltage load shedding device shall not be put to work.
(2) After a serious fault disturbance corresponding to the second-level safety and stability standard occurs in the power system, if corresponding safety and stability measures in the second line of defense works, and the system can remain stable and the voltage is within a reasonable range after it works, the automatic under-voltage load shedding device shall not be put to work; if the under-voltage load shedding is taken as the second line of defense, it is allowed to work.
(3) The shed load amount is sufficient to meet the requirements of system stability and recovery voltage under different faults. At the same time, the load over-shedding shall be avoided.
(4) Set the working voltage and delay time of each round reasonably to correctly reflect the severity of the fault. Each round shall not work beyond its level.
(5) Under-voltage load shedding measures shall be adapted to other measures in the third line of defense to reduce unnecessary load losses and avoid further impact on the power grid.

4. Configuration method of under-voltage load shedding measures

At present, there is no theoretically feasible and engineering practical under-voltage load shedding configuration algorithm, and there are large differences between different power grids in terms of power grid structure and operating conditions. Therefore, at present, the configuration of under-voltage load shedding measures mainly depends on simulation calculation tools, and shall be specifically analyzed by the specific situations.

Under-voltage load shedding measures must meet the requirements of transient voltage stability and medium and long-term voltage stability. In addition, the voltage stability problem after local power grid disconnection also needs special attention. Since the system is usually equipped with some other stability control measures for disconnection, it is necessary to consider the coordination with these measures when configuring under-voltage load shedding measures, and avoid over-shedding and under-shedding, so as not to cause more significant impact on the vulnerable local power grid and lead to power outage accidents of the entire local power grid.

To sum up, the configuration process of under-voltage load shedding measures is summarized as follows.

1. To determine areas with the voltage stability problem

These areas are determined by analysis of simulation results of transient and medium and long-term voltage stability of the whole network.

2. To determine the total amount of sheddable load in the configuration area

The total amount of sheddable load is determined comprehensively according to the load level, load structure and the importance of different loads in the area.

3. To determine the voltage recovery target value after the under-voltage load shedding measures work

The voltage recovery target value is not only an important index to evaluate the rationality of under-voltage load shedding measures, but also a boundary condition to determine each parameter of under-voltage load shedding measures. After the voltage recovery target value is determined, a large number of parameters in subsequent work can be determined based on the target value.

The voltage recovery target value after under-voltage load shedding measures work can be given combined with the following factors:

(1) Stability level of the power grid at the recovery voltage level.
(2) There can be different voltage recovery target values for different fault severity. For example, for an ordinary N-2 fault, a higher value can be given. For very serious cascading faults, the voltage of some nodes in the power grid may be difficult to recover to a higher level. Therefore, when configuring under-voltage load shedding measures, a lower voltage recovery target value may be given

for nodes with the lowest voltage (it is necessary to ensure that the motor is not stalled and a certain stability margin is considered).

4. To determine the preliminary configuration scheme of under-voltage load shedding measures

The preliminary scheme of under-voltage load shedding can be obtained according to a certain algorithm, and can also be given according to the existing configuration experience. For the former, several methods have been proposed, among which *PV* curve method is the more commonly used method. For the latter, it can be given based on the following design methods:

(1) The constant value range of the working voltage of the under-voltage load shedding relay is 8–15% lower than the minimum voltage allowed during normal operation. For systems with large voltage change range, it can also be set according to the voltage change rate.
(2) The time delay of under-voltage load shedding mainly depends on the load characteristics. If the load is mainly the induction motor load and the voltage reduction has a significant influence on it, the velocity of load shedding is fast and the time delay is generally not more than 1.5 s; if the load is mainly a lighting or electric heating device with a constant impedance, the time delay of load shedding can be longer, such as 3–6 s or even longer.
(3) For the load with constant impedance, when the voltage drops to a certain degree and part of the load is shed, the voltage of the system still cannot be restored to the level acceptable for normal operation, and some special rounds need to be added to restore the voltage of the system to the acceptable level.
(4) The load shed by the under-voltage load shedding scheme shall be sufficient, and the load shed under normal circumstances shall not be less than 10%-20% of the load in the area. The same load can be shed for the low-voltage load shedding device and the low-frequency load shedding device. The shedding sequence shall be arranged according to the importance of load.

5. Test the adaptability of the preliminary configuration scheme

The full simulation program is used to analyze the adaptability of the preliminary configuration scheme combined with the requirements of transient voltage stability and medium and long-term voltage stability for low-voltage load shedding measures.

It specifically includes the following three sub-steps:

(1) Test the adaptability of the preliminary scheme to the transient voltage stability of the power grid and modify the preliminary scheme accordingly to form scheme II;
(2) Test the adaptability of scheme II to the medium and long-term voltage stability of the power grid and modifying Scheme II accordingly to form Scheme III;
(3) Test the adaptability of Scheme III to local power grid disconnection and modify scheme III accordingly to form scheme IV.

6. Form the final scheme

The final scheme is formed based on schemes II, III and IV.

7.5 HVDC Regulation

The DC transmission system also needs to consume reactive power equal to about 40–60% active power consumed in DC transmission while supplying active power to the receiving-end AC system. Therefore, it presents unfavorable "characteristics of reactive power load" to the receiving-end AC system. Such "characteristics of reactive power load" are considered factors causing voltage instability.

Practically, owing to the quick regulation characteristics of HVDC transmission system, to adopt proper regulation and control mode can enhance the transient voltage stability through quick control of an AC system. For instance, in case of insufficient reactive support caused by fault disturbance, the reactive power absorbed by DC can be reduced by quickly reducing active power or DC current of the DC transmission line, thereby improving voltage stability of the AC system. If there is no special control mode, when the DC transmission system under constant power control state is subject to low voltage or disturbance, DC current shall increase owing to power control. Therefore, the reactive power absorbed by the current converter shall increase significantly. At this time, switching the control mode from DC system to constant current control mode helps to improve voltage stability.

7.6 Multi-level Voltage Control

Reactive voltage control is mainly to reasonably arrange and sufficiently utilize adjustable capacity of reactive power in the power grid, thereby guaranteeing that voltage of different grid hinge nodes maintains at normal level under normal operation condition and after any accident. In this way, the power grid can operate safely and stably to avoid transmission of a large number of reactive power via long-distance transmission line or multi-level transformers.

Traditional reactive voltage control generally adopts decentralized control. Under such control mode, different voltage control devices (power generator, capacitor bank, and OLTC, etc.) can only acquire local information and independently control local voltage. Such high decentralized control speed not depending on the control center is only to meet local optimization target such as the local reactive voltage within the upper and lower limit range. However, incapability in mutual coordination among controllers may unfavorably impact reactive power distribution and voltage level of main grid.

Centralized voltage control opposite to decentralized control should collect voltage, reactive and other information within the system range. The dispatching

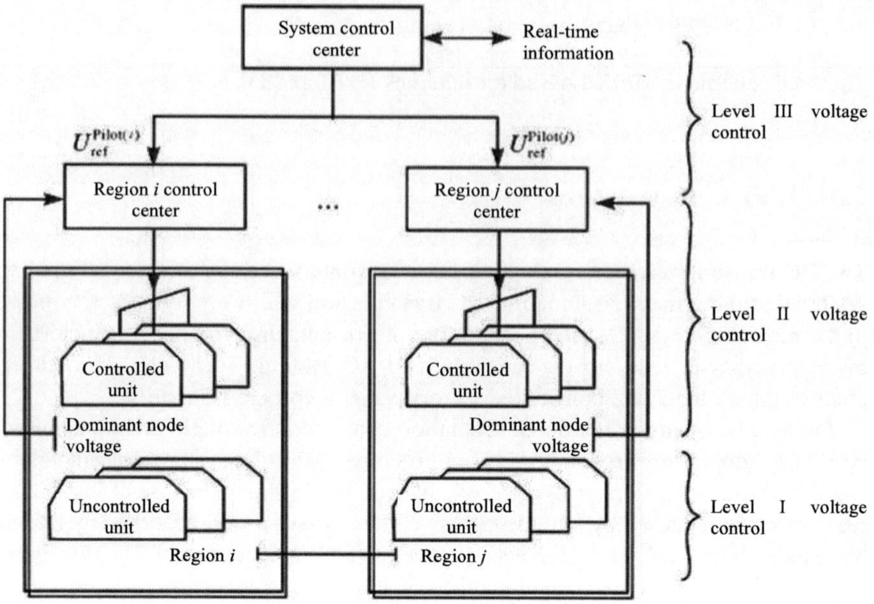

Fig. 7.12 Schematic diagram of voltage control system in hierarchical structure

center should give out a control strategy and issue control signal. Such control mode has higher requirements on reactive measurement accuracy and data communication. Therefore, it is difficult in implementation.

Hierarchical voltage control as a compromise control mode between the two is a better voltage control mode.

In France and other European countries, tertiary voltage control mode is adopted to divide voltage control functions of power systems based on time and space to achieve a hierarchical control structure as shown in Fig. 7.12.

Tertiary voltage control mode is at the top control level over the whole system. The control at this level is implemented by control center of the system, which only requiring the response duration of tens of minutes. Tertiary control systematically coordinates different secondary controls to surely meet the voltage amplitude of dominant nodes in secondary control areas that can guarantee safe restriction for power grid and economic operation of the system with information in the system range.

Secondary voltage control is in the middle level. It is to control a certain area by the control center of different areas at the time constant from tens of seconds to tens of minutes. The main objective is to change the set reference value of primary voltage controller based on the preset control rule and guarantee that voltage of the pilot bus is equal to the set value.

Secondary voltage control as a measure for preventing voltage collapse accident was first presented by Electricite de France (EDF). From single-area test in

1974–1985, the power systems almost all over France are configured with secondary voltage control modes covering 27 control areas, 100 thermal power units and 150 hydropower units, controlling about 30,000 Mvar of reactive power in total. It is demonstrated through operation practice that secondary voltage control simplifies operation control work and better coordinates reactive output among control units. Afterwards, the control mode was also implemented in Italy, Spain and other countries. It is demonstrated from further study that secondary voltage control is an effective method for utilizing the reactive power in the management and control area.

One-level voltage control mode is at the bottom level and arranged in power plants, for users and at different power supply points. Generally, it is the closed-loop control in quick response for controlling local voltage at 1 s to several seconds of response time generally. Control devices compensate for quick and random change of voltage by guaranteeing that the output variables are close to the set value as far as possible.

The main advantage of tertiary voltage control mode is "space-time decoupling". The control area of reactive voltage is divided "spatially" and hinge nodes in different areas are controlled so as to control the area. In this way, reactive resources on the area are fully utilized to avoid reactive flow in wide range. The control effect of reactive voltage is guaranteed through decoupling of different control levels in time.

There are still some disadvantages in the mode adopting power grid division and control hinge nodes. So as to the division, no excessive coupling among different control areas is accepted, otherwise secondary voltage control in a control area shall affect neighboring control areas. In this way, secondary voltage control shall not achieve good control effect. In severe cases, the voltage of neighboring control areas may become instable. Whereas, in selection of a hinge node, the electrical distance between the node and other nodes in the area should be close to each other as much as possible, and sufficiently far from the adjacent control areas so as to avoid unnecessary effect among different areas. Therefore, selection of division and hinge node is sensitive to grid structure. In case of significant change in grid structure, the previous division and hinge node shall be calculated and adjusted again. In addition, secondary voltage control of each control area in the scheme is completed by the control center independently set in the area. The method is not suitable for the power grid under development.

In Italy, secondary voltage regulation strategy is adopted and applied to the hierarchical voltage control to achieve integration between reactive control and voltage control. The microprocessor-based voltage and reactive regulator referred to as REPORT are adopted for the control level of power plant; regional voltage regulator (RVR) is adopted to achieve real-time and closed-loop control through REPORTs and main ultrahigh voltage bus voltage for the control level of regional voltage distribution network. The independent system operator (ISO) of Italy equipped REPORT devices in main power plants and RVR devices in many regional power grid dispatching rooms.

At present, traditional "local" and "manual" voltage control methods are still used extensively, including reactive output distribution of a unit, voltage regulation plan at high-voltage side of a power plant, switching of shunt capacitors and reactors, and application of much effective SVC with high response speed, etc. However, the method has following disadvantages if used in reactive voltage control for the whole power grid:

(1) Reactive output distribution and voltage regulation plan must be based on accurate load prediction, they cannot adapt to dynamic change in power grid;
(2) Assignment and implementation of distribution plans fall behind change in power grid, namely, implementation of control means cannot adapt to the dynamic phenomena of the power grid;
(3) Owing to insufficient coordination between operators of power plants and transformer substations, no synchronization or optimal control can be achieved;
(4) In order to improve the certainty of the control measures, it is required to introduce power grid reactive power, voltage control and monitoring system.

The hierarchical control structure with power grid voltage control as the target for achieving automatic coordination of power grid reactive power supply is referred to as secondary voltage regulation (SVR). The structure has been demonstrated in Italy, Belgium and France, and also widely promoted and applied in Spain. In open electric power markets and revolution of power utilities in Europe, such hierarchical control system has received favorable comments and been improved increasingly. Therefore, it can meet simple and automatic demands of ISO on voltage control.

Promoted by liberalization of electric power market, some power generator manufacturers have equipped reactive power control functions in excitation control system of units, namely, voltage control at high-voltage sides of power plants. In fact, the secondary voltage control system based on automatic voltage control at high-voltage sides of power plants has received attention from North America. Bonneville Hydropower Administration is developing large-area power grid voltage control system, including inter-tripping of power generator and load, regulation of reactive power supply, TCSC/SVC module, voltage regulation at high-voltage side of the power plant and regulation of transformer taps. The development is tied closely with the European SVR plan.

Furthermore, ISO depends on simple, effective and automatic voltage control systems to achieve system voltage control while assuring and improving system security and reliability, power quality and economic efficiency. Automatic coordination of reactive power supply mainly relates to regional reactive standby, including unit capacity, synchronous compensator, shunt capacitor/reactor, SVC and OLTC. Large power grid is divided at first to achieve regional coordination through decentralized voltage control, thereby achieving voltage control target. In each area, in regional dispatching and between power plants and transformer substations, it is required to consider data and control signal exchange required by coordination control. More real-time data exchanges corresponding to power grid dynamics assures better performance and effect of voltage control effect. The economic effect generated by voltage control is closely related to regional coordination and control, and also efficient exchange of data and control signal.

In China, the research on automatic control of reactive voltage within system range has been in a rapid development stage mainly towards optimization control algorithm of reactive voltage, optimization closed-loop control, simulation analysis, and reactive voltage management, etc.

Previously, most reactive voltage control devices operated online in China basically took local reactive voltage control as their target. Based on the nine-zone diagram, the control principle only guarantees that the local reactive voltage controlled within the upper and lower limits may generate adverse effect on reactive power distribution and voltage level in main power grid. In addition, such devices have no joint debugging function in network. Therefore, it cannot achieve reactive voltage optimization and control in the whole grid.

At present, power grids of Fujian, Henan, Jiangsu, Anhui, Liaoning, Hebei and other provinces have already achieved optimization control of reactive voltage. Each system has its own advantages and disadvantages, for example, impractical dynamic zone result and low three-level control robustness, etc. have to be settled urgently in future AVC research. Even so, it shows from AVC systems of different provincial power grids that promotion and application of AVC system do contribute to improvement of system voltage quality and safe and stable operation level, reduction of power grid loss, and also alleviation of operators' workload in frequent reactive regulation.

References

1. Taylor CW (1994) Power system voltage stability. EPRI power system engineering series. McGraw Hill, New York
2. Kundur P (1994) Power system stability and control. EPRI power system engineering series. McGraw Hill, New York
3. Zhou S, Zhu L, Guo X et al (2004) Voltage stability and control of power system. China Electric Power Press, Beijing
4. Wang Q, Zhou S, Zhu L (2003) Improvement of system angle stability through voltage control of high-voltage side. Power Syst Technol 27(6):19–21, 41
5. Cheng L, Sun Y, Jia Y et al (2007) Effect of load compensation on system stability in excitation control of generator. Proc Chinese Soc Electri Eng 27(25):32–37
6. van Cutsem T, Vournas C (1998) Voltage stability of electric power systems. Kluwer Academic Publishers, Boston
7. Ma S, Yi J, Sun H et al (2009) Power system under-voltage load-shedding configuration principle and scheme. Proc CSU-EPSA 33(5):45–49

Chapter 8
Examples of Voltage Stability Analysis

Big power grid operation as the sign of modern power industry brings self-evident economic effect. However, any accident may cause serious consequences. Therefore, safe operation of the big power grid is an important research subject for electric power workers both at home and abroad. Several typical voltage collapse accidents abroad are the profound warning for us in development of our power grid. In accordance with characteristics of interconnected power grid in China, the power grid structure becomes more and more complicated. There are more and more AC power transmission channels, or AC and DC power transmission channels in parallel for transmitting power intensively across different areas in long distances. The difference between peak and valley loads is enlarged with extensive application of new power transmission technology. Therefore, it is more and more difficult to regulate and control voltage; in some interconnected power grids, the risk probability of voltage collapse accident increases owing to coexistence of insufficient dynamic and reactive standby at center of the end load.

Presently, in practical projects, reactive balance analysis, static voltage stability margin analysis, electromechanical transient simulation and other common voltage stability analysis methods are still the typical methods. This chapter introduces how to perform detailed voltage stability analysis for specific power grids through two project cases on voltage stability.

8.1 Case I: Power Grid Voltage Stability Analysis and Research of Fujian

Take the peak load mode in summer and peak load mode in winter of Fujian Power Grid in year 2005 as the research object to comprehensively evaluate static voltage stability level of different ways through scanning and calculation of different static voltage stability index; analyze static voltage stability level of different modes after $N - 1$ and other serious faults of Fujian Power Grid, determine $N - 1$ and serious fault

© Science Press 2021

Y. Tang, *Voltage Stability Analysis of Power System*, Power Systems,
https://doi.org/10.1007/978-981-16-1071-4_8

mode with significant effect on voltage stability; dynamic stability check and calculation of serious fault is performed to simulate the fault modes in a chain reaction and their consequences that may cause voltage stability problems; through simulation and research of different load models, the transient stability characteristics of Fujian Power Grid under different load models is compared to determine effect of different load models on transient stability power limit. Present the countermeasures and comprehensive measures and suggestions for Fujian Power Grid in preventing voltage collapse accident; put forward measures and suggestions for adding reactive compensation and improving voltage stability; analyzing the effect of SVC dynamic reactive compensation on improvement of voltage stability and prevention of voltage collapse accidents; put forward measures for regulation of operation mode and grid structure, power supply configuration measures in weak voltage areas; put forward regulation measures for under-voltage load-shedding scheme.

8.1.1 Main Contents of the Research

Research is performed from following aspects:

1) Analyze reactive balance and existing reactive voltage condition of Fujian Power Grid, put forward reactive compensation suggestions

In combination with the reactive voltage problem exposed in operation and control of Fujian Power Grid, balance analysis is performed for existing reactive compensation devices and capacities, existing reactive voltage regulation modes for Fujian Power Grid are evaluated, and preliminary advices for improving the reactive compensation are put forward.

2) Static voltage stability analysis for Fujian Power Grid under normal mode

Adopt static voltage stability analysis method for peak load mode in summer and peak load mode in winter in the year 2005 to calculate reactive power margin index of load bus, reactive power margin index of 500 kV/220 kV bus, regional active power margin index, and reactive power margin index, etc.; determine key lines, key units and weak links with significant effect on voltage stability through modal analysis and calculation; on the basis, normal mode of Fujian Power Grid is comprehensively evaluated.

3) Static voltage stability analysis for Fujian Power Grid mode after $N-1$ and $N-2$ faults

Static voltage stability analysis is performed on 500 kV heavy-load flow lines after $N-1$ and $N-2$ breakage to determine the fault modes of $N-1$ and $N-2$ with relatively significant effect on voltage stability.

4) Transient voltage stability check and analysis for Fujian Power Grid under multiple and serious fault disturbances

Time-domain simulation is performed for serious fault disturbance mode of Fujian Power Grid to check the voltage instability accident possibility of Fujian Power Grid under major fault disturbance, including disturbance modes of serious faults such as tripping of large units, loss of excitation, and simultaneous tripping of double-circuit line, etc.

5) Presentation of countermeasures and comprehensive measure suggestions in preventing voltage collapse accidents for Fujian Power Grid

Based on static voltage stability analysis result for Fujian Power Grid, put forward measures and suggestions to increase reactive compensation and improve voltage stability, and evaluate the effect of SVC dynamic reactive compensation on improvement of voltage stability and prevention of voltage collapse accidents; in combination with check result of transient voltage stability, put forward emergency voltage control measures for preventing voltage collapse accidents.

8.1.2 Research Conditions and Calculation Principles

1. Basic data and evaluation principle

1) Basic data and operation mode

The research modes are peak load mode in summer and peak load mode in winter in the year 2005, and $N-1$ and multiple accident outage mode based on these two modes.

2) Evaluation principle of voltage stability

Static voltage stability levels of different operation modes are evaluated with active margin of single-load bus and reactive margin of bus, so as to evaluate the receiving capacity of regional power grid based on the active reserve coefficient in the load increase area. At the same time, voltage sensitivities of different buses are reflected by the bus participation factor figured out through modal analysis, so as to find out the relative weak link of Fujian Power Grid.

To solve the active margin index, take a constant power factor as the load increase power factor; to solve regional active reserve coefficient, it is assumed that different bus loads of Fujian Power Grid increase over the same period of previous year based on the basic load. The increased load capacity is balanced by internal units of Fujian Power Grid.

3) Main evaluation index and corresponding unit

Reactive power margin: p.u. (Based on 100 MVA).

Active power margin: p.u. (Based on 100 MVA).

Active power reserve coefficient: %.

Critical voltage: p.u. (Based on reference voltage 525 and 230 kV, etc. of different buses in power flow data).

Voltage sensitivity: kV/Mvar.

Participation factor: dimensionless (relative size makes sense).

2. Technical principles

The main stability standards and principles of power system:

(1) *Technical Guideline on Voltage and Reactive Power of Power System*;
(2) *Technical Guideline of Power System*;
(3) *Guide on Security and Stability for Power System.*

8.1.3 Computational Tool and Method

The computational tool adopted in the research is "PSD power system analysis software" developed by China Electric Power Research Institute, including PSD-VSAP static voltage stability analysis software, PSD-BPA power flow program, PSD-BPA transient stability program and full dynamic simulation program of PSD-FDS power system. Wherein, static voltage stability analysis software mainly includes following modules:

(1) Calculation of voltage stability margin In calculation, linear simulation method is taken as the basis, the superior initial solution of power flow is taken as the initial state, linear mode of power flow in combination with sensitivity algorithm is utilized, change in different nonlinear factors of the power system is taken as the restriction control in the calculation process and reactive output limitation of power generator, etc. are considered to calculate the power margin for single-bus load increase or regional load increase.
(2) Modal analysis method. This method is the application of modular technology in stability analysis of static voltage. Voltage collapse means modular voltage collapse. The most possible instable mode in power grid is found out by resolving the singular value in Jacobian matrix of flow trend. The load bus with larger participation factor is verified as the key load bus. The area with intensive key load buses is a relative weak area in the power grid, and also an effective area for increasing reactive compensation. Original data adopted in the program is the power flow data in BPA format and power flow result data.

8.1.4 Analysis of Voltage and Existing Reactive Condition for Fujian Power Grid

This chapter introduces general conditions of Fujian Power Grid in the year 2005, including grid structure characteristics analysis and power flow analysis.

1. Structure characteristics of Fujian Power Grid

In the year 2005, 500 kV grid of Fujian Power Grid presents a long-distance chain power grid structure. At the same time, a local 500 kV triangular looped network is composed of Houshi and Shuikou power plants as the main force, as shown in Fig. 8.1. Regional grids compose a main grid with 220 kV voltage; with 500 kV/220 kV electromagnetic looped network structure, if considering low external transmission power level under peak load mode, the power flow transfer should not generate serious

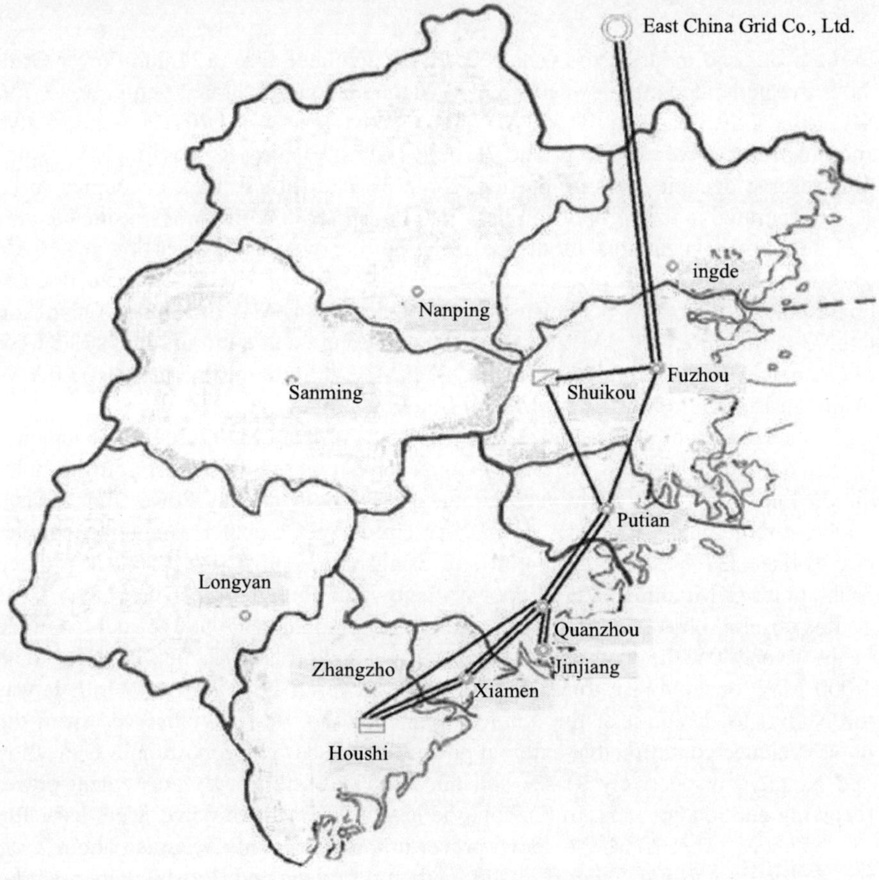

Fig. 8.1 500 kV grid structure schematic of Fujian in year 2005

consequence in case of fault tripping of 500 kV circuit in Quanzhou-Putian line. In case of high external transmission power level, fault tripping of Quanzhou-Putian line may cause large-scale power flow transfer.

2. Installation condition

In the year 2005, the installed capacity of Fujian Power Grid under regulation of top power authority is 11,025 MW. Power supplies from Fujian Power Grid distribute dispersedly. Except that Houshi Power Plant is directly connected to the 500 kV system, units of other power plants are respectively connected to 220 kV or 110 kV voltage grade. It is observed from connection condition of power supply in systems, regional power grids can preferably support power supplies in accordance with disperse connection principle of power supply. However, in Quanzhou and Xiamen, there is relatively insufficient dynamic voltage support capacity owing to high external power demands.

3. Power flow analysis

In the peak load mode of the year 2005, it is calculated that in Fujian Power Grid, the active generated output was 10,832 MW, the reactive generated output was 3,756 Mvar, the active load was 10,666 MW, the reactive load was 4,201 Mvar, the active spinning reserve was 200 MW, and the active transmission loss was 166 MW. Spinning reserve occupied less proportion, about 2%; transmission loss occupies about 1.5% generated output proportion in Fujian Power Grid, which was reasonable.

Under normal peak load mode, there was no line overload. Power flow in 500 kV line: 0 MW of tie line to East China, 2,366.4 MW of Houshi-Xiamen double-circuit line, 1,047.6 MW of Houshi-Quanzhou line, 1,441 MW of Xiamen-Quanzhou double-circuit line, 777.6 MW of Quanzhou-Jinjiang double-circuit line, 273.8 MW of Quanzhou-Putian double-circuit line, 39.7 MW of Shuikou-Putian line, 27.0 MW of Fujian-Putian line and 162.0 MW of Shuikou-Fujian line.

220 kV lines with great power flow: 251.5 MW from each circuit of Huaneng-Qishan double-circuit line, 307.8 MW from each circuit of Shuikou-Minqing double-circuit line, 304.1 MW from each circuit of Meizhouwan Bay Power Plant-Hushi double-circuit line, 247.9 MW from each circuit of Quanzhou-Guanqiao double-circuit line, 231.7 MW from circuit I of Zongshan-Jiaomei line, 250.2 MW from each circuit of Mianhuatan Hydropower- Fengyuan double-circuit line.

It is observed from distribution conditions of regional load under peak load mode in the year 2005, the load level of Fuzhou, Quanzhou and Xiamen is higher than 1,000 MW, occupying a total of 70% load of overall Fujian Power Grid. It was the central load region of the whole Fujian Power Grid. It is observed from the power balance condition that external power occupied high proportion in Quanzhou and Xiamen, respectively 91.4% and 63.9%. Therefore, they were typical power receiving end load centers. In Fuzhou, the load and generated power were basically in a balanced state. 78 MW external power in the region only occupied about 2.8% proportion. Regarding power plants in Fuzhou, Longyan and Houshi, there are relatively sufficient dynamic reactive standby units; however, in Quanzhou and Xiamen, there is nearly no dynamic reactive standby unit. The power generator units have relatively low power factor.

8.1.5 Reactive Power Balance Analysis of Fujian Power Grid

Based on the results of power flow analysis in Sect. 8.1.4 and the current situation of reactive power compensation in Fujian Power Grid, this section focuses on the analysis of the balance in layers and sub-regions of reactive power in Fujian Power Grid, and puts forward measures and suggestions for improving reactive power compensation in Fujian Power Grid.

1. Basic principles of reactive power balance

Natural reactive power load includes reactive power load before power user compensation, reactive power loss of each voltage level transformer and transmission line, reactive power load for power plant, and reactive power consumption of various reactors, etc. Reactive power supply includes actual adjustable reactive capacity of generators, line charging power, and capacitive reactive power capacity in reactive power compensation equipment, etc.

The main principles of reactive power balance are as follows:

1) The power system satisfies the principle of balance in layers and sub-regions of reactive power under normal operation mode

The load distribution of the power grid is influenced by the factors of regional economic development and has high randomness. The selection of power supply points is influenced by many factors such as energy distribution, transportation, environmental protection and so on. The distribution of load and power supply points determines the power flow direction. Power grid planning is done to determine a reasonable transmission grid frame based on load forecasting and power flow direction. Large-capacity and long-distance transmission of active power flows is an inevitable trend in the development of modern large-area interconnected power grids. Due to the development trend of power marketization, power flow changes in power grids are more frequent and irregular. During the operation of the power grid, the long-distance transmission of reactive power not only increases the loss of the transmission grid frame, but also makes the voltage level of the power grid difficult to control. Therefore, in order to maintain the voltage level of each point, reactive power supply and reactive power load must be balanced with each other at any time to achieve the balance in different voltage levels and with each other in each sub-region, thus avoiding the multi-level transformer from transmitting a large amount of reactive power.

Reactive power balance is aimed at different operating points. The output of reactive power load and reactive power supply varies with the operation mode. During the change of operation mode, in order to control the voltage level within the required range, the configuration capacity of reactive power must meet the requirements of upper and lower limits, and the configuration capacity must be selected according to the representative design level year and typical operation mode. Generally, the maximum reactive power load time point and the minimum reactive power load time point are always taken as the balancing objects. The former corresponds to a typical peak load mode. Due to heavy power flow and insufficient reactive power in the system, the voltage drops and compensation of capacitive reactive power needs to be considered; the latter corresponds to a typical small mode. Due to light power flow and unbalanced charging power in the system, the voltage rises and compensation of inductive reactive power needs to be considered.

2) Meet the demand of dynamic reactive power balance and fast reactive power regulation during accidents

After the system encounters a major accident, it is necessary to re-establish a new equilibrium point. With a large number of power flow transfers, the system voltage may drop significantly. During the active power balance process, dynamic reactive power reserve is required to provide voltage support. Therefore, emergency reactive power reserve is essential for the recovery of system voltage after a fault and the prevention of large power grid collapse accidents. After a large disturbance occurs, the system is able to maintain voltage stability and normal power supply after necessary generator shutdown and load shedding measures are taken.

The establishment of dynamic reactive power balance requires reactive power compensation equipment with rapid regulation capability. The generator and synchronous condenser in the system have rapid regulation performance, and part of the capacity is usually reserved as an emergency dynamic reactive power reserve. In addition, SVC, STATCOM and other reactive power compensation equipment are also used as dynamic reactive power reserve in the system.

2. Reactive power balance analysis

During the peak load mode in summer in 2005, the 500 kV system was operated at the voltage of 511.0–530.4 kV and each node voltage was maintained within the qualified examination range.

(1) Compensation and balance of charging power for 500 kV lines

In 2005, Fujian Power Grid had 14 circuits of 500 kV lines with a length of about 1,332 km and 2 circuits of 500 kV inter-provincial tie lines. According to the operating voltage of 525 kV, the total charging power of 500 kV lines in Fujian Power Grid is about 1,505 Mvar, of which the tie-line between Fujian and East China is calculated according to half of the charging power. In 2005, the total capacity of inductive reactive power compensation equipment in Fujian Power Grid 500 kV system was 1,425 Mvar, of which the capacity of HV reactor was 660 Mvar and that of LV reactor was 765 Mvar. The inductive reactive power compensation degree of 500 kV line in Fujian Power Grid is about 95%.

Article 5.1 of the *Technical Guideline on Voltage and Reactive Power of Power System* stipulates that 330–500 kV power grids shall be equipped with HV and LV shunt reactors according to the basic requirements of balance of reactive power in layers on-site to compensate for the charging power of EHV lines. In general, the total capacity of HV and LV shunt reactors is not less than 90% of the line charging power. The capacity distribution of HV and LV shunt reactors is determined through comprehensive research according to the system conditions and their respective characteristics. Meanwhile, Article 2.3.2 of the new *Guide on Security and Stability of Power System* emphasizes that the charging power of 330–500 kV lines should be basically compensated.

Therefore, from the perspective of the balance of reactive power in layers and sub-regions on-site, the total configuration of 500 kV HV and LV shunt reactors in Fujian Power Grid is basically reasonable.

The total capacity of 500 kV capacitive reactive power compensation equipment is 656 Mvar.

It can be seen that the proportion of reactive power injected into the medium voltage side of 500 kV substation in Fujian power grid is not large, and the reactive power is basically balanced. There are relatively more reactive power exchanges in 500 kV substations with heavy loads.

2) Reactive power compensation and balance of 220 kV lines

The capacitive reactive power capacity of 220 kV substation of Fujian Power Grid in 2005 is 2,305 Mvar. Configuration of reactive power compensation equipment in various regions: 370 Mvar in Fuzhou City, 30 Mvar in Ningde City, 90 Mvar in Putian City, 1,030 Mvar in Quanzhou City, 273 Mvar in Xiamen City, 150 Mvar in Zhangzhou City, 130 Mvar in Longyan City, 70 Mvar in Sanming City and 162 Mvar in Nanping City.

3) Division balance of reactive power compensation

The power supply range of Fujian Power Grid is divided into 9 sub-regions, the load power factor in various sub-regions is above 0.92, and the load-side reactive power compensation meets the requirements of the grid for the load-side power factor.

4) Analysis of dynamic reactive power reserve capacity

Compared with the active reserve situation, the system must also have sufficient reactive power reserve capacity for operation reserve and voltage adjustment. The difference is that active power reserve is network-wide, while reactive power reserve is regional. Insufficient active capacity standby causes frequency reduction, and insufficient reactive power reserve causes voltage instability. When there is reactive power disturbance in the system, the dynamic reactive power reserve is quickly transferred to support the system voltage, and other emergency voltage control measures, such as under-voltage load shedding, is coordinated when necessary to restore the system voltage.

Dynamic reactive power reserve generally refers to the reactive power regulation margin of the reactive power compensation equipment that is quickly regulated, the most important of which is the reactive power reserve of the generator. The generator reactive power reserve determined according to the rated capacity is one of the contents to be considered in the static voltage stability evaluation. In areas where the reactive power is relatively lacking, the generator operating power factor is generally low, and the dynamic reactive power reserve level of the system is also low. During the process of large disturbances, the voltage support capability of the system becomes poor, which easily causes voltage instability. Therefore, in some practical voltage safety monitoring systems, the dynamic reactive power reserve of the main unit or key unit is usually regarded as one of the important indexes reflecting voltage safety.

In peak load mode in 2005, the dynamic reactive power reserve of Fujian Power Grid accounted for 38% of the total reactive power load capacity, and the reactive power reserve was relatively sufficient. However, the distribution of reactive power reserve is not balanced, Fuzhou, Longyan, and Nanping have sufficient reserves; Houshi Power Plant also has a reserve of 568 Mvar; in Xiamen and Quanzhou areas that lack power support, the reserve is almost 0, the situation of reactive power reserve also reflects that Xiamen and Quanzhou are the areas with the biggest lack of reactive power in Fujian Power Grid and need to strengthen reactive power compensation.

5) Analysis of reactive power regulation means

Voltage and reactive power regulation means of Fujian Power Grid mainly include generator set voltage regulation, transformer voltage regulation and shunt capacitor bank/reactor bank voltage regulation. The generator set is the best voltage and reactive regulator. The voltage and reactive power regulation capability in near areas can be improved through the rational utilization of generator set. The units in Fujian Power Grid are connected to regional power grids in a decentralized way, so the voltage and reactive power regulation means is more flexible. However, in terms of load center areas such as Quanzhou City and Xiamen City, the voltage support capacity is relatively poor owing to large external power demands. Meanwhile, the reactive power compensation is relatively insufficient. Therefore, it is difficult to control the voltage level in case of high load or weakened network structure.

3. Suggestions on the measures for reactive power compensation and voltage and reactive power regulation in Fujian Power Grid

Through the current situation analysis on voltage and reactive power regulation in Fujian Power Grid, it can be seen that the reactive power balance is basically in line with the principle of the balance in layers and sub-regions and on-site compensation in normal peak load mode. However, due to the lack of power supply and large external power demands in load center areas such as Quanzhou City and Xiamen City, the reactive power compensation capacity is relatively insufficient. It is necessary to receive a certain capacity of reactive power from the upper power grid. Although the requirements for voltage regulation can be met in normal peak load mode, voltage regulation means are few, with a poor adaptability to mode changes. According to the basic principle of reactive power compensation, it is required to increase the power compensation factor in the areas with a small power supply capacity. In order to meet the assessment requirements of reactive power compensation on the load side, it can be considered to enhance the shunt reactive power compensation capacity on the main gird side and reserve more dynamic reactive power compensation capacity to improve the dynamic voltage support capacity of power grid.

8.1.6 Research on Static Voltage Stability of Fujian Power Grid in Peak Load Mode in 2005

The static voltage stability analysis method is adopted to analyze and calculate the static voltage stability in normal peak load mode as well as shutdown modes after $N - 1$ and $N - 2$ faults, so as to comprehensively evaluate the static voltage stability of Fujian Power Grid and determine the weak links of various modes and the fault mode with the most serious influence on the voltage stability according to the evaluation indexes on the static voltage stability.

1. Analysis of static voltage stability in peak load mode in 2005

1) Calculation of dynamic voltage stability margin

Through the calculation of static voltage stability of Fujian Power Grid in peak load mode in 2005, load bus voltage stability indexes in Fujian Power Grid were given: power margin and voltage margin.

Calculation and analysis results show that when the load increases at a constant power factor, the active power margin of 110 kV load bus in Fujian Power Grid was 1.517–6.495 p.u. Only one of power reserve coefficients of 110 kV load buses in Fujian Power Grid is lower than 40%, namely, 38%, while those in other areas are above 40%.

The minimum power reserve coefficient was that of Quanzhou City in peak load mode in 2005, namely, 10.1%, followed by Fuzhou City and Xiamen City with 12.9% and 20.4% respectively.

Reactive power margin of 500 kV bus in peak load mode in 2005 was 4.306–5.207 p.u.

2) Calculation results by modal analysis method

Through the modal analysis of Fujian Power Grid in peak load mode in 2005, key load bus, lines and generator set are given.

The correlation factors of Dongdu load buses in Qingmeng Economic and Technology Development Zone and Shandou Village, Quanzhou City and Xiamen City are ranked first. The load buses with larger correlation factors are located in weak areas of the power grid, which are also the best locations for reactive power compensation.

The participation factors of 220 kV North Suburb-Shuikou, Minqing-Huian, Minqing-Shuikou lines and 500 kV Xiamen-Houshi, Xiamen-Quanzhou lines are the largest. Most of these lines are high power flow lines. The change of their reactive power losses is sensitive, so it is necessary to control the power flow in the operation of the lines.

Meidian, Houshi, Songyu, Nanpu and Huaneng generator sets play an important role in the power grids. It is required to monitor their dynamic reactive power reserves in the operation control and strengthen the reactive power compensation in nearby areas in order to improve the reactive power reserves of the units.

3) Elevation of static voltage stability in peak load mode in 2005

Through power margin scanning calculation and modal analysis calculation above, it can be seen that the Power margin index of Fujian Power Grid was higher in peak load mode in 2005, and active power reserve coefficients in the regions were larger than 10%, which meets the requirements for normal load fluctuation.

It can be seen from the results of modal analysis and calculation that the relatively weak links of Fujian Power Grid are mainly concentrated in Quanzhou City, Xiamen City and Zhangzhou City where reactive power and power supplies are insufficient, which is consistent with the results of reactive power balance analysis.

2. Analysis of the static voltage stability in the mode after $N-1$ fault

Through static voltage stability scanning analysis of 500 kV lines, such as I-circuit Houshi-Xiamen line, Houshi-Quanzhou line, I-circuit Quanzhou-Xiamen line, I-circuit Quanzhou-Jinjiang line, I-circuit Quanzhou-Putian line and Shuikou-Fuzhou line in the mode after $N-1$ fault, it is known that the faults of Houshi-Quanzhou line Houshi-Xiamen line have the most serious influence on Fujian Power Grid. In addition, regional power reserve coefficients are reduced sharply, and the voltage safety of the system cannot be guaranteed. The rest of the 500 kV lines have little influence on the voltage stability in the fault mode and can meet the requirements of voltage safety.

3. Static voltage stability analysis in post-accident mode

It is necessary to consider the fault breaking mode of some 500 kV double-circuit lines and 220 kV double-circuit high power flow lines in post-accident mode. After multiple faults, the structure of power grid frame is weakened, with a larger reactive power loss and weak static voltage stability. Through static voltage stability margin scanning calculation, operators can understand the degree of influence on the static voltage stability of the system after multiple faults and properly formulate corrective measures and emergency control measures.

Through static voltage stability scanning analysis of double-circuit lines, such as Quanzhou-Jinjiang line, Quanzhou-Putian line, Shuikou-Minqing line, Meidian-Hushi line and Miandian-Fengyuan line in the breaking mode, it can be seen that the static voltage stability of the system is greatly affected in the cut-off mode, especially the 220 kV Miandian-Fengyuan line, followed by the 220 kV Shuikou-Minqing line. The influence of 500 kV Quanzhou-Jinjiang and Quanzhou-Putian lines on the voltage stability is more significant than that of Miandian-Fengyuan and Shuikou-Minqing lines due to their small power flow.

8.1.7 Research on Transient Voltage Stability in Fault Disturbance Mode

According to the requirements specified in *Guide on Security and Stability of Power System* of China:

(1) The voltage collapse shall not occur when any element (generator, line and transformer) of the system fails in failure in normal operation mode (including normal maintenance mode);
(2) When external power supply is suddenly lost and a maximum capacity unit trips or loses the excitation in areas with concentrated loads, the voltage collapse shall not occur due to the lack of reactive power.

The fault modes referred to the *Guide on Security and Stability of Power System* mainly refer to a single fault mode, which requires the voltage stability verification be performed. However, it can be seen from a series of large voltage collapse accidents in the world that the proportion of voltage collapse accidents caused by multiple serious fault modes is the largest. Accidental multiple accidents may not occur continuously, but each accident impact is accompanied by a serious power flow transfer and a larger and reactive power loss. Furthermore, the accident impact may be expanded due to the misoperation of safety and stability measures and relay protection and human factors.

When removing the fault of a large capacity generator set in s system in normal operation mode, the voltage collapse shall not occur. Frequency drop problems caused by the lack of a large amount of active power can be solved with active power reserve capacity and by low-frequency load shedding devices, so as to restore the system frequency. Since the fault removal is accompanied with a large power flow transfer, the reactive power loss increases. Moreover, when a generator lacks its reactive power supply, the reactive power of the system is more insufficient. In addition, the functions of load dynamic characteristics and shunt reactive power compensation equipment are reduced when the voltage drops, which has an adverse impact on the voltage stability and may cause voltage collapse in case of serious faults.

After loss of excitation, the generator set not only has no reactive power output, but also absorbs reactive power equivalent to its capacity from the system. If the unit is not disconnected from the power grid, the system needs to provide more than twice of the reserve capacity to meet the reactive power demand. If the power gird has no sufficient dynamic reactive power, the voltage stability may be affected after a large unit loses its excitation. Therefore, if the reactive power reserve capacity of power grid is insufficient, any large units shall be disconnected from the power grid after the loss of excitation. On the contrary, some loads may operate for a short time.

When a double-circuit line on the same channel has a fault in its single-circuit line in high power flow mode, the overload or the misoperation of relay protection in another circuit line caused by power flow transfer may improve the probability of multiple accidents. Several synchronous tripping faults in double-circuit lines have occurred in the Tianjian-Guangzhou DC shunt transmission channel of interconnected power grids in the south of China. Multiple faults have a more significant impact on the system. Although Fujian Power Grid has been developed greatly in recent years, the structure of 500 kV power grid is relatively weak. Multiple faults may lead to voltage collapse in load center areas.

In this section, the transient voltage stability of the system under various fault disturbances is numerically simulated in peak load mode in 2005. The simulated major faults include element fault in each line, loss of excitation of generator set, transformer fault, full shutdown of Houshi Power Plant and multiple faults in AC lines.

1. Single element fault check

The fault types include three phase short-circuit, single generator trip and excitation-loss fault of the lines and buses in 500 and 220 kV voltage level, and the fault state of single 500 kV transformer fault and out of service.

2. Line $N - 1$ fault check

The fault types include three-phase short circuit of the lines and buses in 500 and 220 kV voltage level. The time of fault clearing of tie-line and Shuikou-Putian Line are calculated as 0.1, and 0.09 s for the rest of the 500 kV lines, and 0.12 s for the 220 kV lines. Overload of lines and transformer elements are not considered in fault check.

The calculation results show that when the three-phase short-circuit fault occurs in the 500 kV line including the tie-line and the 220 kV line in Fujian Province, the fast protection can clear the fault correctly and the Fujian Power Grid can maintain stability. The Houshi-Xiamen Line and Houshi-Quanzhou Line is the most serious fault in the transmission section of Houshi Power plant.

3. Generator $N - 1$ and excitation-loss fault

According to the *Guide on Security and Stability of Power System*, for the power system under normal operation mode, if any generator trips or loses excitation, the protection, switch and reclosing shall work correctly, without taking stability control measures; it is necessary to maintain the power system stability operation and normal power supply of power grid, and other elements shall not exceed the specified emergency overload capacity, and have no interlocking trip. The calculation results show that under the peak load mode in winter of 2005, if any generator in Fujian Power Grid trips, the system can maintain steady operation without voltage instability.

The excitation-loss fault is more serious than the generator trip fault, because the occurrence of the fault not only loses the reactive power output of the generator itself, but also needs to absorb the reactive power equivalent to the generator capacity from the system, causing serious reactive power impact on the system. As the calculation results shown, the open-circuit excitation-loss fault of exciting winding occurs in a single generator in Fujian Power Grid, and the system cannot maintain stability due to insufficient dynamic reactive power reserve; the generator unit with excitation loss fault is removed 0.3 s after the fault, so the system can maintain stability and the voltage is maintained at a high level.

4. Transformer $N-1$ fault check

The $N-1$ fault check is performed on the main transformer of the 500 kV system of Fujian Power Grid. A three-phase tripping fault occurred on the transformer HV side at 0 s, and the transformer was out of service after 5 cycles. As the calculation results shown, when $N-1$ fault occurs in 500 kV main transformer of Fujian Power Grid, the system can recover quickly and maintained stability, and the voltage is also kept at a high level.

5. Multiple critical fault check

The fault types include three-phase tripping double-circuit fault of the line, transformer $N-2$ exit, and the full power failure occurred in Houshi Power Plant. The under-voltage load-shedding scheme adopted in this study is shown in Table 8.1, in which the starting value is the per-unit value of bus voltage.

1) Full shutdown of Houshi Power Plant

Houshi Power Plant is directly connected to the 500 kV system of Fujian Power Grid and its power generation capacity (3,600 MW) accounts for about 1/3 of that of Fujian Power Grid. It is an important voltage and reactive power support in the southern area of Fujian Power Grid. If there is a full power failure of 6 units, it would cause a significant impact on the system. As the calculation results shown, the 6 units of Houshi Power Plant without fault trip suffered a sudden reduction of 3,600 MW output, the system could maintain the rotor angle stability, the LV load shedding action could perform 112 MW load shedding, and the system could maintain the voltage stability, which could meet the requirements of the third level safety and stability standard. The bus voltage of the first swing of 500 kV lines in Jinjiang, Quanzhou and Fuzhou were low, which could be recovered to above 0.9 p.u. The lowest frequency of the system was about 49.1 Hz, which could be stabilized at about 49.4 Hz finally. The low-frequency load shedding is not started (the setting value of the first round of action is 49 Hz). After the fault, the tie-line between Fujian and East China and the 500 kV system power flow of Fujian Power Grid were reserved. East China Power Grid supplied power to Fujian Power Grid, and sent 1, 280 MW active power flow to Fujian through the tie-line; and the 500 kV system power flow of Fujian Power Grid also changes from the original South-to-North Power Transmission to

Table 8.1 List of under-voltage load-shedding measures for Fujian power grid

Region	First round			Second round			Third round			Fourth round		
	Starting value	Time delay	Load shedding amount	Starting value	Time delay	Load shedding amount	Starting value	Time delay	Load shedding amount	Starting value	Time delay	Load shedding amount
	/p.u.	/s	/%	/p.u.	/s	/%	/p.u.	/s	/%	/p.u.	/s	/%
Quanzhou	0.83	0.2	6	0.79	0.2	7	0.75	0.2	7	0.71	02	7
Xiamen	0.83	0.2	5	0.79	0.2	5	0.75	0.2	5	0.71	02	5
Zhangzhou	0.83	0.2	9	0.79	0.2	7	0.75	0.2	11	0.71	02	11

North-to-South Power Transmission, and the southern area of Fujian Power Grid was equivalent to the terminal system of East China Power Grid. As shown in the full power failure occurred in Houshi Power Plant, the 500 kV bus voltage, frequency curve and the power flow of the tie-line between Fujian and East China show that the system could maintain stability after the fault.

2) Three-phase tripping double-circuit fault of Houshi-Xiamen Line

In the Houshi-Xiamen Line, a three-phase tripping fault occurred on Houshi side at 0 s, and double-circuit breaking occurred on Houshi-Xiamen Line at 0.09 s. Under the fault mode, the accelerating power of the units of Houshi Power Plant was large, and the system could not maintain stability because of the out of synchronism in the first swing. After the double-circuit breaking of Houshi-Xiamen Line, the output of Houshi Power Plant can only be sent out through the Houshi-Quanzhou Line, and it was obviously unrealistic to transmit 3,600 MV power only through a 500 kV Line. Under the current measures (in case of serious fault, only 2 units could be removed at most in Houshi Power Plant), two 600 MW units of Houshi Power Plant are removed at 0.25 s to reduce the output power of Houshi Power Plant, and the system was still unstable. Considering the limit of feasible existing measures, it was necessary to adopt the concentrated load shedding measures for the fault mode. As the calculation results shown, in addition to the generator shutdown measures, through the measures of 1,600 MW concentrated load shedding in Quanzhou area at 0.3 s, 400 MW concentrated load shedding in Xiamen area, and 103 MW load shedding by under-voltage load-shedding, the system could maintain stability. Under the fault mode, the voltage drop was relatively rapid. If a large number of concentrated load shedding measures are not taken, the under-voltage load-shedding device will be locked and will not play its due role. The instability was reflected as too long LV time in Houshi Power Plant 500 kV bus.

Considering that if the number of generator shutdown units was not limited, as the calculation results shown, through measures of five 600 MW units of Houshi Power Plant load shedding at 0.25 s, and 238 MW load shedding by under-voltage load-shedding action, the system could maintain stability. The instability was reflected as LV time being too long in Houshi Power Plant 500 kV bus.

Considering the measures to speed up the load shedding of units, the calculation results show that after two units of Houshi Power Plant are removed at 0.15 s, the system couldn't maintain stability. Then take the measure of concentrated load shedding. After 700 MW concentrated load shedding in Quanzhou area at 0.3 s, while 460 MW load shedding by under-voltage load-shedding action. The system could maintain stability, and the first swing voltage of Houshi Power Plant 500 kV bus was the lowest. If the shedding of 3 units of Houshi Power Plant could be considered in case of this fault, the limit removal time was 0.19 s, no concentrated load shedding measures are needed, after 684 MW load shedding by under-voltage load-shedding action, the system could maintain stability.

Through the above analysis, it can be seen that the installed capacity of Houshi Power Plant, as the main power source of Fujian Power Grid, accounts for a large proportion of the total system capacity, and the situation of large units with small networks of Fujian Power Grid still exists. When serious faults occur near the exit of Houshi Power Plant, the accelerating power of the unit is large, and the out of synchronism occurs in the first unstable period, which brings certain difficulties for taking stability measures to the system. Under the condition of existing measures, the measures needed to be taken are costly and difficult to implement in case of three-phase tripping double-circuit fault in Houshi-Xiamen Line. It is suggested to speed up the load shedding of units, and to configure the concentrated load shedding measures in case of serious fault in Quanzhou area.

3) Three-phase tripping double-circuit fault of Quanzhou-Xiamen Line

Three-phase tripping fault occurred on the Xiamen side of Xiamen-Quanzhou Line at 0 s, and double-circuit breaking occurred in Quanzhou-Xiamen Line at 0.09 s, and the system could maintain stability. The first swing voltage of Jinjiang and Quanzhou 500 kV buses was the lowest, which could be recovered to about 0.9 p.u. During the first swing, 287 WM load shedding was performed by under-voltage load-shedding action.

4) Three-phase tripping double-circuit fault of Quanzhou-Jinjiang Line

Three-phase tripping fault occurred on Quanzhou side of Quanzhou-Jinjiang Line at 0 s, and double-circuit breaking occurred in Quanzhou-Jinjiang Line at 0.09 s, and the system could maintain stability. The first swing voltage of Jinjiang 500 kV bus was the lowest, but it was not as low as the starting value of under-voltage load-shedding device, and the bus voltage could be recovered to higher than 0.9 p.u.

5) Three-phase tripping double-circuit fault of Quanzhou-Putian Line

Three-phase tripping fault occurred on Quanzhou side of Quanzhou-Putian Line at 0 s, and double-circuit breaking occurred in Quanzhou-Putian Line at 0.09 s, and the system could maintain stability. The first swing voltage of Quanzhou 500 kV bus was the lowest, which could be recovered to above 0.9 p.u. During the first swing, 20 WM load shedding was performed by under-voltage load-shedding action.

6) Two main transformers of Quanzhou in fault

Three-phase tripping fault occurred on the HV side of Quanzhou main transformer at 0 s, the fault was eliminated at 0.1 s, and the two shunt transformers of Quanzhou were in fault state. The system could maintain stability, and the voltage of Xianyuan line was the lowest, but it could still be recovered to about 0.88 p.u. During the first swing, 140 WM load shedding was performed by under-voltage load-shedding action.

7) Two main transformers of Xiamen in fault

Three-phase tripping fault occurred on the HV side of Xiamen main transformer at 0 s, the fault was eliminated at 0.1 s, and the two shunt transformers of Xiamen were in fault state. The system could maintain stability, and the voltage of Xiamen area was the lowest, but it could still be recovered to about 0.9 p.u. During the first and second swing, 72 WM load shedding was performed by under-voltage load-shedding action.

8.1.8 Full Dynamic Simulation

According to the previous analysis, if reasonable measures are taken after a serious fault disturbance, no transient voltage instability will occur in Fujian Power Grid, but considering the actions of slow acting elements such as generator excitation limit unit and transformer tap automatic regulator, the serious fault disturbance may cause dynamic voltage instability in the medium and long-term process. In this section, the full dynamic simulation program of the power system of China Electric Power Research Institute is used to simply simulate the dynamic voltage stability problems that may occur in Fujian Power Grid, taking the full shutdown of Houshi Power Plant under the peak load mode in winter as an example.

1. The main model used in the simulation

The receiving transformer of main load points of Fujian power grid is a LTC transformer, the tap can be adjusted automatically; the load adopts static a model, which is composed of 60% constant power and 40% constant impedance, and the load frequency factor is considered; all power plants simulate the over excitation devices and adopt the OELBPA model. The main parameters of the over excitation model are: after reaching the calorific value limit, the excitation current is limited. If the calorific value cannot be lower than the limit value within 3 s, the generator shutdown occurs.

2. Process of voltage collapse accident

It is assumed that the accident occurs in Houshi Power Plant at 0.2 s, which results in the disconnection of the whole plant and the system. After disconnection, the load point voltage in Xiamen area is low, and the LTC transformer tap was automatically adjusted, resulting in generator shutdown of several power plants in Fujian Power Grid due to over excitation: Songyu Power Plant has over excitation limit at 52.8 s and generator shutdown at 55.8 s; Yongan unit is limited at 57.8 s and has generator shutdown at 60.8 s; Yongan unit is limited at 90.6 s and has generator shutdown at 97.6 s. Thus, the power grid has a power grid collapse accident.

The curve of the generator rotor angle variation is shown in Figs. 8.2 and 8.3. After the accident, the units in the whole network can maintain rotor angle stability. The voltage variation curve of load point (node: Banlan 110, 110 kV) is shown in Fig. 8.4. Before the action of over excitation device of Songyu Power Plant, the load point voltage increases with the variation of tap; after the over excitation device of Songyu

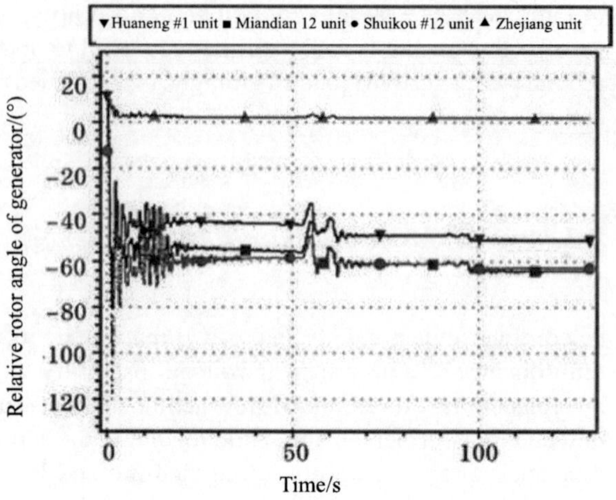

Fig. 8.2 Change of relative rotor angle of generator

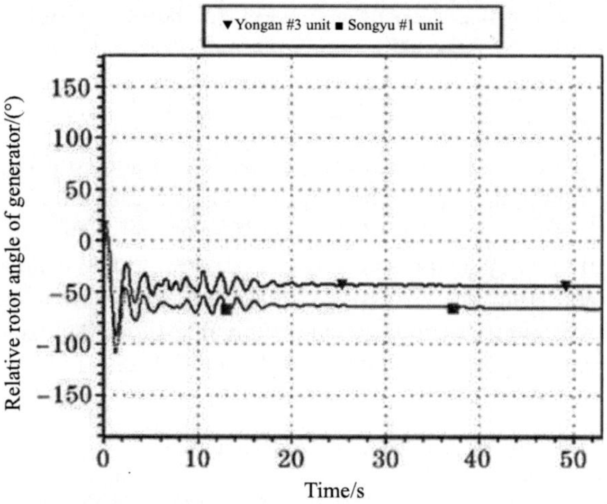

Fig. 8.3 Change of relative rotor angle of generator

Power Plant sends the order of limiting excitation current (52.8 s), the node voltage drops sharply due to the lack of reactive power; after generator shutdown of Songyu Power Plant and other power plants, the node voltage drops further, eventually leading to voltage collapse. The voltage variation curve of the system 500 kV bus is shown in Fig. 8.4, which shows the process of Xiamen bus voltage variation. The curves of excitation current and reactive power output of the power plant suffering generator shutdown due to over excitation are shown in Figs. 8.5 and 8.6. The curve of reactive power variation of the main 220 kV line Xiahe-Banlanshan, which supplies power to the load point (node: Banlan 110, 110 kV) is shown in Fig. 8.7.

The full dynamic simulation results show that the automatic regulation of transformer tap causes some units to trip, which will eventually lead to serious voltage

Fig. 8.4 Busbar voltage

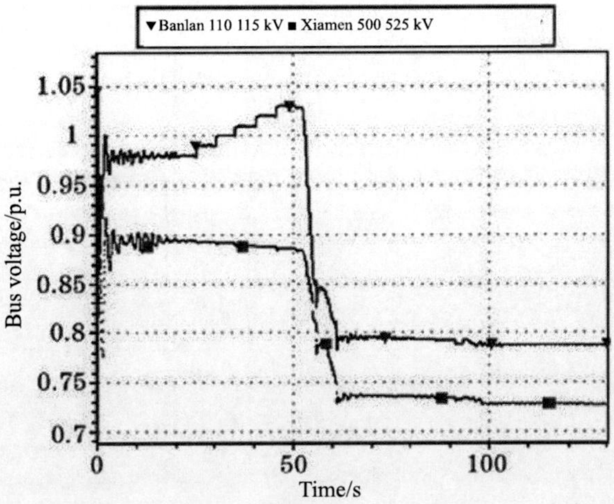

Fig. 8.5 Generator excitation current

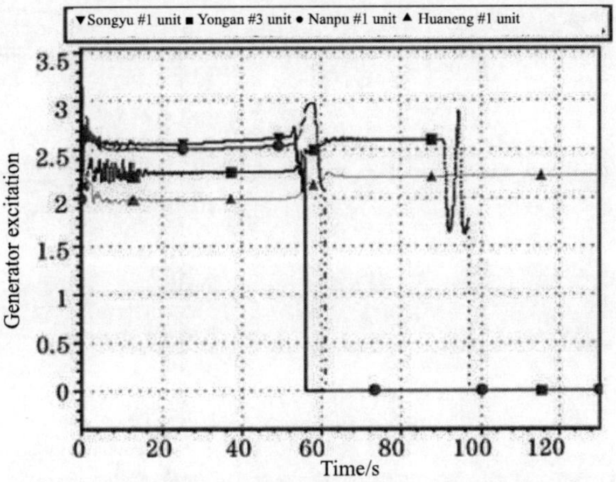

collapse accidents. It should be pointed out that the simulation study is based on the assumption that the transformers at the main load points of Fujian Power Grid are LTC transformers and all power plants are equipped with over excitation devices, which does not necessarily reflect the actual situation of the system. It can also be seen from the simulation that voltage instability may spread to a long time period, and the well-known voltage instability accidents abroad generally have a long development time. Because the transient stability analysis program cannot simulate the actions of slow acting elements such as generator excitation limit and LTC tap adjustment, it may not expose the voltage stability problems in the power grid. It is suggested to further study in depth the model of dynamic regulating elements in the future, such as OLTC, generator excitation current limit, and dynamic recovery characteristics

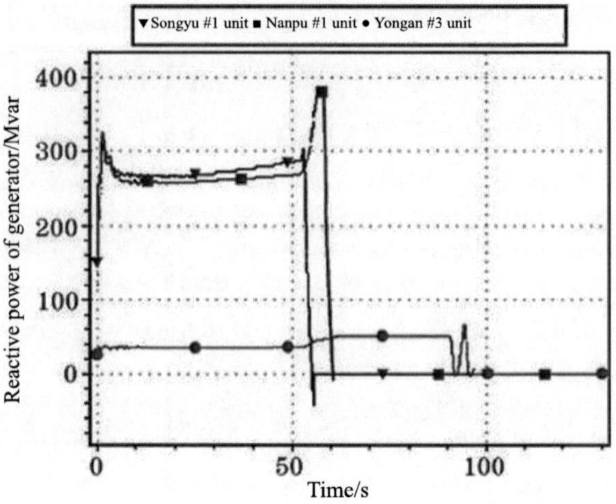

Fig. 8.6 Reactive power of generator

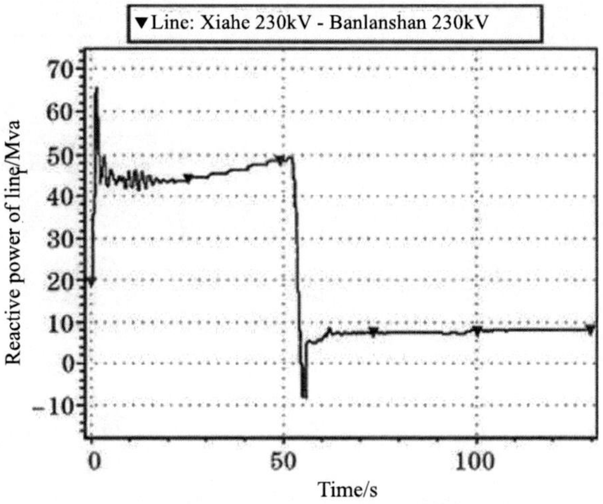

Fig. 8.7 Reactive power of line

of load, to establish a mathematical model more in line with the actual situation of power grid.

This section checks the dynamic voltage stability of single fault and multiple serious faults in Fujian Power Grid. The calculation results show that, for single fault, if the relay protection acts correctly, Fujian Power Grid can maintain stability, and will not cause voltage instability accident. The unit excitation-loss fault is more serious than trip fault, but as long as the excitation-loss unit is removed, the system can still maintain stability; for multiple serious faults, such as the full power failure occurred in Houshi Power Plant, Fujian Power Grid lost more than 30% of the active power. East China Power Grid provides power support to Fujian Power Grid through the tie-line, which is conducive to maintain the stable operation of the power grid;

for three-phase tripping double-circuit fault in Houshi-Xiamen Line, if there are no reasonable measures, there will be rapid instability accidents in Fujian Power Grid, which also shows that the problems of large generator with small grid still exist to some extent; for the double-circuit 500 kV line fault of other channels, no instability accident will occur under peak load mode in winter in 2005. Based on the assumption that all the main load transformers are LTC transformers and all the generators are equipped with over excitation devices, the full dynamic simulation results of Houshi Power Plant full shutdown fault show that the fault will cause a serious voltage collapse accident.

8.1.9 Comprehensive Measures to Improve Voltage Stability of Fujian Power Grid

The measures to improve voltage stability are comprehensive, involving the planning, construction, operation and management of the power grid. In the planning stage, mainly consider the strengthening measures of the power grid structured, and optimize and reasonably configure the power supply at the same time; do well in the reactive power planning and study work to strengthen the reactive power compensation in the central area of the terminal load. In the operation stage, it is necessary to strengthen the online monitoring means of voltage and reactive power and the management of reactive power compensation on the user side, improve the level of dispatching automation, and take preventive correction measures; formulate the emergency voltage control measures, and establish the safety measures to prevent voltage collapse accidents and construct the third line of defense.

Based on the current situation analysis of voltage and reactive power, steady state voltage stability analysis, study conclusion of dynamic voltage stability check and development planning of Fujian Power Grid, the comprehensive measures to improve voltage stability and prevent voltage collapse accidents of Fujian Power Grid at present mainly include the following aspects:

(1) To improve the shunt reactive power compensation capacity of the receiving end load center, mainly including Quanzhou, Fuzhou, Xiamen and Zhangzhou areas, which lack of reactive power compensation in normal large-scale mode and need to receive a certain amount of reactive power outside; in addition, for example, Quanzhou, Xiamen and Zhangzhou areas lack of dynamic power supply support, they need to appropriately improve the compensation capacity.

(2) To improve the dynamic reactive compensation capacity in the receiving-end load center area, since the proportion of external power demands in Quanzhou and Xiamen accounts for 91 and 64% of the total regional load respectively, the dynamic reactive power support capacity is relatively insufficient, which is manifested in the lack of reactive power regulation means under normal and maintenance modes, and the voltage recovery is slow during large disturbances.

(3) To take comprehensive measures to solve voltage collapse accidents caused by serious multiple faults, which mainly depend on the emergency control measures of generator shutdown and load shedding (under-voltage load-shedding or concentrated load shedding), and consider the low-voltage disconnection measures on the other hand.

1. Study on the reactive power compensation measures for Fujian Power Grid

For the reactive power compensation locations, Quanzhou, Xiamen, Fuzhou, Zhangzhou and other areas where the reactive power compensation is relatively insufficient are selected. Different reactive power compensation schemes are studied, the compensation location is mainly selected according to the weak links determined in the modal analysis, considering the capacity limitation of the existing 220 kV substation and the maximum reactive power load capacity, to avoid local over compensation and making local voltage regulation difficult.

The specific compensation location and capacity are given after study. After compensation, the total dynamic reactive power reserve of Fujian Power Grid is greatly increased. The power margin of load bus is improved obviously.

2. Study on the dynamic reactive power compensation measures

When selecting the installation location and capacity of SVC, it is necessary to consider which location is more sensitive to SVC, and it is better to select a location that has a significant supporting effect on voltage. According to the previous calculation results, the voltages of Fuzhou, Quanzhou and Xiamen are the lowest when serious faults occur. After these substations are equipped with SVCs, SVC provides dynamic reactive power support during the transient voltage sags, which is conducive to the recovery of voltage, so as to improve the stability of the system. In addition, the short-circuit capacity of the power grid is relatively high. In this case, SVC must have enough capacity to play an obvious role. Generally, SVC is installed in the third winding of the main transformer, which requires that the third winding of the transformer has enough capacity.

Two SVC configuration schemes are studied and designed, and the fault check is carried out. According to the study results, the scheme of SVC configuration in Fuzhou and Quanzhou is better.

3. Regulation of operation mode

According to the previous calculation results, the phenomenon of large unit with small network still exists in Fujian Power Grid. In the three-phase tripping double-circuit fault of Houshi-Xiamen line, the system can only maintain stability after 2 units generator shutdown in Houshi Power Plant at 0.25 s, 1,600 MW concentrated load shedding in Quanzhou area at 0.3 s, 400 MW concentrated load shedding in Xiamen area, and 103 MW load shedding by under-voltage load shedding action. However, the actual situation of Fujian Power Grid is that at most only 2 units can be removed in Houshi Power Plant when there are serious faults, so the output of Houshi Power

Plant must be limited from the operation mode. As the calculation results shown, limit the output of one unit in Houshi Power Plant to 320 MW, correspondingly reduce the load of Fujian Power Grid by 3% (about 280 MW), in the case of this fault, the generator shutdown of 2 units of Houshi Power Plant can maintain the system stability, with 639 MW load shedding by under-voltage load shedding action, and the first swing voltage of Houshi 500 kV bus is the lowest.

Through comparison, it can be seen that limiting the output of Houshi Power Plant can significantly reduce the load shedding capacity of three-phase tripping double-circuit fault on Houshi-Xiamen Line.

4. Regulation of grid structure

The power of Houshi Power Plant is sent out through Houshi-Xiamen double-circuit and Houshi-Quanzhou lines. More than two-thirds of the output of Houshi Power Plant is sent through Houshi-Xiamen Line, so the three-phase tripping double-circuit fault on Houshi-Xiamen Line is very serious. If the Houshi-Quanzhou Line is strengthened to be double-circuit, after the output power flow from Houshi Power Plant, the power flows of Houshi-Xiamen double-circuit and Houshi-Quanzhou double-circuit are almost the same, accounting for almost half each. No matter whether three-phase tripping double-circuit fault on Houshi-Xiamen Line (230 MW load shedding by under-voltage load-shedding action) or three-phase tripping double-circuit fault on Houshi-Quanzhou Line (no under-voltage load shedding action), the system can maintain stability, and Houshi Power Plant does not need to take generator shutdown measures.

After Houshi-Quanzhou Line is strengthened to be double-circuit, the temporary stability limit of power transmission from Fujian to East China is 1,600 MW, and the most serious fault constraint is the three-phase permanent short-circuit fault on Fuzhou side of Fuzhou-Shuanglong Line. After the fault, a large amount of power in Fujian Power Grid cannot be sent out to East China, and the unit accelerates, resulting in the loss of synchronization between Fujian Power Grid and East China Power Grid. The vibration center is on the tie-line between Fujian and East China. The instability type is Fuzhou 500 kV bus having low voltage for too long.

It can be seen from the above analysis that the scheme of strengthening Houshi-Quanzhou Line into double-circuit has little effect on the delivery limit of Fujian Power Grid. But for the serious fault of Houshi Power Plant exit, the system can maintain stability without even taking generator shutdown measures. So it provides obvious advantages in comparison with the original power grid structure.

5. Configure power supply in areas with weak voltage

Synchronous generators are active power output power supply and reactive power output power supply with the best regulation performance. Through generator excitation regulation, the magnitude of reactive power output can be smoothly changed. The generator unit can output reactive power during the lagging phase operation and absorb reactive power during the leading phase operation. The generator excitation system can exert its overload capacity in the transient process and provide dynamic voltage output to the system through forced excitation and more reactive power output. When there is excess reactive power in the system, reactive power can be absorbed through the leading phase operation to suppress the rise of system voltage.

The proportion of external power in Quanzhou City accounts for 91% of the total load of the area. The proportion of external power is too large, and the dynamic reactive power support capacity is relatively insufficient. Under normal operation and maintenance modes, reactive power regulation means are lacking, and voltage recovery is slow in the process of large fault disturbance. Configuring power supply in Quanzhou City with weak voltage level can improve the voltage output capacity of the receiving-end power grid and provide dynamic reactive power output capacity.

A 300 MW unit is configured in Shandou with relatively weak voltage level. From the results of static voltage stability analysis, the active power reserve coefficient in Quanzhou City has been significantly improved, and the active power reserve coefficient in other areas of Fujian Power Grid has also been improved in various degrees. The results are shown in Table 8.2.

A 300 MW unit is configured in Shandou. In case of three-phase tripping double-circuit fault of Houshi-Xiamen line, the system cannot remain stable if 2 units in Houshi Power Plant are shut down within 0.25 s; if the unit shedding time is accelerated (2 units in Houshi Power Plant are shut down within 0.15 s), the system can be kept stable. Under the normal peak load mode in winter, in case of the fault, the system cannot be kept stable if 2 units in Houshi Power Plant are shut down within 0.15 s.

A 300 MW unit is configured in Shandou. The transient stability limit of Fujian to East China is 1,610 MW. The most serious fault constraint form is the three-phase permanent short-circuit fault on Fujian side of Fujian double-circuit line. The oscillation center is located on the tie-line between Fujian and East China. The instability type is Fuzhou 500 kV bus long-term low voltage.

From the above analysis, it can be seen that configuring power supply in Quanzhou City with weak voltage can improve the active power reserve coefficient of various areas of Fujian Power Grid, especially active power reserve coefficient of Quanzhou City, and also can strengthen the system's ability to resist serious faults.

6. Emergency control measures to prevent voltage collapse in Fujian Power Grid

From the dynamic voltage stability check of multiple serious faults in Fujian Power Grid, it can be seen that if a three-phase tripping double-circuit fault of Houshi-Xiamen line occurred on the outgoing line of Houshi Power Plant, it would lead to a large-area voltage collapse accident in Fujian Power Grid. The main reasons are as follows: Houshi Power Plant is the only 500 kV power supply connected to Fujian Power Grid, and is also the most important voltage output point and active power supply in Quanzhou and Xiamen, accounting for more than 30% of the active power output of Fujian Power Grid. In case of the three-phase tripping double-circuit fault of Houshi-Xiamen line, the accelerating power of units in Houshi Power Plant is relatively large, and the out of synchronism occurs in the first unstable period. After

Table 8.2 Comparison of active power reserve coefficients in different regions (Unit: %)

Mode	Fuzhou City	Putian City	Quanzhou City	Xiamen City	Zhangzhou City	Longyan City	Sanming City	Nanping City	Ningde City
Peak load mode in winter	18.6	50.9	9.8	20.6	24.0	58.1	48.5	60.7	51.6
Provision of power supply	24.2	59.2	14.1	24	26.7	60.1	50.2	63	53.5

the monotonic loss of synchronism of units in Houshi Power Plant, the entire southern Fujian Power Grid loses the most important voltage support. The active power flow of the East China Power Grid can pass through the weak Fujian 500 kV system to supply power to the southern power grid and the southern power grid will become the end system of the entire East China Power Grid. The huge impact of active power flow will increase the reactive power loss of the power grid, thus disable the reactive power balance maintenance. In this process, although the unit works under the strong excitation and outputs more reactive power, and the voltage characteristics of the load will also reduce the reactive power demand on the load side, the reactive power imbalance is too large to re-establish a new stable equilibrium point, and the voltage will continue to drop, eventually destabilizing the whole system.

1) Research on under-voltage load shedding measures

The existing under-voltage load shedding scheme of Fujian Power Grid only configure under-voltage load shedding devices in Quanzhou, Xiamen and Zhangzhou. Referring to the configuration of under-voltage load shedding at home and abroad, and combining with the results of multiple serious fault checks, the under-voltage load shedding setting scheme for Fujian Power Grid is set out in Tables 8.3, 8.4 and 8.5, mainly for adjusting the voltage setting value and load shedding amount in the first round. In Table 8.3, the voltage starting value is still 0.83 p.u. and the time delay is 0.2 s in the first round, and the load shedding amount is mainly adjusted. In Table 8.4, the time delay is 0.2 s and the load shedding amount is unchanged in the first round, and the starting value of voltage is mainly adjusted. In Table 8.5, the time delay is 0.2 s and the voltage setting value is 0.86 p.u. in the first round, and the load shedding amount is adjusted. The following is a study of the above setting scheme.

Original plan: In case of three-phase tripping double-circuit fault of Houshi-Xiamen line, it requires to shut down 2 units in Houshi Power Plant within 0.15 s, shed 700 MW load in Quanzhou in concentrated load shedding, shed 460 MW load by under-voltage load shedding, with a total load shedding amount of 1,160 MW.

Scheme I: If the load shedding amount in the first round increases by 5% in Quanzhou, Xiamen and Zhangzhou, in case of three-phase tripping double-circuit fault of Houshi-Xiamen line, the system will not remain stable if 2 units in Houshi Power Plant are shut down within 0.15 s, 600 MW load is shed in concentrated load shedding, 678 MW load is shed in under-voltage load shedding, with a total load shedding amount of 1,278 MW. The instability type is Houshi Power Plant 500 kV bus long-term low voltage

If the load shedding amount in the first round increases by 7%, in case of three-phase tripping double-circuit fault of Houshi-Xiamen line, the system can remain stable if 2 units in Houshi Power Plant are shut down within 0.15 s, 600 MW load is shed in concentrated load shedding, 691 MW load is shed in under-voltage load shedding, with a total load shedding amount of 1,291 MW.

If the load shedding amount in the first round is set to a larger value, such as 20%, in case of three-phase tripping double-circuit fault of Houshi-Xiamen line, 2 units in Houshi Power Plant are shut down within 0.15 s, 600 MW load is shed in

Table 8.3 Setting Scheme I for under-voltage load shedding

Region		First round			Second round			Third round			Fourth round		
		Starting value/p.u.	Time delay/s	Load shedding amount/%	Starting value/p.u.	Time delay/s	Load shedding amount/%	Starting value/p.u.	Time delay/s	Load shedding amount/%	Starting value/p.u.	Time delay/s	Load shedding amount/%
Sub-scheme I	Quanzhou	0.83	0.2	11	0.79	0.2	7	0.75	0.2	7	0.71	0.2	7
	Xiamen	0.83	0.2	10	0.79	0.2	5	0.75	0.2	5	0.71	0.2	5
	Zhangzhou	0.83	0.2	14	0.79	0.2	7	0.75	0.2	11	0.71	0.2	11
Sub-scheme II	Quanzhou	0.83	0.2	13	0.79	0.2	7	0.75	0.2	7	0.71	0.2	7
	Xiamen	0.83	0.2	12	0.79	0.2	5	0.75	0.2	5	0.71	0.2	5
	Zhangzhou	0.83	0.2	16	0.79	0.2	7	0.75	0.2	11	0.71	0.2	11
Sub-scheme III	Quanzhou	0.83	0.2	20	0.79	0.2	7	0.75	0.2	7	0.71	0.2	7
	Xiamen	0.83	0.2	20	0.79	0.2	5	0.75	0.2	5	0.71	0.2	5
	Zhangzhou	0.83	0.2	20	0.79	0.2	7	0.75	0.2	11	0.71	0.2	11

Table 8.4 Setting Scheme II for under-voltage load shedding

Region	First round			Second round			Third round			Fourth round		
	Starting value/p.u.	Time delay/s	Load shedding amount/%	Starting value/p.u.	Time delay/s	Load shedding amount/%	Starting value/p.u.	Time delay/s	Load shedding amount/%	Starting value/p.u.	Time delay/s	Load shedding amount/%
Quanzhou	0.86	0.2	6	0.83	0.2	7	0.79	0.2	7	0.75	0.2	7
Xiamen	0.86	0.2	5	0.83	0.2	5	0.79	0.2	5	0.75	0.2	5
Zhangzhou	0.86	0.2	9	0.83	0.2	7	0.79	0.2	11	0.75	0.2	11

Table 8.5 Setting Scheme III for under-voltage load shedding

Region	First round			Second round			Third round			Fourth round		
	Starting value/p.u.	Time delay/s	Load shedding amount/%	Starting value/p.u.	Time delay/s	Load shedding amount/%	Starting value/p.u.	Time delay/s	Load shedding amount/%	Starting value/p.u.	Time delay/s	Load shedding amount/%
Quanzhou	0.86	0.2	13	0.83	0.2	7	0.79	0.2	7	0.75	0.2	7
Xiamen	0.86	0.2	12	0.83	0.2	5	0.79	0.2	5	0.75	0.2	5
Zhangzhou	0.86	0.2	16	0.83	0.2	7	0.79	0.2	11	0.75	0.2	11

concentrated load shedding and 775 MW load is shed in under-voltage load shedding, with a total load shedding amount of 1,375 MW.

Scheme II: If the working voltage is set to 0.86 p.u. and the load shedding amount remains unchanged in the first round, in case of three-phase tripping double-circuit fault of Houshi-Xiamen line, the system can remain stable if 2 units in Houshi Power Plant are shut down within 0.15 s, 600 MW load is shed in concentrated load shedding, 627 MW load is shed in under-voltage load shedding, with a total load shedding amount of 1,227 MW.

Scheme III: Adjust the working voltage setting to 0.86 p.u in the first round, and the load shedding amount is increased by 7% in the first round, in case of three-phase tripping double-circuit fault of Houshi-Xiamen line, the system can remain stable if 2 units in Houshi Power Plant are shut down within 0.15 s, 500 MW load is shed in concentrated load shedding, 823 MW load is shed in under-voltage load shedding, with a total load shedding amount of 1,323 MW.

From the above analysis, it can be seen that increasing the load shedding proportion and the setting a value of the working voltage in the first round have no advantages in terms of the total load shedding amount required to maintain the stability of the system in case of serious faults. However, properly adjusting the load shedding proportion and the setting value of the working voltage in the first round can reduce the required concentrated load shedding amount.

2) Discussion on under-voltage disconnection measures

Under-voltage disconnection measure is a safety and stability measure as the third line of defense. In the process of large disturbances, the voltage will oscillate and fall greatly, which is often the precursor of voltage collapse accidents. Preventive under-voltage disconnection measures shall be taken to realize disconnection of regional power grids, to prevent further expansion of the accident, and to prevent whole-network collapse.

Under-voltage disconnection measures have certain risks and the setting of a fixed value of the device is relatively difficult. This is mainly because the dynamic instability characteristics of different fault modes are different, the voltage instability process is often expandable and the dynamic voltage instability changes at the critical point are abrupt. If disconnection is too early, the operating benefit and mutual support of the power grid may be lost. If disconnection is too late, the system stability may not be kept. In an actual project, it is necessary to set and check according to the specific situation to ensure the reliability of the device.

Considering that the center line of Fujian Power Grid is 500 kV line and its main grid structure is composed of 220 kV lines, its oscillation center is constantly changing due to different modes and faults. Judging from the characteristics of grid structure, there is no condition for local disconnection.

(3) Emergency control measures to prevent voltage collapse in Fujian Power Grid

Under the peak load mode in winter of 2005, a serious fault at the outlet of Houshi Power Plant will have a serious impact on the system. If emergency load shedding measures are not taken, the power flow passing through in East China Power Grid will have a significant impact on Fujian Power Grid. At present, the emergency load shedding measures of Fujian Power Grid include low-cycle load shedding measures and under-voltage load shedding measures. After the low-cycle and under-voltage load shedding devices are considered, simulation calculation is carried out. Under this fault mode, the system can be kept stable only when 2 units in Houshi Power Plant are shut down, 1,600 MW load in Quanzhou and 400 MW load in Xiamen are shed in concentrated load shedding, and 103 MW load is shed in the under-voltage load shedding, and the low-frequency load shedding does not work.

To sum up, for the most serious three-phase tripping double-circuit fault of Houshi-Xiamen line of Fujian Power Grid, the measures that can be taken to keep the system stable include the following aspects.

(1) Limit the output of Houshi Power Plant: By limiting the output of one unit of Houshi Power Plant to 320 MW and accordingly shedding 3% load (about 280 MW) of Fujian Power Grid, the system can be kept stable by adopting the measure of shutdown of 2 units of Houshi Power Plant with 0.25 s in case of this fault.
(2) Increase the number of inter-tripping units in case of fault: The system can be kept stable if 5 600 MW units of Houshi Power Plant are shut down at 0.25 s and 238 MW load is shed in the under-voltage load shedding; or 3 units of Houshi Power Plant are shut down with the limit shedding time of 0.19 s, 684 MW load is shed in under-voltage load shedding.
(3) Accelerate the unit shutdown time in case of fault: The system can be kept stable if 2 units of Houshi Power Plant are shut down with 0.15 s, 700 MW load in Quanzhou are shed at 0.3 s in concentrated load shedding and 460 MW load is shed in under-voltage load shedding.
(4) Adjust grid structure: The Houshi-Quanzhou line is strengthened to be a double-circuit; in case of serious fault of the outlet of Houshi Power Plant, the system can be kept stable even without adopting tripping measures.
(5) Configure power supply in Quanzhou: 300 MW units will be configured in Shandou. In case of serious fault of the outlet of Houshi Power Plant, the system can be kept stable if 2 units of Houshi Power Plant are shut down at 0.15 s.
(6) Impact of dynamic reactive power compensation scheme on transient stability: Configuring SVC can reduce the amount of the concentrated load shedding required to a certain extent, but due to the limited amount of configurable SVC, it has little impact on system transient stability.
(7) Adjustment of under-voltage load shedding measures: Adjusting the load shedding proportion and the setting value of working voltage in the first round has no advantage in terms of the total load shedding amount required to keep the system stable in case of serious faults, but can reduce the amount of the concentrated load shedding required.

8.1.10 Conclusion

Taking the peak load mode in summer and peak load mode in winter in 2005 as the research object, the static voltage stability level of different modes is comprehensively evaluated through scanning and calculation of different static voltage stability indexes; the static voltage stability level of various serious post-fault modes of Fujian Power Grid is calculated and analyzed; various cascading fault modes and their consequences that may lead to voltage stability problems are simulated through dynamic stability checking calculation of serious faults; research on emergency control measures to prevent voltage collapse in Fujian Power Grid are made. The main research conclusions are summarized as follows.

1) Reactive power balance analysis of Fujian Power Grid

From the analysis of reactive power balance in Fujian Power Grid, it can be seen that reactive power balance basically conforms to the principles of balance in layers and zones and local compensation. However, due to the lack of power supply and large external power demands in load center areas such as Quanzhou and Xiamen, the reactive power compensation capacity is relatively insufficient, and it is necessary to receive a certain capacity of reactive power. Voltage regulation means are few, with a poor adaptability to mode changes.

2) Static voltage stability analysis

(1) According to the static voltage stability analysis results of the peak load mode in summer and peak load mode in winter in 2005, since the dynamic reactive power reserve is close to 40% of the total reactive load capacity, the reactive power reserve is relatively sufficient, the bus reactive power margin index of Fujian Power Grid is also relatively high, and the minimum reactive power margin of 500 kV bus is 4.3 p.u. (reference power is 100 MVA); from the single load bus active power margin index, the minimum active power reserve coefficient is 38% and the power reserve coefficient of each area is above 10%, which meet the requirement of normal load fluctuation. However, the distribution of reactive power reserve is not balanced. The dynamic reactive power reserves in Fuzhou, Longyan and Nanping are relatively sufficient, while the reactive power reserves in the areas with a lack of power supply support, such as Xiamen and Quanzhou are almost zero. The situation of reactive power reserve can also reflect that Xiamen and Quanzhou are the areas with least reactive power in Fujian Power Grid.

(2) According to the static voltage stability index of Fujian Power Grid of the single and multiple serious post-accident modes corresponding to peak load mode in summer and peak load mode in winter of 2005, the reactive power margin and active power margin of each bus of the post-fault mode have decreased. Among the modes after a single fault, the modes of Houshi-Quanzhou line maintenance, Houshi-Xiamen line I-circuit maintenance and Jinjiang main transformer decommissioning are relatively typical. The reactive power margin of 500 kV bus has decreased by more than 80% compared

to the normal mode, and the regional power reserve coefficient is also below 3%. Among the modes after multiple serious faults, the modes of decommissioning of Xiamen shunt transformer, Quanzhou shunt transformer, Shuikou-Minqing line double-circuit disconnection, Quanzhou-Jinjiang line double-circuit disconnection, and Quanzhou-Putian double-circuit disconnection are relatively typical. The reactive power margin of 500 kV bus has decreased by more than 70% compared to the normal mode, and the regional power reserve coefficient is also below 5%.

3) Dynamic voltage stability checking

According to the dynamic voltage stability checking calculation results of single fault and multiple serious faults in Fujian Power Grid, for single fault, quick protection can work normally to ensure the stability of Fujian Power Grid without causing voltage instability accidents. The excitation-loss fault of the unit is more serious than the tripping of the unit, but the system can still remain stable as long as the excitation-loss unit is shut down on time. For multiple serious faults, such as three-phase tripping double-circuit fault of Houshi-Xiamen line, if no reasonable measures are taken, Fujian Power Grid will have large-area voltage instability accidents, the failure of other double-circuit 500 kV lines would not have instability accident under peak load mode in 2005. Based on the assumption that all the main load transformers are LTC transformers and all the generators are equipped with over excitation devices, the full dynamic simulation results of Houshi Power Plant full shutdown fault show that the fault would cause a serious voltage collapse accident. It is suggested that in the future further study in the future on the model of dynamic regulating elements, such as OLTC, generator excitation current limit and dynamic recovery characteristics of load be carried out, it should be studied in depth to establish a mathematical model more in line with the actual situation of power grid.

4) Measures to prevent voltage collapse in Fujian Power Grid

An in-depth study on three-phase tripping double-circuit fault of Houshi-Xiamen line that may cause voltage collapse accidents in Fujian Power Grid is done, measures to limit output of Houshi Power Plant and static and dynamic reactive power compensation schemes for Fujian Power Grid are proposed, and regulation measures of under-voltage load shedding schemes are studied. From the perspective of grid structure adjustment, the measure that the Houshi-Quanzhou line is strengthened to be a double-circuit is studied. The results show that this measure can greatly improve the system's ability to resist serious faults. In case of a serious fault of the outlet of Houshi Power Plant, the system can be kept stable even without adopting tripping measures. The measures of power supply configuration in areas with weak voltage are studied. The results show that the configuration of power supply in Quanzhou can improve the regional power reserve coefficient of each area, and at the same time, can enhance the ability of the system to resist serious faults.

8.2 Example II: Study and Comprehensive Evaluation on Voltage Stability of Chongqing Power Grid

In this section, based on the 2008 mode of Chongqing power grid, the current situation of reactive power balance in Chongqing power grid is comprehensively evaluated through power flow analysis and voltage and reactive power balance analysis. The static voltage stability level of Chongqing power grid under normal mode, maintenance mode, and the mode after multiple serious faults is comprehensively evaluated through the calculation and analysis of various static voltage stability indexes. The transient voltage stability level of Chongqing power grid is evaluated through transient voltage stability analysis. Considering many factors such as serious faults, cascading faults, SVC devices and the load of induction motors with low voltage self-release function, the simulation and analysis for long-term voltage stability of Chongqing power grid is carried out and the possibility of long-term voltage collapse of Chongqing power grid and preventive control measures are discussed. The dynamic voltage stability margin of Chongqing power grid is given through the full simulation study of continuous load growth, and the static voltage stability margin and dynamic voltage stability margin of Chongqing power grid are compared and analyzed; the influence of different load models on voltage stability of Chongqing power grid is studied through voltage stability analysis of typical load models.

8.2.1 *Conditions and Principles*

1. Principle of research

In this research, the following stability standards and principles of power system is followed:

(1) *Technical Guideline on Voltage and Reactive Power of Power System*;
(2) *Technical Guideline of Power System*;
(3) *Guide on Security and Stability for Power System.*

2. Research tool

PSD-BPA power flow and transient stability program of "PSD Power Tools" of China Electric Power Research Institute are adopted as the main tools for this study, mainly including:

PSD-BPA Power Flow Calculation Program (Version 4.0);

PSD-BPA Stability Calculation Program (Version 5.70);

PSD-FDS Full Dynamic Simulation of Power System Program;

PSD-PCS Power System Digital Platform (Version 2.0);

PSD-CLIQUE Program for Drawing of Power Flow Diagram in Format of Geographical Wiring Diagram;

Table 8.6 Stability calculation load characteristics of chongqing power grid

Power grid	Induction motor		Constant impedance		Frequency factor	
	Active/%	Reactive/%	Active/%	Reactive/%	D*P*/d*f*	D*Q*/d*f*
Chongqing	40	40	60	60	0	0

Table 8.7 Low-frequency load shedding allocation of chongqing power grid

	Round 1	Round 2	Round 3	Round 4	Round 5	Round 6	Round 1	Round 2
Rated frequency/Hz	49.0	48.8	48.6	48.4	48.2	48.0	49.0	48.5
Delay/s	0.2	0.2	0.2	0.2	0.2	0.2	20	15
Clearing load percentage/%	6.1	6.1	6.1	6.1	6.1	6.1	4.2	4.2

PSD-MYCHART Stability Curve Drawing Tool.

3. Main element models

1) Generator and its regulating system

The generator adopts E''_d and E''_q change models, and simulates exciter, PSS, prime mover and governor.

2) Load model

The induction motor and constant impedance model are adopted, and its load characteristics are shown in Table 8.6.

4. Low-frequency load shedding allocation scheme

See Table 8.7 for low-frequency load shedding allocation of Chongqing power grid in 2008.

The total allocation of low-frequency load shedding in Chongqing power grid is about 500 MW, with 4 rounds. It is distributed in Bashan, Huangjingbao, Tianjia, Meihuashan, Youting, and Liangting.

5. Fault shooting time

See Table 8.8 for the fault clearing time of simulation calculation.

Table 8.8 Time for fault clearing in Chongqing power grid

Fault type	Voltage level/kV	Fault form	Fault clearing time	Fault consequence
Line fault	500	Three-phase short-circuit	Near-end: 0.09 s Far-end: 0.1 s	Clearing of 1 line
		Three-phase short-circuit	Near-end: 0.09 s Far-end: 0.1 s	Clearing of 2–3 lines
Faults of transformer	500	Three-phase short-circuit	0.10 s	Clearing of 1–2 transformers
Faults of line, single-phase switch fault	500	Three-phase short-circuit	0.10 s for two phases 0.25 s for three phases	Clearing of 1–2 lines or 1 line and 1 transformer
Engine fault	Loss of excitation		Generator tripping in 4 s	Clearing of loss-of-field generator set
	Generator tripping			Clearing of 1–2 large units
	Loss of large capacity power plant			Clearing of all units in power plant

8.2.2 Analysis on Voltage and Reactive Power Balance of Chongqing Power Grid in Basic Operation Mode in 2008

1. Structure characteristics and installation conditions of Chongqing power grid in 2008
1) Characteristics of grid frame

In 2008, Chongqing's 500 kV backbone network was in the transition from a 日-shape single looped network to a 日-shape double looped network, as shown in Fig. 8.8.

The mode of wiring of 500 kV/220 kV power grid in normal operation is as follows:

500 kV Honghou-Banqiao double-circuit line and Huangyan-Wanxian No. 1 line are connected to Sichuan power grid (forming Sichuan-Chongqing section), and under normal circumstances, there is no electromagnetic looped network between Sichuan and Chongqing. After Huangyan-Wanxian No. 2 line is put into operation in 2008, quadruple-circuit 500 kV lines are used for networking operation between Sichuan and Chongqing. At the same time, Chongqing power grid is connected to Hubei power grid through a total of four circuits including 500 kV Wanzhou-Longquan double-circuit line and Zhangjiaba-Enshi double-circuit line (forming the Chongqing and Hubei section).

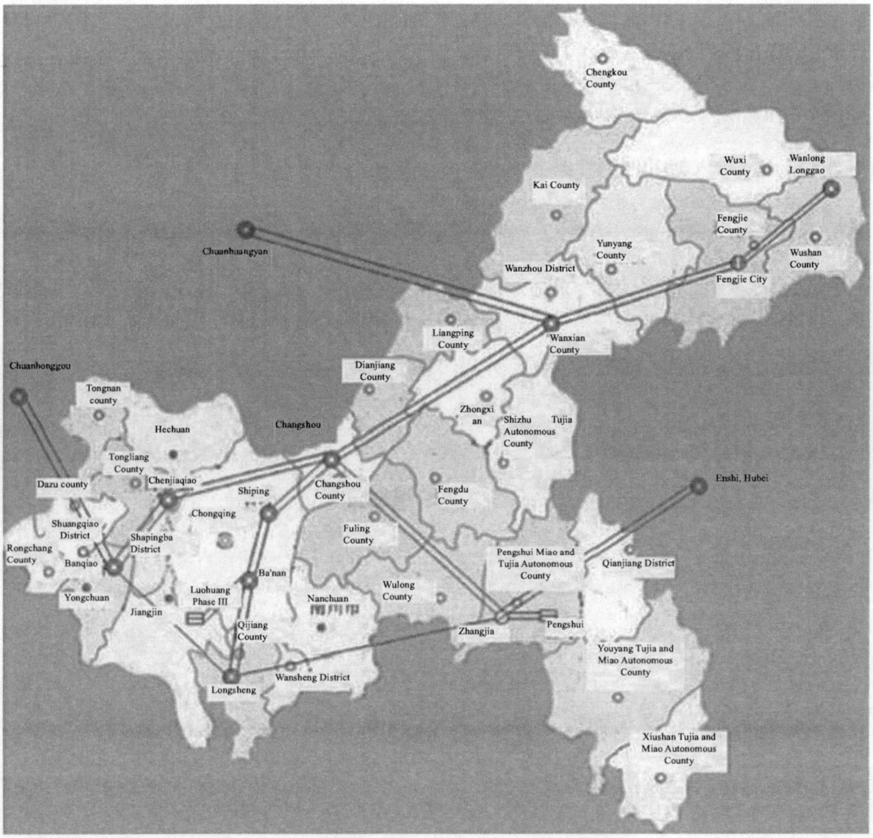

Fig. 8.8 Structure diagram of Chongqing 500 kV power grid in 2008

Some units of Guizhou Xishui Power Plant can be connected to Chongqing power grid through 220 kV Xishui-Longjing Line and Xishui-Tongzi-Qijiang line. In addition, part of the load in Qinan area can also be supplied by Guizhou power grid through 220 kV Xishui-Tongzi-Qijiang line.

In Chongqing power grid, a 500 kV/220 kV electromagnetic looped network is maintained between Chenjiaqiao, Banqiao and Longsheng station (hereinafter referred to as ("Chenjiaqiao-Banqiao-Longsheng"), and a 500 kV/220 kV electromagnetic looped network is maintained between Shiping and Banan station (hereinafter referred to as "Shiping-Banan"). An electromagnetic open ring is adopted between "Chenjiaqiao-Banqiao-Longsheng" and "Shiping-Banan", and between "Shiping-Banan" and Changshou.

Under normal circumstances, according to the requirements for power supply by sections after the electromagnetic ring network is disconnected, the tie-line switches of 220-kV power grid in different power supply sections should be disconnected. Among them, the Zhujiaba side switch of Zhuping line should be disconnected for standby, the connecting switch between Phase I and Phase II of Luohuang power plant should be disconnected for standby, and the Dazhulin side switch of Zhujie line should be disconnected for standby. The other 220 kV lines of Chongqing power grid should be put into operation.

(2) Installation conditions

By the end of 2007, the total installed capacity of Chongqing power grid was 6,473.5 MW, of which the total installed capacity for thermal power was 5,770 MW, accounting for 89%; the total installed capacity for hydropower was 703.5 MW, accounting for 11%.

2. Basic principles of reactive power balance

Natural reactive power load includes reactive power load before power user compensation, reactive power loss of each voltage level transformer and transmission line, reactive power load for power plant, and reactive power consumption of various reactors, etc. Reactive power supply includes actual adjustable reactive capacity of generator, line charging power, and capacitive reactive power capacity in reactive power compensation equipment, etc.

The main principles of reactive power balance are described in Sect. 8.1.5.

3. Analysis of power flow and voltage and reactive power balance in typical mode in 2008

1) Analysis of power flow and reactive power balance in the mode of wet season peak load

(1) Power flow analysis

In 2008, the active output was 5,440.0 MW, the reactive output was 650.3 Mvar, the active load was 7,124.8 MW, the reactive load was 1,881.7 Mvar, the active spinning reserve was 150 MW, and the active power loss was 170.4 MW during the mode of wet season peak load. The spinning reserve accounts for about 2.8% of the power load of Chongqing power grid; the power loss accounts for about 3.1% of the power load of Chongqing power grid.

Under this mode, Chongqing power grid receives 4,590.4 MW from Sichuan, of which 1,841.2 MW is sent via Huangyan-Wanxian 500 kV double-circuit line and 2,749.2 MW is sent via Honggou-Banqiao 500 kV double-circuit line. Chongqing power grid sends 2,806.9 MW to Hubei power grid, of which 1,962.8 MW is sent via Wanxian-Longquan 500 kV double-circuit line and 844.1 MW via Zhangjiaba-Enshi 500 kV double-circuit line.

The results of power flow calculation show that the distribution of power flow in Chongqing power grid is basically reasonable, without overloaded lines and transformers.

The main transformer of Chenjiaqiao is overloaded (2×637.1 MW, capacity—load ratio of 1.15), which cannot meet the requirements of static $N-1$ safety analysis. The main grid in Chongqing is divided into eastern and western parts.

The load is mainly concentrated in the west, and the western load is connected by Chenjiaqiao station and Banqiao station, most of which is supplied by Chenjiaqiao main transformer. In addition, the Honggou-Banqiao 500 kV double-circuit line in Sichuan-Chongqing section is overloaded.

(2) Analysis of voltage and reactive power balance

In 2008, the 500 kV system voltage of Chongqing power grid was kept at 519.6–529.8 kV during the mode of wet season peak load, with the lower voltage corresponding to the bus of Banqiao substation and the higher voltage corresponding to the bus of Fengjie series compensation station. The voltage of 220 kV substation was 225.5–238.7 kV, with the lower voltage corresponding to the bus of Chengkou and Shuanggui substation, and the higher voltage corresponding to the bus of the high-voltage side of the outlet transformer of Shuanghuai power plant, reaching the voltage for normal operation.

According to the *Technical Guideline on Voltage and Reactive Power of Power System*, when the 220 kV substation is at the maximum load of the main transformer, the power factor at the secondary side should be kept at 0.95–1, and the overall load side power factor of Chongqing power grid is high, meeting the requirements.

(3) Analysis of dynamic reactive power reserve capacity

The dynamic reactive power reserve capacity of Chongqing power grid basically comes from generators. Judging from the power factor of units, the power factor of units connected to 500 kV system is above 0.99; only the power factor of Chongdian units connected to 220 kV system is slightly lower than 0.95, and the active output of Chongqing Grid units has not reached its maximum output. On the whole, Chongqing power grid has sufficient dynamic reactive power reserve.

(4) Summary

The reactive power of Chongqing power grid in the mode of wet season peak load in 2008 is basically in line with the principles of balance in layers and zones and on-site compensation. Besides Wanxian and Changshou main transformers, the power factor at the medium voltage side of 500 kV step-down transformer is kept above 0.95. The power factor at the medium voltage side of Wanxian and Changshou main transformers is very low, but the active and reactive power transmitted is relatively low. In this case, it is more meaningful to use the absolute reactive power transmission to check whether the principle of balance in layers and zones is met. According to this principle, Wanxian and Changshou transformer regions also meet the principle of balance in layers and zones. The power factor of the main units is kept above 0.97 except that the Chongqing Grid unit is slightly lower than 0.95, and the dynamic reactive power reserve is sufficient. Due to the light load in Pengshui, Qianjiang, Youyang and Xiushan in southeast Chongqing, in the mode of wet season peak load, Jiangkou power plant's leading phase operation is applied to absorb excess line charging power to adjust the voltage of the 220 kV network, increasing reactive power by the main transformer of Changshou substation. If the reactive power compensation on the low voltage side of the main transformer of 220 kV Pengshui substation is withdrawn and the leading phase operation depth of Jiangkou power plant is reduced, reactive power input by the main transformer of Changshou substation can be reduced, and the voltage of the long-chain network of Changshou substation to Xiushan substation can be kept within the allowable range.

2) Analysis of power flow and reactive power balance in the mode of dry season peak load

(1) Power flow analysis

In 2008, the active output was 4630.0 MW, the reactive output was 571.1 Mvar, the active load was 6244.3 MW, the reactive load was 1860.4 Mvar, the active spinning reserve was 450 MW, and the active power loss was 119.7 MW under the mode of dry season peak load. The spinning reserve accounts for about 9.7% of the power load of Chongqing power grid; the power loss accounts for about 2.6% of the power load of Chongqing power grid.

Under this mode, Chongqing power grid receives 3830.4 MW from Sichuan, of which 1778.8 MW is sent via Huangyan-Wanxian 500 kV double-circuit line and 2051.6 MW is sent via Honggou-Banqiao 500 kV double-circuit line. Chongqing power grid sends 2142.8 MW to Hubei power grid, of which 1336.6 MW is sent via Wanxian-Longquan 500 kV double-circuit line and 806.2 MW via Zhangjiaba-Enshi 500 kV double-circuit line.

The results of power flow calculation show that the distribution of power flow in Chongqing power grid is basically reasonable, without overloaded lines. There are no 500 and 220 kV heavy load lines. In 500 kV substations, the load of the main transformer in Chenjiaqiao is reduced (2×555.3 MW, capacity—load ratio of 1.30) compared with that in wet season peak load mode, but it still cannot meet the requirements of static $N-1$ safety analysis.

(2) Analysis of voltage and reactive power balance

The 500 kV system voltage of Chongqing power grid was kept at 518.9–534.6 kV during the mode of dry season peak load, with the lower voltage corresponding to the bus of Chenjiaqiao substation and the higher voltage corresponding to the bus of Fengjie series compensation station. The voltage of 220 kV substation was 224.6–235.5 kV, with the lower voltage corresponding to the bus of Bashan substation and the higher voltage corresponding to the high-voltage side bus of Luohuang Phase II power plant, reaching the voltage for normal operation.

The average load power factor of the whole Chongqing Grid is 0.958, which meets the requirements.

(3) Analysis of dynamic reactive power reserve capacity

Judging from the power factor of units, the power factor of units connected to 500 kV system is above 0.98; the power factor of most units connected to the 220 kV system is larger than 0.98, while that of the units of Chongqing Grid, Yongchuan and Luohuang Phase I is slightly lower than 0.95, and the dynamic reactive power reserve of the entire power grid is sufficient.

(4) Summary

The reactive power of Chongqing power grid in the mode of dry season peak load in 2008 is basically in line with the principles of balance in layers and zones and on-site compensation. Only the power factor by the medium voltage side of step-down transformer in Changshou 500 kV substation is low, however, due to the small reactive power, the loss is limited. Most of the main power units in the power grid have power factors above 0.98, only the Chongqing Grid, Yongchuan and Luohuang Phase I have power factors slightly lower than 0.95, and the whole power grid has sufficient dynamic reactive power reserve.

3) Analysis of power flow and reactive power balance in the wet season valley load mode

(1) Power flow analysis

In 2008, the network topology of Chongqing power grid in the wet season valley load mode was basically the same as that of Chongqing power grid in the wet season peak load mode. The active output was 2055.0 MW, the reactive output was 362.2 Mvar, the active load was 2886.6 MW, the reactive load was 964.5 Mvar, the active spinning reserve was 2440 MW, and the active power loss was 27.6 MW under this mode. The spinning reserve accounts for about 118.7% of the power load of Chongqing power grid; the power loss accounts for about 1.3% of the power load of Chongqing power grid.

Under this mode, Chongqing power grid receives 1677.4 MW from Sichuan, of which 780.6 MW is sent via Huangyan-Wanxian 500 kV double-circuit line and 896.8 MW is sent via Honggou-Banqiao 500 kV double-circuit line. Chongqing power grid sends 827.2 MW to Hubei power grid, of which 718.4 MW is sent via Wanxian-Longquan 500 kV double-circuit line and 108.8 MW via Zhangjiaba-Enshi 500 kV double-circuit line.

The results of power flow calculation show that the distribution of power flow in Chongqing power grid is basically reasonable. There are no overloaded lines or heavy load lines, and the load of main transformer in 500 kV substation is relatively light.

(2) Analysis of voltage and reactive power balance

The 500 kV system voltage of Chongqing power grid was kept at 522.6–541.4 kV under the wet season valley load mode, with the lower voltage corresponding to the bus of Banqiao substation and the higher voltage corresponding to the bus of Fengjie series compensation station. The voltage of 220 kV substation was 224.5–234.0 kV, with the lower voltage corresponding to the bus of Wulong substation, and the higher voltage corresponding to the bus of the high-voltage side of the outlet transformer of Shuanghuai power plant, reaching the voltage for normal operation.

Under the mode of wet season valley load, the load power factor is lower, which is conducive to absorbing the excess charging power of 500 kV system and suppressing the excessive voltage.

(3) Analysis of dynamic reactive power reserve capacity

Judging from the power factor of the units, the power factor of the units connected to the 500 kV system is above 0.95. Among the units connected to the 220 kV system, Jiangkou power plant performs leading phase operation to control the voltage in southeast Chongqing; the power factor of Hengtai, Anwen, Jiulong, Luohuang Phase III, and Pengshui power plant is larger than 0.95; since the active output of Shuanghuai, Yongchuan, Chongqing and Luohuang Phase I power plant is far smaller than the rated value, the power factor is relatively low, the actual reactive output is far less than the maximum value, and the reactive power reserve is relatively large. In general, the reactive power reserve of Chongqing power grid is sufficient.

(4) Summary

The reactive power of Chongqing power grid in the wet season valley load mode in 2008 is basically in line with the principles of balance in layers and zones and on-site compensation. The power factor on the medium voltage side of 500 kV step-down transformer is kept above 0.95 except for Changshou (the total amount of active power and reactive power is small) and Longsheng main transformer (a set of 60 Mvar low-reactance device can be removed). The reactive power reserve of the main units is sufficient. Due to the light load in Pengshui, Qianjiang, Youyang and Xiushan in the southeast of Chongqing, Jiangkou power plant needs to absorb the excess line charging power through leading phase operation.

4) Analysis of power flow and reactive power balance in the dry season valley load mode

(1) Power flow analysis

In 2008, the active output was 2440.0 MW, the reactive output was 116.7 Mvar, the active load was 2250.4 MW, the reactive load was 1066.7 Mvar, the active spinning reserve was 1400 MW, and the active power loss was 22.6 MW during the mode of dry season valley load. The spinning reserve accounts for about 57.4% of the power load of Chongqing power grid; the power loss accounts for about 0.9% of the power load of Chongqing power grid.

Under this mode, Chongqing power grid delivers 212.9 MW to Sichuan, of which 170.5 MW is delivered via Huangyan-Wanxian 500 kV double-circuit line and 383.4 MW is delivered via Honggou-Banqiao 500 kV double-circuit line. Chongqing power grid receives 44.6 MW from Hubei power grid, of which 390.3 MW is sent via Wanxian-Longquan 500 kV double-circuit line and 345.7 MW via Zhangjiaba-Enshi 500 kV double-circuit line.

The results of power flow calculation show that the distribution of power flow in Chongqing power grid is basically reasonable. There are no overloaded lines or heavy load lines, and the load of main transformer in 500 kV substation is relatively light.

(2) Analysis of voltage and reactive power balance

The 500 kV system voltage of Chongqing power grid was kept at 526.1–544.4 kV under the dry season valley load mode, with the lower voltage corresponding to the bus of Banqiao substation and the higher voltage corresponding to the bus of Fengjie series compensation station. The voltage of 220 kV substation was 220.4–238.4 kV, with the lower voltage corresponding to the bus of Xiushan substation, and the higher voltage corresponding to the bus of the high-voltage side of the outlet transformer of Baihe power plant, reaching the voltage for normal operation.

The average load power factor of the power grid is 0.904, which is low.

(3) Analysis of dynamic reactive power reserve capacity

Judging from the power factor of units, most units in Chongqing power grid have a high power factor and sufficient dynamic reactive power reserve. Some units with low power factors do not have large actual reactive outputs, and the reason for the lower power factor is that the active spinning reserve is large. In order to control the voltage in each area, the leading phase operation is adopted for Anwen, Luohuang Phase III and Pengshui units. In a word, Chongqing power grid has sufficient dynamic reactive power reserve under this mode.

(4) Summary

The reactive power of Chongqing power grid in the dry season valley load mode in 2008 is basically in line with the principles of balance in layers and zones and on-site compensation. Most of the power factors on the medium voltage side of 500 kV step-down transformer are kept above 0.96, only those of Banan and Longsheng main transformers are low, and the reactive power compensation capacity can be appropriately increased by connecting the 220 kV power grid. The dynamic reactive power reserve of the power grid is sufficient.

4. Evaluation on and measures for voltage and reactive power balance in typical mode in 2008

Judging from the above analysis of voltage and reactive power balance in typical modes, Chongqing power grid basically meets the requirements of the balance of reactive power in layers and zones, but there are also some problems in the arrangement of operation modes. The following is a summary based on the specific conditions of each area:

(1) Chongqing Xicheng District is the main load center of Chongqing power grid, with the load accounting for more than half of the total load of Chongqing power grid. Under various operation modes, the reactive power in Xicheng District basically meets the principle of balance in layers and zones, and the node voltages of 220 kV and above meet the operation requirements.

(2) In southeast Chongqing, the load side power factor under four typical operation modes is lower than 0.9, and the load side power factor does not meet the requirements during the mode of wet season peak load and the mode of dry season peak load. However, since southeast Chongqing is dominated by a long-chain transmission network and the load is very light under various modes, the low power factor on the load side is beneficial to absorb the excess charging power of the line. In addition, the power factor of Changshou transformer is lower under the modes of wet season peak load, dry season peak load and wet season valley load, which is mainly due to the fact that the active power exchanged between the 220 kV power grid connected to the Changshou substation and the main grid is very small under these modes. Therefore, although the total reactive power is small, the power factor is lower. In these modes, the principle of balance of reactive power in layers and zones is basically met due to the fact that the actual absolute reactive power on the medium voltage side of the Changshou substation is relatively small.

(3) The northeast Chongqing has relatively small total active and reactive loads, and it is supported by the voltage of Baihe power plant. The power exchange with other areas is relatively small under various modes, which basically meets the requirements of power balance in layers and zones.

(4) The load side power factor is relatively high under the peak load mode in Dongcheng District of Chongqing, and the reactive power is well balanced in layers and zones. In valley load mode, the load side power factor is slightly lower than 0.95, which is beneficial to absorb the excess charging power of the line. The whole area basically meets the requirements of power balance in layers and zones.

8.2.3 Research on Static Voltage Stability of Chongqing Power Grid

In this section, the static voltage stability analysis method is used to analyze and calculate the static voltage stability of Chongqing power grid in 2008 in the wet season peak load mode, the modes after $N-1$ fault and other serious faults, to comprehensively evaluate the static voltage stability of Chongqing power grid, and to determine the weak links of voltage stability of Chongqing power grid through various static voltage stability evaluation indexes.

1. Analysis of static voltage stability under basic mode
1) Calculation results of dynamic voltage stability margin

Stability margin index includes the active margin index of single-load bus, regional power reserve coefficient index and reactive margin index of buses.

According to the scanning of active margin of single-load bus, when the load increases based on a constant power factor, the active margin of 110 kV load bus in Chongqing power grid is 0.614–5.008 p u., The lower one corresponds to Fuling bus in southeast Chongqing (mainly because the load power factor connected to Fuling bus is very low, the same below), and the higher one corresponds to Pengshui bus in southeast Chongqing. The active reserve coefficient of 110 kV load bus is 20–110%, with the lower corresponding to Bashan bus in Xicheng district and the higher corresponding to Pengshui bus in southeast Chongqing district.

According to the scanning of regional power reserve coefficient, the regional power reserve coefficient of each area of Chongqing power grid is quite different, among them, the southeast Chongqing is the highest, reaching 46.2%; the northeast Chongqing and Dongcheng District are next, with 29.6% and 32.7%, respectively; the coefficient of Xicheng district is the lowest, with only 12.8%, and Xicheng district is the area with the weakest static voltage stability in Chongqing power grid.

According to the scanning of bus reactive power margin, the reactive power margin of 110 kV load bus in Chongqing power grid is 0.6–1.684 p.u., The lower one corresponds to Chengkou bus in northeast Chongqing, and the higher one corresponds to Wansheng bus in Xicheng district.

2) Calculation results by modal analysis method

Modal analysis method can be used to find out the key buses, key lines and key generators related to voltage stability in power grid. Through the modal analysis of Chongqing Power Grid in wet season peak load mode in 2008, key load bus, lines and generator set are given.

Among the first 20 buses, load bus participation factors are arranged as Xicheng District (45%), Northeast Chongqing (20%), Dongcheng District (20%) and southeast Chongqing (15%). The load bus with large correlation factor is the weak voltage stability area in the power grid and is also the best location for reactive power compensation. As Xicheng District occupies about half of the first 20 buses with large correlation factors, it can be seen that this area is the weakest static voltage stability area in Chongqing power grid—this conclusion is consistent with the conclusion of regional power reserve coefficient.

The 500 kV lines with large participation factors include Banqiao-Longsheng, Luohuang-Banan, Shiping-Banan, and Shiping-Changshou, etc.; 220 kV lines with large participation factors include Fuling-Wulong, Fuling-Changshou, and Pengshui-Qianjiang, etc., and all these are key lines.

The calculation results of generator unit participation factors show that Luohuang 1 (Phase III) is a relatively vital generator unit. It is necessary to detect its dynamic reactive power reserve during operation control, and at the same time strengthen reactive power compensation in its vicinity to improve the reactive power reserve of the unit.

2. Analysis of the static voltage stability in the mode after $N-1$ fault

In this section, through static voltage stability analysis of the mode after $N-1$ fault, a single fault mode that has a more significant impact on the voltage stability of Chongqing power grid is studied.

In the mode after $N - 1$ fault, the power margin index of Chongqing power grid is reduced, especially the active reserve coefficient of Xicheng district is often lower than 10% (only 5.7% when the Chenjiaqiao #1 main transformer is broken), and the active reserve coefficient is low. In addition, the active power reserve coefficient and reactive power margin of the single-load buses in Xicheng district are relatively low.

3. Analysis of the static voltage stability in $N - 2$ post-fault mode

In this section, through static voltage stability analysis of the serious post-fault mode, a serious fault mode that has a more significant impact on the voltage stability of Chongqing power grid is studied.

In $N - 2$ post-fault mode, the power margin indexes in Xicheng District of Chongqing power grid are further reduced, but in general, under the $N - 2$ fault mode, the power supply areas of Chongqing power grid still have high stability margin and the overall voltage stability is high.

8.2.4 Study on Transient Voltage Stability of Chongqing Power Grid in Basic Operation Mode in 2008

The transient voltage stability of the system under various fault disturbances is simulated and analyzed in the wet season peak load mode and the dry season peak load mode adopted in 2008. The simulated major fault modes include the line $N - 1$ fault, generator set $N - 1$ and loss of excitation fault, transformer $N - 1$ fault, full shutdown of power plant, shunt transformer fault, and multiple fault modes of the AC line.

1. Research on transient voltage stability in wet season peak load mode
1) Analysis of transient voltage stability in the case of single element fault

The fault types include three phase short-circuit, single generator trip and excitation-loss fault of the lines in 500 kV voltage level, and the fault state of single 500 kV transformer fault and out of service.

(1) Line $N - 1$ fault simulation analysis

Each line with a 500 kV voltage fault type has a three-phase short circuit fault each, and the fault clearing time of 500 kV lines is taken as 0.09 s for the near fault point side and 0.1 s for the far fault point side; all 500 kV lines in Sichuan-Chongqing power grid are considered to be used as the fault lines, and overloading of lines and transformer elements is not considered during fault checking.

The calculation results of the three-phase short-circuit fault of 500 kV lines show that after the three-phase short-circuit fault occurs in 500 kV lines of Sichuan-Chongqing power grid, the fault can be cleared by quick protection correctly, and Chongqing power grid can remain stable.

2) Generator $N - 1$ and excitation-loss fault

According to the *Guide on Security and Stability of Power System*, for the power system under normal operation mode, if any generator trips or loses excitation, the protection, switch and reclosing shall work correctly, without taking stability control measures; it is necessary to maintain the power system stability operation and normal power supply of power grid, and other elements shall not exceed the specified emergency overload capacity, and have no interlocking trip. The calculation results show that under the wet season peak load mode in 2008, if any generator in Sichuan-Chongqing power grid trips, the system can maintain steady operation without voltage instability.

The excitation-loss fault is more serious than the generator trip fault, because the occurrence of the fault not only loses the reactive power output of the generator itself, but also needs to absorb the reactive power equivalent to the generator capacity from the system, causing a serious reactive power impact on the system. The calculation results show that after the open-circuit excitation-loss fault of exciting winding occurs in a single generator in Sichuan-Chongqing power grid, the loss of excitation unit is removed at 3.0 s, the system can be kept stable, the under-voltage load shedding does not operate, and the voltage remains at a high level.

3) Transformer $N - 1$ fault.

The $N - 1$ fault analysis is performed on the main transformer of the 500 kV system of Chongqing and Sichuan grids. A three-phase tripping fault occurred on the transformer HV side at 0 s, and the transformer is out of service after 5 cycles. As the calculation results have shown, when $N - 1$ fault occurs in 500 kV main transformer of Chongqing and Sichuan grids, the system can recover quickly and maintain stability, there is no under-voltage load shedding action and the voltage is also kept in a high level. However, in case of $N - 1$ fault of Chenjiaqiao main transformer, the other main transformer will be severely overloaded.

2) Analysis of transient voltage stability in the case of multiple serious faults
(1) 220 kV line three-phase tripping double-circuit fault

It is assumed that there is a three-phase tripping fault on the 220 kV line in Chongqing power grid, and the double-circuit line trips after 6 cycles. The calculation results show that:

① Chongqing power grid can maintain stability and the system voltage can quickly return to normal level when such faults occur.
② In the event of three-phase tripping double-circuit faults of Changshou-Fuling, Chenjiaqiao-University Town, Fuling-Wulong, Pengshui-Qianjiang, Pengshui-Wulong, Youyang-Qianjiang and Fengjie-Wanxian, some 220 kV systems will be disconnected from the main grid. The main grid will remain stable and the voltage will quickly recover.

③ Some 220 kV three-phase tripping double-circuit faults will lead to disconnection of some power plants. In the event of three-phase tripping double-circuit fault of Shuanghuai-Huayuan, Hengtai-Wansheng, Jiangkou-Wulong and Laisu-Yongchuan, Shuanghuaishu power plant, Hengtai power plant, Jiangkou power plant and Yongchuan power plant will be out of operation respectively; in the case of three-phase tripping double-circuit fault of Wanxian-Baihe power plant, Baihe power plant and some 220 kV lines will be disconnected from the main grid.

(2) 500 kV line three-phase tripping double-circuit fault

It is assumed that there is a three-phase tripping fault of 500 kV line, and the double-circuit line trips in 0.09 s at the near fault point side and 0.1 s at the far fault point side.

When this kind of fault occurs in Sichuan-Chongqing power grid, the system will lose stability under partial fault. In case of a fault on each side of Chongqing Banqiao-Sichuan Honggou, Chongqing Wanxian-Sichuan Huangyan, Chongqing Wanxian-Hubei Longquan, Sichuan Honggou-Sichuan Puti, Sichuan Ertan-Sichuan Puti and Sichuan Tanjia-Sichuan Nanchong Tanjia, the system will lose stability, and corresponding clearing of generator and load shall be made to maintain the stability of the system. In addition, in the event of three-phase tripping double-circuit faults of Sichuan Jiulong-Sichuan Shimian, Sichuan Maoxian-Sichuan Tanjia, Sichuan Ya'an-Sichuan Shimian, and Sichuan Ertan-Sichuan Shiban, due to the grid frame, some power grids will be disconnected from the main grid, the main grid will not operate at low voltage load shedding, the voltage will quickly recover, and no voltage instability will occur.

(3) 500 kV shunt transformer fault

It is assumed that there is a three-phase tripping fault on the high-voltage side of 500 kV transformer, both transformers of this station trip after 5 cycles. In such a fault, Chongqing main grid can remain stable and the system voltage can quickly return to normal level without voltage instability.

When the Wanxian double shunt-transformer was in fault, the voltage of the power grid with 220 kV and below connected to the Wanxian transformer dropped significantly. The reason is that: the low voltage at the fault time causes the reactive power output compensated by shunt capacitors of Chengkou, Fengjie, and Shuanggui transformers greatly reduced. Due to the lack of other reactive power sources in this area, the bus voltage of Chengkou and other substations cannot return to the normal level.

When the Changshou double shunt-transformer was in fault, the voltage of the power grid with 220 kV and below connected to the Changshou transformer dropped significantly. The reason for the large drop in voltage is that due to the long-chain transmission line used from Changshou to Xiushan, the reactive power transmission is difficult. At the same time, the recovery characteristics of motor load after the fault led to an increase in reactive power demand. After the interaction, the bus voltages of

Wulong, Pengshui, Youyang and Xiushan transformer decreased successively, and the recovery voltage of 220 kV bus of Xiushan transformer was below 0.75 p.u.

(4) Full shutdown of the main power plants

The power plant full shutdown fault analysis was carried out for the main power plant units in Sichuan-Chongqing power grid. After the full shutdown fault of the main power plant units in Sichuan-Chongqing power grid occurred, the system remained stable, and low-frequency load shedding and under-voltage load shedding were not performed.

(5) 500 kV line three-phase short-circuit single-phase switch does not work

Assuming that there is a three-phase short circuit fault on the 500 kV line, the opposite switch trips after 0.1 s, the single-phase switch on this side does not work and trips for two phases, and the reserve performs protection after 0.34 s by cutting off this line and the line/transformer on the same 3/2 connection.

The calculation shows that the system cannot remain stable after the three-phase short circuit and single-phase switch fault occurs in some 500 kV lines. Among them, the single-phase switch fault on Wanxian side of Wanxian-Longquan Line 1, Wanxian-Longquan Line 2 and Changshou-Wanxian Line 1 will cause instability of Sichuan-Chongqing power grid relative to Central China power grid; Banqiao side faults of Longsheng-Banqiao Line 1, Honggou-Banqiao Line 1 and Honggou-Banqiao Line 2 will also cause the system to lose stability. After clearing generators, the system can remain stable.

(6) 500 kV transmission channel is broken

Assuming that there is a three-phase fault of a single 500 kV transmission line of 0 s Sichuan-Chongqing transmission channel, all lines on the transmission channel trips off or trips into 3-circuit lines after 0.1 s (Sichuan-Chongqing and Chongqing-Hubei channels are dominated by two 500 kV lines with 4-circuit lines).

When the 500 kV transmission channel is broken, the system will lose stability due to the breaking of any side of the other transmission channels except for the breaking of the Chongqing Wanxian-Hubei Longquan transmission channel and the breaking of the Sichuan Honggou-Chongqing Banqiao transmission channel, which can maintain the stability of the Chongqing power grid. Therefore, the clearing of generators and loads is needed.

2. Research on transient voltage stability in dry season peak load mode

In 2008 and under the dry season peak load mode, the 500 kV grid frame of Chongqing power grid was still in the transition from a 日-shape single looped network to a 日-shape double looped network. After the second main transformer was built in 500 kV substations such as Longsheng and Banan, the power supply capacity and reliability of 500 kV substations were further increased.

Since August 2008, the second main transformer of Longsheng and Banan substations and Zhangjiaba #1 main transformer have been put into operation one after another, and Zhangjiaba has also been transformed from a switching station to a substation. At the end of the year, the second-circuit line from Zhangjiaba to Longsheng was put into operation, and the transmission capacity of Chongqing 500 kV power grid was further improved. The change of 500 kV frame of Sichuan power grid adjacent to Chongqing power grid was also relatively small. The main change was that the Dongpo-Jianshan line has increased from one circuit to two circuits, and Jianshan-Ya'an two-circuit line was withdrawn.

In addition, under the dry season peak load mode in 2008, the networking mode in North and Central China was changed: the 1,000 kV single-circuit AC transmission line from southeast Shanxi to Jingmen via Nanyang has been put into operation, and the connection between North and Central China power grids was been further strengthened.

1) Analysis of transient voltage stability in the case of single element fault

The fault types include the disengagement due to three-phase short-circuit of the lines and single generator trip and excitation-loss fault at 500-kV voltage level, and the voltage fault of a single 500-kV transformer.

(1) Line $N - 1$ fault

Each line with a 500 kV voltage fault type has a three-phase short-circuit respectively, and the fault clearing time of 500 kV lines is taken as 0.09 s for the near fault point side and 0.1 s for the far fault point side; all lines in Sichuan-Chongqing power grid are considered to be used as the fault lines, and overloading of lines and transformer elements is not considered during fault checking.

The calculation results of the three-phase short-circuit fault of 500 kV lines show that after the three-phase short-circuit fault occurs in 500 kV lines of Sichuan-Chongqing power grid, the fault can be correctly cleared by quick protection, and Chongqing power grid can remain stable.

(2) Generator $N - 1$ and excitation-loss fault

The calculation results show that under the dry season peak load mode in 2008, if any generator in Sichuan-Chongqing power grid trips, the system can maintain steady operation without voltage instability.

The excitation-loss fault is more serious than the generator trip fault, because the occurrence of the fault not only loses the reactive power output of the generator itself, but also needs to absorb the reactive power equivalent to the generator capacity from the system, causing serious reactive power impact on the system. The calculation results show that after the open-circuit excitation-loss fault of exciting winding occurs in a single generator in Sichuan-Chongqing power grid, the loss of excitation unit is removed at 3.0 s, the system can be kept stable, and the voltage remains at a high level.

(3) Transformer $N-1$ fault

The $N-1$ fault analysis is performed on the main transformer of the 500 kV system of Chongqing and Sichuan grids. A three-phase tripping fault occurred on the transformer HV side at 0 s, and the transformer was out of service after 5 cycles. As the calculation results show, when $N-1$ fault occurs in 500 kV main transformer of Chongqing and Sichuan grids, the system can recover quickly and maintain stability, there is no under-voltage load shedding action and the voltage is also kept in a high level. However, in case of $N-1$ fault of Chenjiaqiao main transformer, the other main transformer will be severely overloaded; in case of $N-1$ fault of the main transformer of Puti substation, the power grid connected to the Puti substation is disconnected from the main grid.

2) Analysis of transient voltage stability in the case of multiple serious faults

(1) 220 kV line three-phase tripping double-circuit fault

It is assumed that there is a three-phase tripping fault on the 220 kV line, and the double-circuit line trips after 6 cycles. The calculation results show that:

① Chongqing power grid can maintain stability and the system voltage can quickly return to the normal level when such faults occur.

② In the event of three-phase tripping double-circuit faults of Changshou-Fuling, Chenjiaqiao-University Town, Fuling-Wulong, Pengshui-Qianjiang, Pengshui-Wulong, some grids will be disconnected from the main grid. The main grid will remain stable and the voltage will quickly recover.

③ Some double-circuit line tripping caused by faults will cause some power plants to be disconnected from the system. In the event of three-phase tripping double-circuit faults of Shuanghuai-Huayuan, Hengtai-Wansheng, Jiangkou-Wulong and Laisu-Yongchuan, Shuanghuaishu power plant, Hengtai power plant, Jiangkou power plant and Yongchuan power plant will be out of operation respectively; in the case of three-phase tripping double-circuit fault of Wanxian-Baihe power plant, Baihe power plant and some 220 kV lines will be disconnected from the main grid.

(2) 500 kV line three-phase tripping double-circuit fault

It is assumed that there is a three-phase tripping fault on the 500 kV line, and the double-circuit line trips in 0.09 s for the near fault point side and 0.1 s for the far fault point side.

In most cases, when such faults occur in Sichuan-Chongqing power grid, Chongqing power grid can remain stable and the system voltage can quickly return to the normal level. However, a fault on either side of Sichuan Ya'an-Sichuan Shimian and Sichuan Guang'an-Sichuan Huangyan will lead to rotor angle instability of North China power grid relative to Central China power grid; the Ertan side fault of Sichuan Ertan-Sichuan Puti will lead to rotor angle instability of Ertan units; in the event of three-phase tripping double-circuit faults of Sichuan Ertan-Sichuan Shiban and Sichuan Jiulong-Sichuan Shimian, due to the grid frame, some power grids will be disconnected from the main grid, the main grid low-frequency load shedding does not work, the voltage will quickly recover, and no voltage instability accident will occur.

(3) 500 kV shunt transformer fault

It is assumed that there is a three-phase tripping fault on the high-voltage side of 500 kV transformer, both transformers of this station trip after 0.1 s.

In case of such faults, Chongqing's main grid can remain stable, the system voltage can quickly return to the normal level, and there will be no voltage instability accident. In case of 500 kV shunt transformer fault at Changshou substation, the power grid of area with 220 kV voltage connected with the substation will be disconnected; when the Jiulong, Shimian and Ya'an substations fail, the regional power plants connected to them will be out of operation.

(4) Full shutdown of the main power plants

Full shutdown fault of the main units of Sichuan-Chongqing power grid are analyzed. After the full shutdown fault occurs in some power plants, the system cannot remain stable: the full shutdown fault of Luohuang, Ertan and Guang'an power plants will lead to rotor angle instability of the North China power grid relative to Central China power grid.

(5) 500 kV line three-phase short-circuit single-phase switch does not work

Assuming that there is a three-phase short circuit fault of 500 kV line, the opposite switch trips after 0.1 s, the single-phase switch on this side does not work and trips for two phases, and the reserve performs protection after 0.34 s by cutting off this line and the line/transformer on the same 3/2 connection.

The system can remain stable after the three-phase short-circuit fault and single-phase switch fault occur in 500 kV line; however, the fault of Chenjiaqiao side of Banqiao-Chenjiaqiao Line 2 will lead to the overload of Chenjiaqiao #2 main transformer.

(6) 500 kV transmission channel is broken

Assuming that there is a three-phase fault of a single 500 kV transmission line of 0 s on Sichuan-Chongqing transmission channel, all lines on the transmission channel trips off or trips into 3-circuit lines after 0.1 s (Sichuan-Chongqing and Chongqing-Hubei channels are dominated by two 500 kV lines with 4-circuit lines). It can be seen that the system can remain stable when the Chongqing-Hubei transmission channel is broken; when a three-circuit line breaking occurs in Sichuan-Chongqing transmission channel, the system cannot remain stable.

3. Comparative analysis of transient voltage stability in wet season peak load mode and dry season peak load mode

As far as the frame of Chongqing power grid is concerned, there were two main changes during the dry season peak load mode compared with that during the wet season peak load mode in 2008: 1. The 500 kV grid frame was still in the transition from a 日-shape single looped network to a 日-shape double looped network, and the power transmission and supply capacity was further improved; 2. The 1,000 kV single-circuit AC transmission line from southeast Shanxi to Jingmen via Nanyang has been put into operation, which has changed the networking mode in North and Central China. The above changes not only strengthened the frame of the power grid, which was conducive to improving the transient stability level of the system, but also led to some new situations, mainly including the following four faults.

1) 500 kV line three-phase tripping double-circuit fault

In wet season peak load mode, when each side of Chongqing Banqiao-Sichuan Honggou, Chongqing Wanxian-Sichuan Huangyan, Chongqing Wanxian-Hubei Longquan, Sichuan Honggou-Sichuan Puti, Sichuan Ertan-Sichuan Puti and Tanjia side of Sichuan Tanjia-Sichuan Nanchong fail, the system will lose stability, and clearing of generator and load shall be done to keep the system stable; in the dry season peak load mode, due to the strengthening of the grid frame and the change of load level, in most cases, if this fault occurs in Sichuan and Chongqing power grid, Chongqing power grid can maintain stability; however, the faults on any side of Sichuan Ya 'an-Sichuan Shimian and Sichuan Guang'an-Sichuan Huangyan will lead to new motor angle instability of the North China power grid relative to Central China power grid.

2) Full shutdown of the main power plants

In wet season peak load mode, the system can remain stable when the shutdown fault of units of main power plants occurs in Sichuan and Chongqing power grid occur; in the dry season peak load mode, the fall shutdown faults of Luohuang (including Phase I, II, and III), Ertan and Guang'an power plants will lead to the motor angle instability of the North China power grid relative to Central China power grid.

3) 500 kV line three-phase short-circuit single-phase switch does not work

In wet season peak load mode, the single-phase switch fault on Wanxian side of Wanxian-Longquan Line 1, Wanxian-Longquan Line 2 and Changshou-Wanxian Line 1 will cause instability of Sichuan-Chongqing power grid relative to Central China power grid, Banqiao side faults of Longsheng-Banqiao Line 1, Honggou-Banqiao Line 1 and Honggou-Banqiao Line 2 will also cause the system to lose stability; in the dry season peak load mode, the system can remain stable after the three-phase short-circuit and single-phase switch fault occur in 500 kV line.

4) 500 kV transmission channel is broken

In wet season peak load mode, the system will lose stability if a three-circuit line break occurs on either side of Sichuan-Chongqing transmission channel and Chongqing-Hubei channel; in the dry season peak load mode, due to the commissioning of UHV transmission channels in Central and Northern China and the changes at the load level of the power grid, the power grid of Chongqing can be kept stable in case of a fault of transmission channels, except for the breaking of three-circuit lines of Sichuan-Chongqing channel, which will cause system instability.

4. Evaluation of the transient stability of Chongqing power grid

This section checks the transient voltage stability of Chongqing power grid under single fault and multiple serious faults, and the following conclusions can be obtained.

1) Wet season peak load mode
(1) Single element fault
① The system can be stable in case of $N-1$ fault of 500 kV line.
② In case of $N-1$ fault of any generator and relay protection acts correctly, Chongqing power grid can remain stable without causing voltage instability.
③ In case of $N-1$ fault of 500 kV main transformer, the system can maintain stability, under-voltage load-shedding does not work, and the voltage is maintained at a high level. However, in case of $N-1$ fault of Chenjiaqiao main transformer, the other main transformer will be severely overloaded.
(2) Multiple critical faults

   ① In the case of three-phase tripping double-circuit fault of the 220 kV line, the system can maintain stability and the voltage recovery level is high. Among them, in the case of three-phase tripping double-circuit fault of Changshou-Fuling, Chenjiaqiao-University Town, Fuling-Wulong, Pengshui-Qianjiang, Pengshui-Wulong will cause some 220 kV systems to be disconnected from the main grid; in the case of three-phase tripping double-circuit fault of Shuanghuai-Huayuan, the units of Shuanghuaishu power plant will be disconnected from the system; in the case of three-phase tripping double-circuit fault of Wanxian-Baihe power plant, the units of Baihe power plant will be disconnected from the system.
   ② In the case of three-phase tripping double-circuit fault of 500 kV line and when partial faults occur, the system will lose stability. In case of a fault on each side of Chongqing Banqiao-Sichuan Honggou, Chongqing Wanxian-Sichuan Huangyan, Chongqing Wanxian-Hubei Longquan, Sichuan Honggou-Sichuan Puti, Sichuan Ertan-Sichuan Puti and Sichuan Tanjia-Sichuan Nanchong Tanjia, the system will lose stability; in addition, when there are three-phase tripping double-circuit faults in Sichuan Ertan-Sichuan Shiban, Sichuan Jiulong-Sichuan Shimian, Sichuan Maoxian-Sichuan Tanjia, Sichuan Ya 'an-Sichuan Shimian, some power grids will be disconnected from the main grid due to the grid frame.
   ③ If 500 kV shunt transformers fail at the same time, Chongqing main grid can remain stable and the system voltage can quickly return to the normal level without voltage instability.
   ④ In the case of full shutdown fault of units of the main power plant, the system can maintain stability, and the low-frequency and under-voltage load-shedding does not need to operate.

⑤ When three-phase short-circuit fault and single-phase switch fault occurs in the 500 kV line, the system cannot be kept stable sometimes. The single-phase switch fault of operation failure on Wanxian side of Wanxian-Longquan Line 1, Wanxian-Longquan Line 2, and Changshou-Wanxian Line 1 will cause instability of Sichuan-Chongqing power grid relative to Central China power grid. Banqiao side faults of Longsheng-Banqiao Line 1, Honggou-Banqiao Line 1 and Honggou-Banqiao Line 2 will also cause the system to lose stability. After clearing generators, the system can remain stable.

⑥ The 500 kV transmission channel breaking fault refers to a situation where lines with more than three circuits are disconnected on the Sichuan-Chongqing or Chongqing-Hubei channels. If all the lines on the Sichuan-Chongqing or Chongqing-Hubei channels are broken at the same time, the main grids of Sichuan-Chongqing and Central China can be kept stable respectively, otherwise stability control measures need to be taken.

2) Dry season peak load mode

(1) Single element fault

① The system can be stable in case of $N-1$ fault of 500 kV line.

② In case of $N-1$ fault of any generator and if relay protection acts correctly, Chongqing power grid can remain stable without causing voltage instability.

③ In case of $N-1$ fault of 500 kV main transformer, the system can maintain stability, under-voltage load-shedding does not need to be used, and the voltage is maintained at a high level. However, in case of $N-1$ fault of Chenjiaqiao main transformer, the other main transformer will be severely overloaded; in case of $N-1$ fault the main transformer of Puti substation, the power grid connected to the Puti substation is disconnected from the main grid.

(2) Multiple critical faults

① In the case of three-phase tripping double-circuit fault of the 220 kV line, the system can maintain stability. In the case of three-phase tripping double-circuit fault of Changshou-Fuling, Chenjiaqiao-University Town, Fuling-Wulong, Pengshui-Qianjiang, Pengshui-Wulong will cause some grids to be disconnected from the main grid; in the case of three-phase tripping double-circuit fault of Shuanghuai-Huayuan, the units of Shuanghuaishu power plant will be disconnected from the system; in the case of three-phase tripping double-circuit fault of Wanxian-Baihe power plant, the units of Baihe power plant will be disconnected from the system.

② In the case of three-phase tripping double-circuit fault of 500 kV transmission line, the system cannot remain stable after some faults occur. Fault on either side of Sichuan Ya'an-Sichuan Shimian and Sichuan Guang'an-Sichuan Huangyan will lead to rotor angle instability of North China power grid relative to Central China power grid; the Ertan side fault of Sichuan Ertan-Sichuan Puti will lead to rotor angle instability of Ertan units; in the event of three-phase tripping double-circuit faults of Sichuan Ertan-Sichuan Shiban, Sichuan

Jiulong-Sichuan Shimian and Sichuan Maoxian-Sichuan Tanjia, due to the grid frame, some power grids will be disconnected from the main grid.

③ If 500 kV shunt transformers fail at the same time, Chongqing main grid can remain stable and the system voltage can quickly return to the normal level without voltage instability. Among them, when Changshou substation fails, the power grid of area with 220 kV voltage connected to the substation will be disconnected; when the Jiulong, Shimian and Ya'an substations fail, the regional power plants connected with them will be out of operation.

① The system cannot keep stable when the shutdown fault of units of main power plants occurs; the fall shutdown faults of Luohuang (including Phase I, II, and III), Ertan and Guang'an power plants will lead to the motor angle instability of the North China power grid relative to Central China power grid.

② The system can remain stable after the three-phase short-circuit fault and single-phase switch fault occur in most 500 kV lines. However, the fault on the Chenjiaqiao side of Chenjiaqiao-Changshou Line 1 will lead to motor angle instability of units in Chongqing Power Plant and Jiulong Power Plant, and the fault of Chenjiaqiao side of Banqiao-Chenjiaqiao Line 2 will lead to an overload of Chenjiaqiao #2 main transformer.

③ In the case of breaking of the 500 kV transmission channel: When the Chongqing-Hubei transmission channel breaks, the system can maintain stability; when a three-circuit line breaking occurs in Sichuan-Chongqing transmission channel, the system cannot remain stable and corresponding security and stability control measures need to be taken.

Through the above analysis, it can be seen that there is no transient stability and dynamic stability problem for the $N-1$ fault of each section in the Chongqing power grid in 2008. The transient stability problem caused by the $N-2$ fault of Sichuan-Chongqing and Chongqing-Hubei sections can be solved by the stability control device for Sichuan-Eastern China power transmission. A few $N-2$ faults in other sections are mainly thermal stability problems (such as faults of main transformer in Chenjiaqiao). On a whole, the frame of Chongqing power grid in 2008 is relatively strong and it has a strong ability to resist the impact of serious faults. With reasonable safety control measures, the system can remain stable, the voltage can be recovered back to a higher level, and the possibility of transient voltage instability is relatively small, which meets the requirements of the three-level security and stability standard of *Guide on Security and Stability of Power System.*

8.2.5 Simulation Study on Medium and Long-Term Voltage Stability of Chongqing Power Grid

All calculation results in this section are based on data of Chongqing power grid obtained in 2008 in wet season peak load mode. In order to approximately simulate the on-load tap changer function of transformers with lower voltage levels, 220 kV transformers are set with on-load tap changer in the calculation so as to evaluate the influence of OLTC operation characteristics on long-term voltage stability.

1. Evaluation of long-term voltage stability in the case of single fault

The fault types include disengagement due to three-phase short-circuit of the lines, single generator trip and excitation-loss fault at 500 kV voltage level, and the single 500 kV transformer fault.

1) 500 kV line $N-1$ fault simulation analysis

Each line with a 500 kV voltage fault type has a three-phase short-circuit fault respectively, and the fault clearing time of 500 kV lines is taken as 0.09 s for the near fault point side and 0.1 s for the far fault point side; all 500 kV lines in Chongqing power grid are considered to be used as the fault lines, and overloading of lines and transformer elements is not considered during fault checking.

The calculation results show that after the three-phase short-circuit fault occurs in 500 kV lines of Chongqing power grid, the fault can be correctly cleared by quick protection, and Chongqing power grid can remain stable.

2) Generator $N-1$ and excitation-loss fault

Under the wet season peak load mode in 2008, if any generator in Chongqing power grid trips, the system can maintain steady operation without voltage instability.

3) 500 kV transformer $N-1$ fault

The $N-1$ fault analysis is performed on the main transformer of the 500 kV system of Chongqing and Sichuan power grids. A three-phase tripping fault occurred on the transformer HV side at 0 s, and the transformer is out of service after 5 cycles. As the calculation results shown, when $N-1$ fault occurs in 500 kV main transformer of Chongqing power grid, the system can recover quickly and maintain stability, there is no under-voltage load shedding action and the voltage is also kept in a high level. However, in case of $N-1$ fault of Chenjiaqiao main transformer, the other main transformer will be severely overloaded.

2. Evaluation of long-term voltage stability in the case of multiple serious faults
1) 220 kV line three-phase tripping double-circuit fault

It is assumed that there is a three-phase tripping fault of 220 kV line in Chongqing power grid, and the double-circuit line trips after 6 cycles. The full simulation results show that the system can keep stable.

2) 500 kV line three-phase tripping double-circuit fault

It is assumed that there is a three-phase tripping fault of the 500 kV line, and the double-circuit line trips in 0.09 s at the near fault point side and 0.1 s at the far fault point side. The results show that, when this kind of fault occurs in Chongqing power grid, the system will lose stability under partial fault. In case of fault on each side of Chongqing Banqiao-Sichuan Honggou, Chongqing Wanxian-Sichuan Huangyan, Chongqing Wanxian-Hubei Longquan, Sichuan Honggou-Sichuan Puti, Sichuan Ertan-Sichuan puti and Sichuan Tanjia-Sichuan Nanchong Tanjia, the system rotor angle will lose stability, and corresponding clearing of generator and load shall be made to maintain the stability of the system. In addition, in the event of three-phase tripping double-circuit faults of Sichuan Jiulong-Sichuan Shimian, Sichuan Maoxian-Sichuan Tanjia, Sichuan Ya'an-Sichuan Shimian, and Sichuan Ertan-Sichuan Shiban, due to the grid frame, some power grids will be disconnected from the main grid, the main grid will not operate at low voltage load shedding, the voltage will quickly recover, and no voltage instability will occur.

3) Decommissioning of 500 kV shunt transformer

The following is the calculation of the decommissioning of the 500 kV substation of Chongqing power grid. The fault type is as follows: three-phase short circuit fault occurs at the high voltage side of the 0 cycle transformer, and all transformers exit at 5 cycles. The simulation results show that the transient motor angle becomes unstable after the decommissioning of transformers of Wanxian and Banqiao substations; after the decommissioning of transformer of Changshou substation, the main grid remains stable, and the areas under 220 kV are disconnected from the main grid and the regional voltage becomes unstable; long-term voltage instability occurs after the decommissioning of transformers of Chenjiaqiao substation; the system remains stable after the decommissioning of transformers of other substations.

Table 8.9 shows the sequence of OLTC and generator over-excitation limiting actions after the decommissioning of Chongqing Chenjiaqiao double main transformer.

There are many lines and heavy loads in the power supply area of Chenjiaqiao substation. After the decommissioning of the Chenjiaqiao double main transformer, as a large part of the power source is lost in the 220 kV area connected to the transformer, the generator excitation current of Chongqing power plant and Jiulong power plant located in the area is increased to increase the voltage. During this period, OLTC kept operating to maintain the voltage, which drops slightly in the area. However, since the generators of Chongqing Power Plant and Jiulong Power Plant were in an over-excitation state for a long time, at 300 s, the generator over-excitation limiting link of Chongqing Power Plant started to operate, which led to the continuous increase of excitation current in Jiulong Power Plant. At 332 s, the over-excitation limiting link of Jiulong Power Plant started to operate, causing the over-excitation limiting link of Luohuang Phase I Power Plant to operate at 352 s. During this period, the voltage in Chenjiaqiao dropped rapidly and became unstable.

Table 8.9 Sequence of OLTC and over-excitation limiting actions after decommissioning of Chongqing Chenjiaqiao double main transformer

Fault time/s	OLTC action location	Over-excitation limiting generator	Effect on voltage
0			The voltage of Chongqing Jinjiayan dropped from 1.05–0.99 p.u.
30.10	Chongqing Baishubao		At 30 s, OLTC started to operate in the surrounding area of Chenjiaqiao, and the voltage of Jinjiayan peaked at 1.08 p.u., then slowly dropped to 0.97 p.u.; At 35 s, the low voltage in the Jinjiayan area led to OLTC's second round of action, with the voltage rising to 1.06 p.u. and then slowly falling to 0.97 p.u.; After that, OLTC continued to operate for several rounds, but the voltage still could not return to that in pre-fault level and finally stabilized at 0.978 p.u.
	Chongqing University Town		
	Chongqing Liangting		
	Chongqing Meihuashan		
	Chongqing Tianjia		
	Chongqing Yuhuangguan		
30.11	Chongqing Dazhulin		
	Chongqing Huayuan		
	Chongqing Liangfengya		
	Chongqing Shuangshan		
	Chongqing Shuinian		
30.12	Chongqing Bashan		
	Chongqing Huangjingbao		
	Chongqing Longjing		
30.13	Chongqing Jinjiayan		
	Chongqing Jinlong		
	Chongqing Malanya		
	Chongqing Zoumayang		
30.14	Chongqing Laisu		
30.15	Chongqing Chadian		
30.16	Chongqing Youting		
35.11	Chongqing Meihuashan		

(continued)

Table 8.9 (continued)

Fault time/s	OLTC action location	Over-excitation limiting generator	Effect on voltage
100.12	Chongqing Qijiang		
298.18		Chongqing power plant	The voltage drops rapidly to 0.85 p.u.
300.98		Chongqing 1	
328.19	Chongqing Wansheng		In 300–332 s, the voltage drops slowly to 0.79 p.u.
332.05		Chongqing Jiulong	The voltage drops rapidly to 0.71 p.u.
			In 332–352 s, OLTC continues to operate, but the voltage still slowly drops to 0.69 p.u.
352.08		Chongqing Luohuang 81	The voltage drops rapidly to 0.45 p.u.
		Chongqing Luohuang 82	
362.06	Chongqing Cuiyun Chongqing Daxi Chongqing Renhe		Due to the effect of OLTC, the voltage increased slightly and finally stabilized at 0.47 p.u.
362.07	Chongqing Jiguanshi		
383.08	Chongqing Dongxincun Chongqing Zhujia		

After that, although OLTC continued to operate, due to serious lack of active power, the voltage in Chenjiaqiao became unstable.

4) Disconnection of a power transmission channel

Assuming that there is a three-phase fault of a single 500 kV transmission line on the 0 s Sichuan-Chongqing transmission channel, and all the lines on the transmission channel trip after 0.1 s (Sichuan-Chongqing and Chongqing-Hubei channels are dominated by two 500 kV lines with 4-circuit lines), the simulation results show that the Chongqing power grid can remain stable when the transmission channels from Chongqing to Hubei and Sichuan to Chongqing are disconnected, among which, when the transmission channel from Sichuan to Chongqing is disconnected, the Sichuan grid is disconnected from the Central China power grid, and the units in Sichuan need to be cleared.

5) Results of simulation of decommissioning of a power plant

The long-term simulation analysis of power plant decommissioning of Chongqing power grid shows that the system can still maintain stable for a long time after the decommissioning of a power plant in Chongqing power grid.

3. Influence of SVC on long-term voltage stability of Chongqing power grid

Through analyzing the influence of SVC (including three SVCs in Chenjiaqiao, Wanxian and Sichuan Honggou) on the long-term stability of Chongqing power grid through decommissioning of power plant and disconnection of channel, the analysis is based on data of Chongqing power grid obtained in 2008 in wet season peak load mode.

The basic function of SVC is to absorb reactive power that can be continuously regulated from or to power grid, so as to maintain the voltage at the installation point constant and facilitate the reactive power balance of the power grid. After voltage fluctuation occurs in the system, SVC can regulate the system voltage to make the stability margin of the system voltage larger, which is conducive to voltage stability. In addition, in the power system, generators run in parallel through power transmission lines. Under the action of small disturbance, the generator rotor electrical angles will swing relatively, and the power on the power transmission lines will also oscillate correspondingly. If the power system lacks necessary positive damping, the oscillation time will be prolonged and it will even lead to amplitude oscillation, and stability will be lost. SVC plays a positive role in improving the stability of the system operation, and improving the damping oscillation characteristics of the system is one of its main functions.

Combined with the actual fault simulation of Chongqing power grid and the above analysis, it can be seen that SVC greatly improves the damping after the system fault, effectively reduces the system oscillation time after the fault, and enables the system to stop the oscillation faster and maintain stability; when SVC is not used, the system voltage and generator motor angle oscillate, which is not conducive to voltage stability.

4. Long-term simulation analysis and study in the case of multiple cascading faults
1) Cascading fault mode I

The cascading fault mode I refers to the decommissioning of Chongqing Chongdian and Jiulong power plant, and the disconnection of Chenjiaqiao-Shuangshan and Chenjiaqiao-Baishubao in succession: after the decommissioning of Chongqing Chongdian and Jiulong at 0 s, the Chenjiaqiao-Shuangshan and Chenjiaqiao-Baishubao lines are overloaded, and the Chenjiaqiao-Shuangshan lines trips after 10 s, the Chenjiaqiao—Baishubao single circuit line trips after 20 s; after the Chenjiaqiao-Baishubao double-circuit line trips after 30 s, since the Jinjiayan and Shuinian substations are relatively remote power receiving areas, and the loads in the Jinjiayan and Shuinian are 340 MW and 370 MW respectively, which are relatively heavy, due to the loss of power supply from Chongqing Chongdian and Jiulong power plant, voltage instability occurs.

At 30 s after the fault occurs, since Jinjiayan and Shuinian have low voltage, OLTC starts to operate to increase the voltage. However, due to insufficient reactive power in the area, transmission of line active power is difficult, although the voltage is increased after OLTC operation, it falls slowly immediately, causing OLTC to operate again, and causing voltage to rise and fall repeatedly; at 103 s, the units of Luohuang Phase I Power Plant are in an over-excitation working state for a long time, and the excitation limiting link starts to operate, resulting in further active power deficiency, rapid voltage drop and further increase of excitation current in Chongqing Yongchuan Power Plant; at 253 s, the over-excitation of Yongchuan power plant results in the voltage instability of Jinjiayan and Shuinian.

2) Cascading fault mode II

Cascading fault mode II refers to the decommissioning of Chongqing Chongdian and Jiulong, and the disconnection of Shuangshan-Shuinian. That is, after the decommissioning of Chongqing Chongdian and Jiulong at 0 s, the power flow on the Shuangshan-Shuinian line is relatively heavy, the single-circuit line is disconnected at 10 s, and the double-circuit line is disconnected at 20 s.

The transmission capacity of the Shuangshan-Shuinian 220 kV double-circuit line is 407.4 MW under the current mode. After the decommissioning of Chongqing Power Plant and Jiulong Power Plant, the Shuangshan-Shuinian line is heavily loaded. After the line trips, Shuinian and Jinjiayan have lost most of their active power sources. The voltage drops rapidly. OLTC in the surrounding area is constantly in action. Although the voltage could be kept constant, the time is very short. The voltage continues to drop rapidly and finally drops to 0.52 p.u.

3) Cascading fault mode III

Cascading fault mode III refers to the disconnection of single-circuit line of Shuangshan-Shuinian and Chenjiaqiao-Dazhulin under the maintenance mode of one unit of Chongqing Chongdian, that is, the three phases of Shuangshan-Shuinian single-circuit line are disconnected after the short circuit in 0 s under the maintenance of one unit of Chongqing Chongdian; after 10 s, the Shuangshan-Shuinian double-circuit line is disconnected; after 20 s, the Chenjiaqiao-Dazhulin single-circuit line is disconnected.

At 40 s after the fault, OLTC in the Shuinian starts to operate, but due to insufficient active power in the area, the voltage still slowly drops to 0.94 p.u.; at 155 s, the voltage of another unit of Chongqing Chongdian rapidly drops to 0.82 p.u. due to the effect of over-excitation protection. At 207 s, the excitation current further increased due to the decommissioning of Chongqing Chongdian, and the unit over-excitation protection of Chongqing Jiulong power plant is carried out, and the voltage drops rapidly to 0.64 p.u. After that, the voltage slowly recovered to 0.68 p.u. due to the action of OLTC.

5. Medium and long-term voltage stability analysis in the case of continuous increase of load

Slow and continuous load increase is common in power systems. Due to low load increase rate, it is not easily detected for a period of time, but after continuous increase for a period of time, system voltage instability could easily occur. Therefore, dispatchers need to understand the system voltage stability margin under slow and continuous load increase.

The traditional static load margin is based on the continuation method. Starting from the current working point, the power flow is solved in turn with the increasing load until it passes through the critical point, thus obtaining the node PV curve to obtain the load margin. However, the influence of dynamic elements on voltage stability is ignored when solving the margin based on power flow, and the results are conservative. Therefore, after research and analysis, based on PSD-FDS Full Dynamic Simulation of Power System Program, dynamic voltage stability margin index is proposed. The dynamic voltage stability margin index is used to calculate the margin of the power system from the voltage collapse point when the load increases slowly at a certain rate. The dynamic voltage stability margin takes into account the influence of OLTC, generator over-excitation limit and other dynamic elements in the system, and can calculate the dynamic margin of the system under various load increase modes such as single node, partition and whole grid, providing operators with a more intuitive measure of the distance from the current operating point of the system to the voltage collapse point. The distance from the operating point of the system to the collapse point and the dynamic margin are linearly related. Therefore, the dynamic margin index is of great practical guiding significance.

1) Analysis of voltage stability in the case of continuous load increase in the Chongqing power grid under no fault

This section analyzes the system voltage stability margin when the load of the whole power grid and the region of Chongqing power grid continuously increases without fault, compares with the static margin, and evaluates the voltage stability of Chongqing power grid.

To calculate the continuous load increase, the system load increase rate needs to be determined first. In this section, the load increase rate is 1%/s, i.e., the increase of loads at all nodes is 1% per second in the initial mode. The calculation results are given in Table 8.10.

Table 8.10 Dynamic voltage stability margin in the case of continuous growth of long-term load in Chongqing power grid under no fault

Mode of increase	Rate of rise/(%/s)	Dynamic voltage stability margin/%
Whole grid	1.0	24.3
Northeast Chongqing	1.0	63.6
Southeast Chongqing	1.0	62.9
Xicheng district of Chongqing	1.0	26.5
Dongcheng district of Chongqing	1.0	47.7

In the process of continuous load increase of the whole grid without fault, the voltage gradually decreases, causing OLTC to start to operate. After a period of time, the voltage starts to oscillate slightly. The units of Chongqing Power Plant, Luohuang Phase I Power Plant and Jiulong Power Plant are in a strong excitation state in turn, and then over-excitation protection is performed at 178 s, 196 s and 204 s respectively, resulting in a rapid voltage drop in some areas.

During the process of continuous load increase in northeast Chongqing without fault, the regional voltage is unstable, and Baihe power plant carries out over-excitation protection at 854 s.

In the process of continuous load increase in southeast Chongqing under no fault, the over-excitation protection of Jiangkou power plant at 1,031 s causes the voltage in some areas to drop rapidly and oscillate sharply, eventually leading to voltage instability.

In the process of continuous load increase in Chongqing Xicheng district without fault, the voltage decreases, causing OLTC to start to operate. After a period of time, the voltage starts to oscillate slightly. The units of Chongqing Power Plant, Luohuang Phase I Power Plant, Jiulong Power Plant and Yongchuan Power Plant take turns being in a strong excitation state, and then over-excitation protection is performed at 354 s, 385 s, 426 s, and 480 s respectively, resulting in a rapid voltage drop in some areas.

In the process of continuous load increase in Dongcheng district of Chongqing without fault, the voltage rapidly decreases and the regional voltage is unstable due to the over-excitation protection of the generators in Phase III and Phase I of Chongqing Luohuang power plant at 125 and 267 s.

2) Analysis on voltage stability in the case of continuous load increase in the mode of breaking of Chongqing Chenjiaqiao-Chongqing Banqiao single-circuit line

This section analyzes the voltage stability when the load of Chongqing power grid continues to increase under the mode of breaking of Chongqing Chenjiaqiao-Chongqing Banqiao single-circuit line. The results are shown in Table 8.11. Chongqing Chenjiaqiao-Chongqing Banqiao line is the line with the maximum power

Table 8.11 Dynamic voltage stability margin in the case of continuous load increase of the chongqing power grid in the mode of breaking of Chongqing Chenjiaqiao-Chongqing Banqiao single-circuit line

Mode of increase	Rate of rise/(%/s)	Dynamic voltage stability margin/%
Whole grid	1.0	10.4
Northeast Chongqing	1.0	17.8
Southeast Chongqing	1.0	29.7
Xicheng district of Chongqing	1.0	11.7
Dongcheng district of Chongqing	1.0	22.3

flow in the Chongqing power grid under the wet season peak load mode in 2008. After the line is broken and the power flow is redistributed, the system voltage stability has also changed greatly, which needs attention.

Firstly, the voltage stability of Chongqing power grid under the continuous load increase after the line is disconnected is calculated. According to the results, the bus voltage of Chongqing power grid gradually decreases during the load increase. The units of Chongqing Luohuang Phase I power plant carry out over-excitation protection at 159 s, the units of Chongqing power plant carry out over-excitation protection at 168 s, the Chongqing Jiulong power plant and Chongqing Luohuang Phase II power plant carry out over-excitation protection at 186 s and 189 s respectively, resulting in a serious drop in the voltage of the Chongqing power grid.

Under this mode, due to the power flow transfer, the power supply from Chongqing Luohuang Phase I power plant to Chenjiaqiao is increased, which makes the key faulty elements of Chongqing power grid change from Chongqing power plant without fault to Chongqing Luohuang Phase I power plant. The load margin of the Chongqing power grid is slightly different from that without fault.

The voltage stability of Chongqing power grid is calculated respectively when the load of each district is continuously increasing under the mode of breaking of Chongqing Chenjiaqiao-Chongqing Banqiao single-circuit line. The results are as follows:

In the case of load increase in the northeast Chongqing, the voltage in the northeast Chongqing gradually fails to meet the operation requirements, and the Chongqing Baihe power plant carries out over-excitation protection at 1,309 s.

In the case of load increase in southeast Chongqing, Chongqing Jiangkou power plant carries out over-excitation protection at 891 s, the region loses the voltage support and the voltage oscillates and becomes unstable.

In the case of load increase in Chongqing Xicheng district, the voltage drop causes OLTC to start to operate and the voltage begin to oscillate slightly. Chongqing power plant, Jiulong power plant and Chongqing Luohuang phase I units take turn being in a strong excitation state, and over-excitation protection is performed at 420 s, 460 s and 480 s respectively, resulting in a rapid voltage drop in some areas.

In the process of load increase in Dongcheng district of Chongqing, the voltage rapidly decreases and the regional voltage is unstable due to the over-excitation protection of the generators in Phase III of Chongqing Luohuang power plant at 350 s.

3) Evaluation of voltage stability in the case of long-term continuous load increase

This section studies the impact of continuous load increase in Chongqing power grid under wet season peak load mode in 2008 on long-term voltage stability. The results show that:

(1) The dynamic voltage stability margin calculated by full simulation based on continuous load increase is slightly different from the static voltage stability margin calculated by static voltage stability method, which is caused by the

difference in calculation conditions. However, the general trend of the results calculated by the two methods is basically the same.

(2) Chongqing's long-term voltage stability is weak in Chongqing's Xicheng district, and its continuous load increase margin is much smaller than those of other districts, which is determined by the load distribution of Chongqing power grid.

(3) The voltage stability problem could easily occur in Chongqing Shuangshan, Chongqing Bashan, Chongqing Shuinian, Chongqing Jinjiayan and other areas after the load continues to increase.

(4) The long-term voltage stability problem caused by the continuous load increase is mainly due to the increase of generator excitation current after the continuous load increase, which eventually causes over-excitation limiting action. OLTC's action characteristics also have a more significant impact on it. When over-excitation occurs in Chongqing Power Plant, Chongqing Jiulong Power Plant and Chongqing Luohuang Phase I Power Plant, the regional voltage will be seriously reduced and voltage stability problems will easily occur; Chongqing power plant and Chongqing Jiulong power plant provide strong voltage supporting to Chongqing Xicheng district. Tripping of any unit will cause a chain reaction and make voltage collapse in Chongqing Xicheng district.

6. Study and evaluation of long-term voltage stability

In this section, the long-term voltage stability of Chongqing power grid is studied based on data of Chongqing power grid obtained in 2008 in wet season peak load mode. Through serious fault scanning and cascading fault analysis, the faults of long-term voltage instability of Chongqing power grid are determined; through calculation and comparative analysis of some faults, the influence of SVC on voltage stability of Chongqing power grid is studied; by analyzing the long-term voltage stability when the load continues to increase, the dynamic stability margin in the case of the continuous load increase is proposed.

1) Evaluation of long-term voltage stability under single and multiple faults

For single faults in the Chongqing power grid, relay protection acts correctly, and the system can remain stable without causing voltage instability. The system is stable in channel disconnection and power plant shutdown. In the decommissioning of main transformer of 500-kV substation, and after the decommissioning of main transformers of Wanxian, Banqiao, Chenjiaqiao and Changshou substations, the regional voltage becomes unstable; the system remains stable after the decommissioning of main transformer of Banan substation.

2) Evaluation of long-term voltage stability under cascading faults

According to the actual situation of the Chongqing power grid, the voltage collapse accidents possibly caused by cascading faults are calculated. The results show that the potential cascading faults of Chongqing power grid are mainly distributed in Banqiao of Chongqing Xicheng district and the areas connected to Chenjiaqiao main transformer. The cause of voltage instability caused by cascading faults is the loss of some active power sources in the area, resulting in the increase of power flow, voltage drop, OLTC successive actions, and the increase of excitation current of regional generators eventually leading to the cascading actions of over-excitation protection. In addition, according to the simulation, all the cascading faults that may cause local voltage collapse are related to the decommissioning of Chongqing and Jiulong power plant (the decommissioning of other power plants in the region will not cause voltage collapse), which greatly supports the voltage safety in the region.

3) Evaluation of influence of SVC on voltage stability of Chongqing power grid

The SVC greatly improves the damping after system fault and effectively reduces the time of system oscillation after a fault; when SVC is not used, the system voltage and generator motor angle will oscillate for a long time, which shows that SVC is of positive significance for maintaining the stability of Chongqing power grid after a fault.

4) Analysis and evaluation of long-term voltage stability in the case of continuous increase of the load

According to the dynamic voltage stability margin index when the load of Chongqing power grid continues to increase under the wet season peak load mode in 2008, the voltage stability problem of Chongqing power grid is mainly concentrated in Chongqing Xicheng district, in this district, Chongqing power plant, Chongqing Jiulong power plant and Chongqing Luohuang phase I power plant are in the center of heavy load, and the voltage instability accidents are mostly caused by chain reactions caused by over-excitation protections of one or more units in the three power plants, so these power plants are of great significance to maintaining voltage stability in the region.

8.2.6 Conclusions and Recommendations

Based on the typical mode in 2008, the voltage stability of Chongqing power grid is studied intensively by using a variety of mature analysis methods. The completed studies mainly include: reactive voltage balance analysis under typical operation mode, static voltage stability analysis under the wet season peak load mode, *the modes after* $N-1$ fault and other serious faults, and transient voltage stability analysis under various single element faults and multiple serious faults; in addition, based on the data obtained in 2008 in the wet season peak load mode, the study on the long-term voltage stability of Chongqing power grid is carried out in an exploratory way, and the influences of factors such as continuous load increase, SVC devices and cascading faults on the long-term voltage stability of Chongqing power grid are analyzed and evaluated. The following is a summary of the main conclusions and recommendations.

1) Voltage and reactive power balance

In 2008, Chongqing power grid basically meets the requirements of reactive power balance in layers and zones under wet season peak load mode, dry season peak load mode, wet season valley load mode and dry season valley load mode. Under various operation modes, the reactive power in Xicheng district basically meets the principle of balance in layers and zones, and the node voltages of 220 kV and above meet the operation requirements; the load-side power factor of southeast Chongqing is lower than 0.9 under four typical operation modes, and the load-side power factor does not meet the requirements under the wet season peak load mode and dry season peak load mode; in addition, the power factor of the Changshou transformer is relatively low under wet season peak load mode, dry season peak load mode and wet season valley load mode, but the total reactive power is not large in general. The northeast Chongqing has relatively small total active and reactive loads, and it is supported by the voltage of Baihe power plant. The power exchange with other areas is relatively small under various modes, which basically meets the requirements of power balance in layers and zones. The load-side power factor of Chongqing Dongcheng district is higher under the peak load mode, and the reactive power is well balanced in layers and zones; in valley load mode, the power factor on the load side is slightly lower than 0.95, which is helpful to absorb the excess charging power of the line, and the whole area basically meets the requirements of power balance in layers and zones.

2) Static voltage stability

The regional active reserve coefficient of the Chongqing power grid is unbalanced under the wet season peak load mode in 2008. The static voltage stability level in Xicheng District as the load center of the Chongqing power grid is relatively low. However, in general, there is no obvious weak area in the Chongqing power grid and the participation factors of buses, lines and generators in each area are relatively average.

In the mode after $N-1$ fault, the power margin index of Chongqing power grid is reduced, especially the active reserve coefficient of Xicheng district is often lower than 10%, and the active reserve coefficient is low. In addition, the active power reserve coefficient and reactive power margin of single-load bus in Xicheng district are relatively low.

In the mode after $N-2$ fault, the power margin indexes in Xicheng District of Chongqing power grid are further reduced and the voltage stability level is relatively low.

(1) Xicheng District is the load center of Chongqing power grid, with the load accounting for more than half of the total load of Chongqing power grid. In wet season peak load mode, the reactive power reserve in Xicheng district is relatively low, so it is necessary to strengthen the construction and management of reactive power.

3) Transient voltage stability

In the wet season peak load and wet season peak load modes in 2008, there is no transient stability and dynamic stability problem for the $N - 1$ fault of each section of Chongqing power grid. The transient stability problem caused by the $N - 2$ fault of Sichuan-Chongqing and Chongqing-Hubei sections can be solved by the stability control device for Sichuan-Eastern China power transmission. A few $N - 2$ faults in other sections are mainly thermal stability problems (such as faults of main transformer in Chenjiaqiao). As a whole, the frame of Chongqing power grid in 2008 is relatively strong and it has a strong ability to resist the impact of serious faults. With reasonable safety control measures, the system can remain stable, the voltage can be recovered back to a higher level, and the possibility of transient voltage instability is relatively small, which meets the requirements of the three-level security and stability standard of *Guide on Security and Stability of Power System.*

4) Medium and long-term voltage stability

Under wet season peak load mode in 2008, in case of a single fault or some serious faults in the Chongqing power grid, the system can be stable as long as the relay protection works normally; if voltage instability occurs in the system under some serious faults and cascading faults, the system can remain stable after disconnecting some units and loads; the long time relatively weak area in Chongqing power grid is still Xicheng district. In the dynamic margin index, Xicheng district has the lowest margin, especially the heavy load areas of Banqiao and Chenjiaqiao, which are prone to voltage stability problems under serious faults. As a whole, the structure of Chongqing power grid is relatively reasonable and the system is strong and has a strong ability to resist the impact of faults.

References

1. China Electric Power Research Institute (2005) Power grid voltage stability analysis and research of Fujian. Technical Report of China Electric Power Research Institute, Beijing
2. China Electric Power Research Institute (2008) Further study and comprehensive evaluation on full time domain voltage stability of chongqing power grid. Technical Report of China Electric Power Research Institute, Beijing